Host-Guest Polymer Complexes

Host-Guest Polymer Complexes

Special Issue Editors

Alan Edward Tonelli
Ganesh Narayanan
Alper Gurarslan

MDPI • Basel • Beijing • Wuhan • Barcelona • Belgrade

MDPI

Special Issue Editors
Alan Edward Tonelli
North Carolina State University
USA

Ganesh Narayanan
North Carolina State University
USA

Alper Gurarslan
Istanbul Technical University
Turkey

Editorial Office
MDPI
St. Alban-Anlage 66
Basel, Switzerland

This is a reprint of articles from the Special Issue published online in the open access journal *Polymers* (ISSN 2073-4360) from 2017 to 2018 (available at: http://www.mdpi.com/journal/polymers/special_issues/host_guest_polymer_complexes)

For citation purposes, cite each article independently as indicated on the article page online and as indicated below:

LastName, A.A.; LastName, B.B.; LastName, C.C. Article Title. *Journal Name* **Year**, *Article Number*, Page Range.

ISBN 978-3-03897-194-8 (Pbk)
ISBN 978-3-03897-195-5 (PDF)

Contents

About the Special Issue Editors

Alan Edward Tonelli, Professor Alan Edward Tonelli received B.S. (with distinction) at University of Kansas Chemical Engineering in 1964, and Ph. D. (with P. J. Flory) at Stanford Polymer Chemistry in 1968. From September 2013, he has been the INVISTA Professor of Fiber & Polymer Chemistry, TECS. He has published over 400 publications, including over 120 in *Macromolecules*, two books ("NMR Spectroscopy and Polymer Microstructure: The Conformational Connection", Wiley, 1989; "Polymers From the Inside Out; An Introduction to Macromolecules", Wiley, 2001) and over 25 book chapters. His research interests include the conformations, configurations, and structures of synthetic and biological polymers, their determination by NMR, and establishing their effects on the physical properties of polymers.

Ganesh Narayanan, Ph.D., After obtaining his PhD under the guidance of Prof Tonelli, Dr. Narayanan pursued postdoctoral fellowship in the Department of Orthopaedic Surgery at UConn Health (with Prof. Laurencin), and he is currently pursuing his second postdoctoral fellowship in chemistry at Mississippi State University (with Prof Smith). His research interests include: synthesis, characterization, and modification of polymers, musculoskeletal tissue regeneration, and composites.

Alper Gurarslan, Ph.D., After graduating with a first honour degree from Textile Engineering Department of Usak University, Turkey, Dr. Alper Gurarslan pursued his graduate studies at North Carolina State University, USA. He received his master's degree in Textile Chemistry with a minor in Chemical Engineering and his PhD degree in Fiber and Polymer Science with a co-major in Materials Science and Engineering. He completed his postdoctoral studies at Mechanical and Aerospace Engineering where he worked on flexible electronics made of 2-Dimensional materials.

polymers

MDPI

Editorial

Host–Guest Polymer Complexes

Alan E. Tonelli [1,*], Ganesh Narayanan [1] and Alper Gurarslan [2]

[1] Fiber & Polymer Science Program College of Textiles, North Carolina State University, Campus Box 8301, 2401 Research Drive, Raleigh, NC 27695-8301, USA; gnaraya@ncsu.edu

[2] Faculty of Textile Technologies and Design, Istanbul Technical University, Inonu Cad. No 65 Gumussuyu, Beyoglu, Istanbul 34437, Turkey; gurarslan@itu.edu.tr

* Correspondence: atonelli@ncsu.edu; Tel.: +1-919-515-6588

Received: 20 July 2018; Accepted: 3 August 2018; Published: 13 August 2018

The discovery of host–guest complexation has resulted in significant advancements in various facets of materials science, including: drug delivery, tissue engineering, etc. Various host molecules such as cyclodextrins (CD), urea, thiourea, cyclo-triphosphazines, pillar[6]arene, cucurbit[6]uril, and calix[4]arene have been identified as undergoing inclusion complexation with a range of guest molecules. Among these host molecules, cyclodextrins are well-known for their complexation capabilities with a variety of guest molecules. Through such complexation, the stability of the guest against thermo-chemical, environmental, light, and air degradation is ensured. In addition, the widespread availability of CD-derivatives (randomly methylated, 2-hydroxypropylated, etc.) has partly contributed to their widespread use in applications extending from textiles to pharmacologics to nuclear waste removal. Another benefit of forming such complexes is that they improve the polymeric properties (for example, mechanical, rheological, and thermal properties) by coalescing the guest polymers from their CD inclusion complexes.

While the host–guest complexation occurs predominantly via non-covalent interactions, the underlying mechanism differs among the studied guest molecules, with predominant complexation occurring via hydrogen and ionic bonding, van der Waals and hydrophobic forces/interactions. Even though the mechanisms differ between the chosen host/guest molecules, a common criterion for successful complexation remains lowered Gibbs free energy, which can occur either by an enthalpy- or an entropy-driven process. The thermodynamics of the host–guest complexation and associated binding constants are typically characterized using spectroscopy-based techniques such as nuclear magnetic resonance, UV-vis, mass spectrometry, fluorescence-based spectroscopy or calorimetric techniques such as differential scanning and isothermal titration calorimetry [1]. With such a wide range of possibilities in terms of guest and host molecules in host–guest chemistry, the purpose of this special Issue is to provide the latest advancements as well as novel applications of these host–guest compounds.

This Special Issue consists of 15 articles, including four comprehensive review articles, written by experts in the field. The first eleven articles are dedicated to the synthesis and characterization of host–guest containing compounds for the fabrication of devices, in particular for biomedical applications. The next four articles are comprehensive review articles discussing the state-of-art synthesis, characterization, and fabrication of various host–guest complexes, including CD–polymer, amylose–polymer, polysaccharide–metal ion, and urea–polymer inclusion complexes. In this series, the first research article reports the synthesis of novel host–guest supramolecular hydrogels comprised of pNIPAm microgels bearing poly[acrylic acid] with attached β–CDs as the host and adamantane–dextrans as the guest molecules. By forming in situ complexation, hydrogels were synthesized that showed sol–gel transitions at physiologically relevant temperatures (37 to 41 °C) [2].

One of the advantages of such a system is in the field of the controlled release/retention of small molecules, mediated in part by the concerned host and the guest molecules. Likewise, the second study reports the fabrication of host–guest aerogels containing fluorescently active quantum dots. In the same way that the interaction with CD protects the small molecules from degradation, this study demonstrated the fluorescence-retention potential of quantum dots upon complexation/interaction with β–CD [3].

A major challenge in tissue engineering remains the lack of oxygen and frequent bacterial infection during tissue regeneration. By complexation, the third and fourth articles in this Special Issue address these two major challenges in tissue regeneration [4,5]. By encapsulating oxygen into the cavities of α–CD and α–CD polymers, in vitro cell studies demonstrated a marked reduction in hypoxia–reoxygenation injury, leading to favorable cell growth and lowered cell mortality [4]. Similar to the approach reported by the first two studies [2,3], the fourth study reported the fabrication of a polyrotaxane (PR) containing α–CD and poly [ethylene glycol] containing silver sulfadiazine. When encapsulated in the PR matrix, silver sulfadiazine demonstrated not only superior antibacterial activity against *E. coli* and *S. aureus*, but also resistance to light-induced degradation [5]. The fifth article reports a novel technique to impart anti-bacterial activity by surface grafting CD containing triclosan to a woven polypropylene mesh, resulting in the sustained release of triclosan, making this medical device suitable in clinical settings [6].

The next three articles reported the synthesis and electrostatic-force-induced complexation of anionic polyelectrolytes and dendrigraft poly(L-Lysine) [7], poly (γ-glutamic acid) and ethyl lauryl alginate [8], pillar[6]arene containing polyelectrolytes [9]. As shown by these studies, one of the advantages of utilizing polyelectrolytes is in the utilization of a layer-by-layer approach for the fabrication of thin films, which could be subsequently formulated for various applications, including: antibacterial films/patches.

In the past two decades, our group (https://textiles.ncsu.edu/blog/team/alan-tonelli/) has reported a facile technique to render incompatible polymers more compatible by threading through the CD cavity and subsequently removing the CD, resulting in improved compatibility between otherwise incompatible blends. In addition, over the years, we have extended this technique to urea–polymer inclusion complexes as well. Similar to this established approach, Liu et al. report an improved ability to crystallize and compatibility between PLLA/PDLA by coalescing from a common inclusion complex [10]. In a similar vein, but without using host molecules for complexation, the next article discusses directional alignment of polyfluorene copolymers at patterned solid-liquid interfaces [11]. The final research article in this Special Issue reports a novel complexation between the collagen (host) and polyphosphates (guest). This is particularly crucial because collagen forms the basic structural unit of human anatomy and polyphosphates are superior metabolic fuel for the synthesis and maintenance of extracellular tissue [12].

The next section of this Special Issue contains review articles from expert researchers on the synthesis, characterization and fabrication of state-of-the-art devices based on inclusion complexation. The first review, from our group, is based on the fabrication of aliphatic polyester nanofibers containing uncomplexed and complexed CDs or inclusion complexes. In this review, we report various possibilities for improving the topographical, mechanical and cell-adhesive properties of aliphatic polyester nanofibers via an electrospinning process [13]. The next two articles discuss the synthesis and characterization of novel inclusion complexes between amylose and polymers, and polysaccharides and metal ions [14,15]. Unlike most of the articles reported in this Special Issue, which are based on CDs, complexation with polysaccharides is unique and hard to accomplish; this review provides a useful starting resource for further research. The final review article, also from our group, discusses the complexation between host (CD and urea) and guest polymeric molecules. This review discusses the morphological and thermal changes observed in the homo- and co-polymer guests upon coalescence, as well as changes in their crystallizability [16].

Overall, we anticipate this Special Issue to be a ready reference for those interested in the synthesis, characterization and fabrication of inclusion complexes between guest polymers and host molecules. We hope these reviews will encourage advanced research on host–guest inclusion complexes. As the editors of this Special Issue, we acknowledge and offer our sincere thanks to the authors and the reviewers for their invaluable contribution to this Special issue.

Funding: This research received no external funding.

Conflicts of Interest: The authors declare no conflicts of interest.

References

1. Narayanan, G.; Boy, R.; Gupta, B.S.; Tonelli, A.E. Analytical techniques for characterizing cyclodextrins and their inclusion complexes with large and small molecular weight guest molecules. *Polym. Test.* **2017**, *62*, 402–439. [CrossRef]
2. Antoniuk, I.; Kaczmarek, D.; Kardos, A.; Varga, I.; Amiel, C. Supramolecular Hydrogel Based on pNIPAm Microgels Connected via Host–Guest Interactions. *Polymers* **2018**, *10*, 566. [CrossRef]
3. Liang, X.-Y.; Wang, L.; Chang, Z.-Y.; Ding, L.-S.; Li, B.-J.; Zhang, S. Reusable Xerogel Containing Quantum Dots with High Fluorescence Retention. *Polymers* **2018**, *10*, 310. [CrossRef]
4. Femminò, S.; Penna, C.; Bessone, F.; Caldera, F.; Dhakar, N.; Cau, D.; Pagliaro, P.; Cavalli, R.; Trotta, F. α-Cyclodextrin and α-Cyclodextrin Polymers as Oxygen Nanocarriers to Limit Hypoxia/Reoxygenation Injury: Implications from an In Vitro Model. *Polymers* **2018**, *10*, 211. [CrossRef]
5. Liu, S.; Zhong, C.; Wang, W.; Jia, Y.; Wang, L.; Ren, L. α-Cyclodextrins Polyrotaxane Loading Silver Sulfadiazine. *Polymers* **2018**, *10*, 190. [CrossRef]
6. Sanbhal, N.; Mao, Y.; Sun, G.; Li, Y.; Peerzada, M.; Wang, L. Preparation and Characterization of Antibacterial Polypropylene Meshes with Covalently Incorporated β-Cyclodextrins and Captured Antimicrobial Agent for Hernia Repair. *Polymers* **2018**, *10*, 58. [CrossRef]
7. Lounis, F.M.; Chamieh, J.; Leclercq, L.; Gonzalez, P.; Rossi, J.-C.; Cottet, H. Effect of Dendrigraft Generation on the Interaction between Anionic Polyelectrolytes and Dendrigraft Poly(L-Lysine). *Polymers* **2018**, *10*, 45. [CrossRef]
8. Gamarra-Montes, A.; Missagia, B.; Morató, J.; Muñoz-Guerra, S. Antibacterial Films Made of Ionic Complexes of Poly(γ-glutamic acid) and Ethyl Lauroyl Arginate. *Polymers* **2018**, *10*, 21. [CrossRef]
9. Nicolas, H.; Yuan, B.; Xu, J.; Zhang, X.; Schönhoff, M. pH-Responsive Host–Guest Complexation in Pillar[6]arene-Containing Polyelectrolyte Multilayer Films. *Polymers* **2017**, *9*, 719. [CrossRef]
10. Liu, P.; Chen, X.-T.; Ye, H.-M. Enhancing Stereocomplexation Ability of Polylactide by Coalescing from Its Inclusion Complex with Urea. *Polymers* **2017**, *9*, 592. [CrossRef]
11. Pan, X.; Li, H.; Zhang, X. Directional Alignment of Polyfluorene Copolymers at Patterned Solid-Liquid Interfaces. *Polymers* **2017**, *9*, 356. [CrossRef]
12. Müller, W.E.G.; Relkovic, D.; Ackermann, M.; Wang, S.; Neufurth, M.; Radicevic, A.P.; Ushijima, H.; Schröder, H.-C.; Wang, X. Enhancement of Wound Healing in Normal and Diabetic Mice by Topical Application of Amorphous Polyphosphate. Superior Effect of a Host–Guest Composite Material Composed of Collagen (Host) and Polyphosphate (Guest). *Polymers* **2017**, *9*, 300. [CrossRef]
13. Narayanan, G.; Shen, J.; Boy, R.; Gupta, B.S.; Tonelli, A.E. Aliphatic Polyester Nanofibers Functionalized with Cyclodextrins and Cyclodextrin-Guest Inclusion Complexes. *Polymers* **2018**, *10*, 428. [CrossRef]
14. Orio, S.; Yamamoto, K.; Kadokawa, J.-I. Preparation and Material Application of Amylose-Polymer Inclusion Complexes by Enzymatic Polymerization Approach. *Polymers* **2017**, *9*, 729. [CrossRef]
15. Wang, C.; Gao, X.; Chen, Z.; Chen, Y.; Chen, H. Preparation, Characterization and Application of Polysaccharide-Based Metallic Nanoparticles: A Review. *Polymers* **2017**, *9*, 689. [CrossRef]
16. Gurarslan, A.; Joijode, A.; Shen, J.; Narayanan, G.; Antony, G.J.; Li, S.; Caydamli, Y.; Tonelli, A.E. Reorganizing Polymer Chains with Cyclodextrins. *Polymers* **2017**, *9*, 673. [CrossRef]

polymers MDPI

Review

Reorganizing Polymer Chains with Cyclodextrins

Alper Gurarslan, Abhay Joijode, Jialong Shen, Ganesh Narayanan, Gerry J. Antony, Shanshan Li, Yavuz Caydamli and Alan E. Tonelli *

Fiber & Polymer Science Program, College of Textiles, North Carolina State University,
Raleigh, NC 27606-8301, USA; gurarslan@itu.edu.tr (A.G.); asjoijod@ncsu.edu (A.J.); jshen3@ncsu.edu (J.S.);
gnaraya@ncsu.edu (G.N.); gajohn@ncsu.edu (G.J.A.); sli31@ncsu.edu (S.L.); ycaydam@ncsu.edu (Y.C.)
* Correspondence: atonelli@ncsu.edu; Tel.: +1-919-515-6588

Received: 22 October 2017; Accepted: 22 November 2017; Published: 4 December 2017

Abstract: During the past several years, we have been utilizing cyclodextrins (CDs) to nanostructure polymers into bulk samples whose chain organizations, properties, and behaviors are quite distinct from neat bulk samples obtained from their solutions and melts. We first form non-covalently bonded inclusion complexes (ICs) between CD hosts and guest polymers, where the guest chains are highly extended and separately occupy the narrow channels (~0.5–1.0 nm in diameter) formed by the columnar arrangement of CDs in the IC crystals. Careful removal of the host crystalline CD lattice from the polymer-CD-IC crystals leads to coalescence of the guest polymer chains into bulk samples, which we have repeatedly observed to behave distinctly from those produced from their solutions or melts. While amorphous polymers coalesced from their CD-ICs evidence significantly higher glass-transition temperatures, T_gs, polymers that crystallize generally show higher melting and crystallization temperatures (T_ms, T_cs), and some-times different crystalline polymorphs, when they are coalesced from their CD-ICs. Formation of CD-ICs containing two or more guest homopolymers or with block copolymers can result in coalesced samples which exhibit intimate mixing between their common homopolymer chains or between the blocks of the copolymer. On a more practically relevant level, the distinct organizations and behaviors observed for polymer samples coalesced from their CD-ICs are found to be stable to extended annealing at temperatures above their T_gs and T_ms. We believe this is a consequence of the structural organization of the crystalline polymer-CD-ICs, where the guest polymer chains included in host-IC crystals are separated and confined to occupy the narrow channels formed by the host CDs during IC crystallization. Substantial degrees of the extended and un-entangled natures of the IC-included chains are apparently retained upon coalescence, and are resistant to high temperature annealing. Following the careful removal of the host CD lattice from each randomly oriented IC crystal, the guest polymer chains now occupying a much-reduced volume may be somewhat "nematically" oriented, resulting in a collection of randomly oriented "nematic" regions of largely extended and un-entangled coalesced guest chains. The suggested randomly oriented nematic domain organization of guest polymers might explain why even at high temperatures their transformation to randomly-coiling, interpenetrated, and entangled melts might be difficult. In addition, the behaviors and uses of polymers coalesced from their CD-ICs are briefly described and summarized here, and we attempted to draw conclusions from and relationships between their behaviors and the unique chain organizations and conformations achieved upon coalescence.

Keywords: polymers; cyclodextrins; inclusion compounds; coalescence

---

1. Introduction

Unlike atomic and small molecule solids, the behaviors and properties of materials made from polymers can be significantly altered during their processing, because they are closely related to the organizations, structures, and morphologies of their constituent chains. The conformations and arrangements of their inherently flexible long chains are amenable to modifications through processing, so materials made from the same polymer can behave very distinctly when different means are used to process them. One need only compare the behaviors of garbage bags and gel-spun Spectra® fibers both made from poly(ethylene) (PE). Even though the "same" polymer is used in both applications, the highly oriented and crystalline PE chains in Spectra PE produce extremely strong fibers and may be fabricated into lightweight armor, while molded articles, such as melt-blown PE garbage bags, with randomly-coiled and entangled PE chains, are not nearly as strong, but have a much greater elasticity. The widely different means used to process PE Spectra fibers and PE garbage bags produce widely different organizations, structures, and morphologies of their polymer chains and resultant properties.

Here we describe a means for nano-processing polymers into solids exhibiting unique properties. First, we form non-covalently bonded inclusion compounds (ICs) between guest polymers and host cyclodextrins (CDs) [1,2]. The minimum cross-sectional areas of guest polymers adopting highly extended conformations determines which CD can include them. This is followed by coalescing the guest polymer chains into a bulk solid sample by carefully removing the host CDs. In Figure 1, CDs are shown to form ICs with guest polymers, which are included and reside in the very narrow CD-IC channels (~0.5–1.0 nm in diameter) formed by their crystalline host lattice [3]. There, the guest chains are stretched to high extension and isolated from other chains. It was hoped that, through the careful removal of the crystalline host CD lattice, the resulting coalesced polymer chains (c-polymers) would retain a significant degree of their included extended and un-entangled natures, so they would be organized in a manner quite different from samples processed from their solutions or melts, where they randomly coil and entangle. As we will now demonstrate, this has indeed been found to be the case [2–73]. Their behaviors and properties differ significantly from, and are improved with respect to those of ordinarily processed samples.

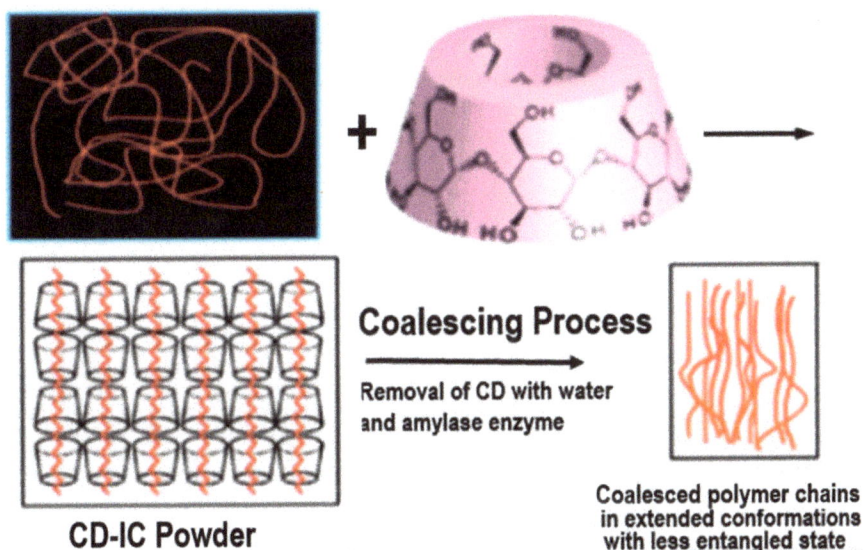

CD-IC Powder

Coalescing Process

Removal of CD with water and amylase enzyme

Coalesced polymer chains in extended conformations with less entangled state

Figure 1. Formation of and coalescence of a polymer sample from its crystalline cyclodextrin complex. Figure adapted with permission from Reference [2], Copyright 2009 Springer-Verlag London 2009.

Some of the results presented here were previously summarized [74] and their experimental details can be found in the original papers [2–73].

2. Polymers Reorganized through Complexation with and Coalescence from Their CD-ICs

2.1. Amorphous Polymers

Figure 2 presents the 2nd heating DSC scans of as-received (asr) and coalesced (c) samples of the amorphous polymers atactic poly (vinyl acetate) (PVAc) and atactic poly(methyl methacrylate) (PMMA) [70]. These coalesced samples were in fact obtained from their urea (U)-ICs, not their CD-ICs. However, we have demonstrated [43,64,67,70] that c-polymers obtained from their CD- and U-ICs behave in virtually identical manners. Compare the T_g of c-PVAc obtained from its γ-CD-IC [43] in Figure 3 to that of the c-PVAc obtained from its U-IC shown in Figure 2. Note, in both cases, the T_g of the c-PVAc is raised more than 10 °C above the T_g of asr-PVAc. The T_g observed for c-PMMA is similarly elevated above that of asr-PMMA. It should be noted that in their CD- and U-ICs the PMMA and PVAc are isolated from each other and consequently do not experience a T_g or show any other change in heat capacity until the IC samples degrade above 250 °C.

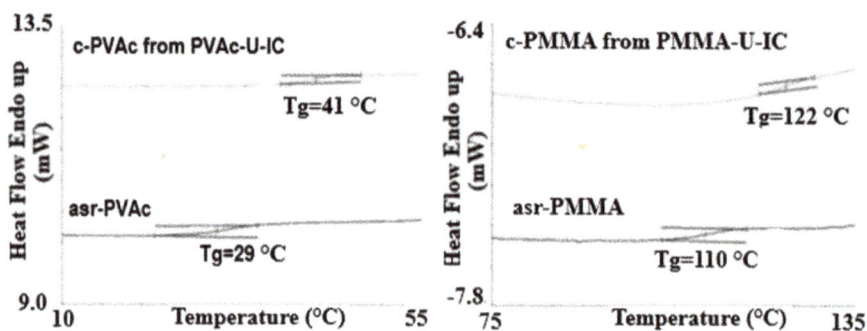

Figure 2. Differential Scanning Calorimetry (DSC) observed glass transitions in amorphous polymers as-received (asr) and coalesced (c) from their U-ICs. Figure adapted with permission from Reference [70], Copyright 2013 Wiley Periodicals, Inc.

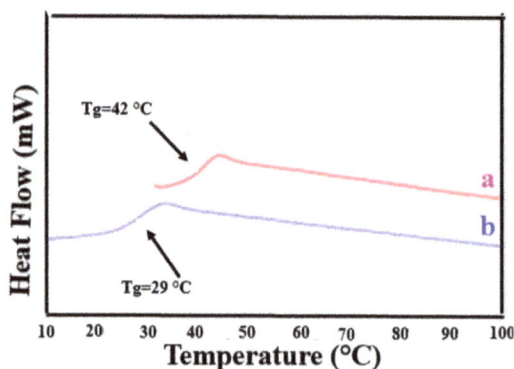

Figure 3. DSC thermograms of the second heating scan of: (a) c-PVAc from its γ-CD-IC; and (b) asr-PVAc. Figure adapted with permission from Reference [43]. Copyright 2005 Elsevier Ltd.

Table 1 summarizes density measurements performed on asr- and c-PVAc samples both below and above their glass-transition temperatures. Note that PVAc coalesced from either its γ-CD- or U-IC is denser than asr-PVAc, both below and above their T_gs. This increase in density is likely also due to the more extended conformations and closer packing of chains in c-PVAcs suggested in Figure 1.

Table 1. Measured Densities for as-received and coalesced PVAcs. Adapted with permission from Reference [72]. Copyright Joijode et al. North Carolina State University 2014.

Sample	Density at 25 °C (g/cm^3) (below T_g)	Density at 58 °C (g/cm^3) (above T_g)
asr-PVAc	1.093	1.040
c-PVAc (γ-CD)	1.156	1.077
c-PVAc (urea)	1.154	1.076

If we use an excess of guest polymer when forming an IC with host CDs, a non-stoichiometric (n-s)-polymer-CD-IC is formed, where portions of the guest chains are un-included and dangle outside the crystalline host CD lattice (see Figure 4). As evidenced in Figure 5, the T_gs of the un-included portions of PVAc and PMMA chains in their (n-s)-γ-CD-ICs are significantly elevated above their neat bulk asr-samples. In fact, even though they were never included in the narrow channels created by the γ-CD-IC crystalline lattice, the un-included chains in (n-s)-CD-ICs may have T_gs even higher than those of samples completely coalesced from either their stoichiometric- or (n-s)-CD-ICs. This has been suggested to be a result of the high density and high extension of the brush-like un-included chain portions emanating from the CD crystal surfaces of (n-s)-CD-ICs (see Figure 4 and subsequently Figure 17) [57,61].

We may also begin to blend and mix the un-included portions of the PVAc and PMMA chains by forming common (n-s)-γ-CD-ICs containing both polymers. The T_gs of the un-included portions of PVAc and PMMA chains in their common (n-s)-γ-CD-ICs are presented in Figure 6. There, it is very clear, by comparison to the T_gs observed in Figure 5, that the common 3:1 (n-s)-γ-CD-IC made using a 1:1 PVAc: PMMA molar ratio results in very little mixing between the un-included portions of the PVAc and PMMA chains. On the other hand, when a 2:1 molar ratio of PVAc: PMMA was used to make their common (n-s)-γ-CD-IC, substantial mixing of the un-included portions of PVAc and PMMA chains was generated, because the T_gs of segregated un-included PVAc and PMMA chains are not observed. Instead, three distinct T_gs are observed, and they may be attributed to PVAc-rich and PMMA-rich phases, and a well-mixed PVAc/PMMA phase.

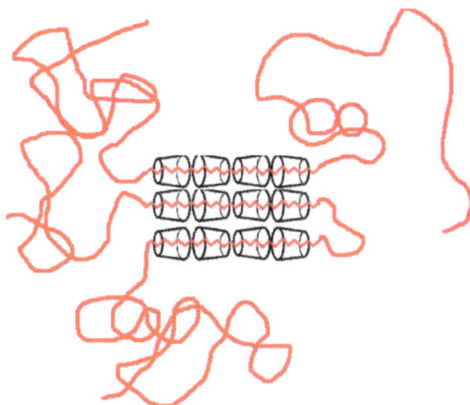

Figure 4. Schematic depiction of a (n-s)-polymer-CD-IC sample. Adapted with permission from Reference [57] Mohan, A.; Joyner, X.; Kotek, R.; Tonelli, A.E. *Macromolecules* **2009**, *42*, 8983–8991. Copyright 2009 American Chemical Society.

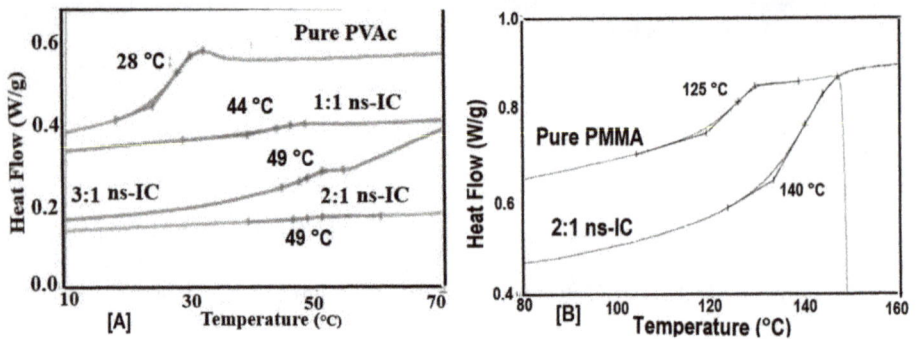

Figure 5. Heating DSC scans of (**a**) asr-PVAc and (**b**) asr-PMMA and their (n-s)-c-CD-ICs. The 3:1 and 2:1 ratios signify the amounts of PVAc and PMMA chains in excess of their stoichiometric amounts. Figure adapted with permission from Reference [70], Copyright 2013 Wiley Periodicals, Inc.

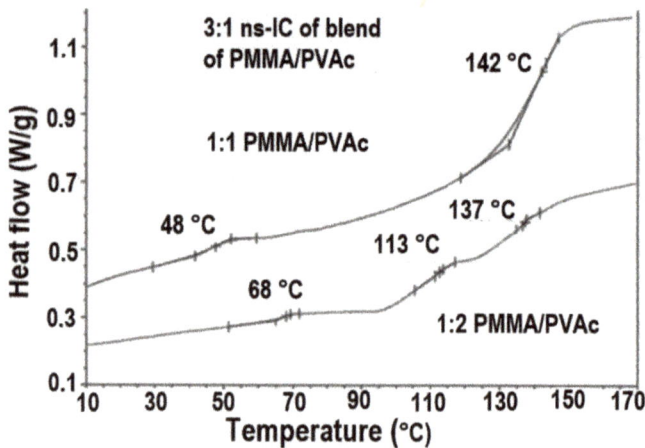

Figure 6. Heating DSC scans of (n-s)-common-PVAc/PMMA-c-CD-ICs. Figure adapted with permission from Reference [70], Copyright 2013 Wiley Periodicals, Inc.

The blending of inherently immiscible amorphous polymers can also be achieved through formation of and coalescence from their fully covered chain included common stoichiometric ICs, as illustrated in Figure 7 for PVAc/PMMA blends coalesced from their U-ICs [72]. They are not completely mixed and appear to possess three phases: PVAc and PMMA rich phases with T_gs similar to those of their neat coalesced samples and well-mixed samples with intermediate T_gs. It is interesting to notice that the T_gs of the well mixed phases in their coalesced blends seem to be proportional to their molar compositions.

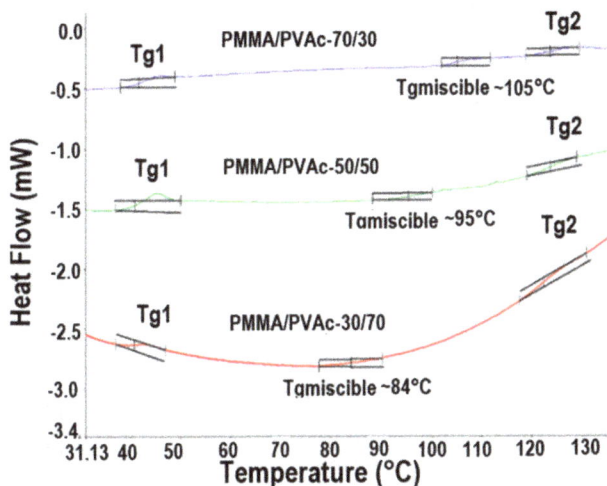

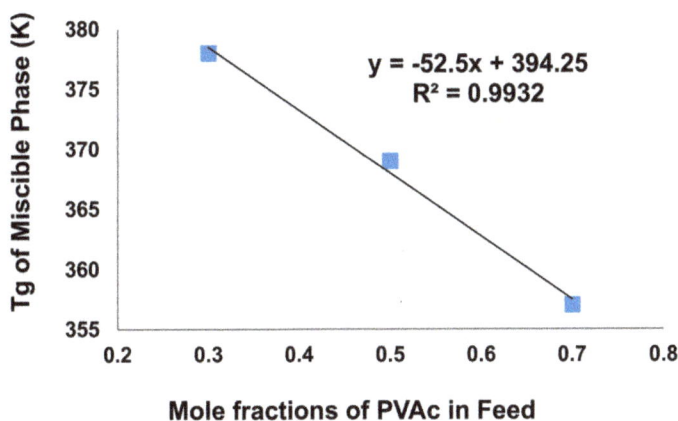

Figure 7. The DSC observed T_gs observed from PVAc/PMMA blends achieved upon coalescence of their common stoichiometric U-ICs (**top**); and plotted versus their composition (**bottom**) [72].

2.2. Semi-Crystalline Polymers

Polymers that can crystallize also form ICs with CD and U hosts. Figure 8 presents the DSC cooling scans observed at different cooling rates for melts of asr-poly(ε-caprolactone) (asr-PCL) and c-PCL, the latter obtained from its α-CD-IC. Though both PCL samples crystallize at decreasingly lower temperatures the faster their melts are cooled, the c-PCL melt crystallizes at significantly higher temperatures than the asr-PCL melt regardless of the cooling rate. asr-PCL and c-PCL, with the latter obtained from its urea-inclusion compound (U-IC), crystallize from their melts upon cooling at $-20\,°C/min$, respectively, at 11.7 and 33.3 °C [67], while as seen below in Figure 8, the same PCL coalesced from its IC formed with host α-CD crystallized at $T_c = 36.8\,°C$ from the melt upon cooling at $-20\,°C/min$ [64]. In addition, crystalline exotherms observed on cooling c-PCL melts are substantially narrower than those shown by asr-PCL. Upon cooling from the melt, not only does the c-PCL begin to crystallize sooner than asr-PCL, but its crystallization ceases much more abruptly than that of asr-PCL.

Both features cause the semi-crystalline morphology c-PCL to be finer and more uniform, i.e., more homogeneous, as can be seen in Figure 9, and lead to improved mechanical and other properties.

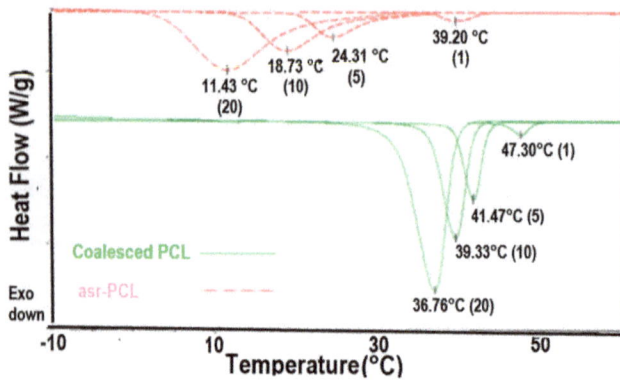

Figure 8. Melt-crystallization curves of asr-s and c-PCL observed at 20, 10, 5 and 1 °C/min cooling rates by DSC. Figure adapted with permission from Reference [64], Copyright 2011 Elsevier Ltd.

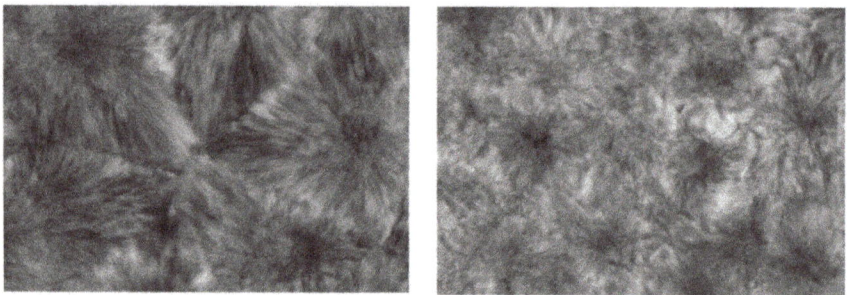

Figure 9. Optical microscopy images (500×), crossed polarizers, $\frac{1}{4}$ λ plate) of: melt pressed films of asr-PCL (**left**); and c-PCL obtained from its U-IC (**right**). Unpublished research from Prof. Tonelli's research group.

In Figure 10, the melt rheology of asr- and c-PCL samples measured at T = 100 °C are compared and reveal a dramatic difference. Notice the ~100-fold decrease in the viscosity of the c-PCL melt, possibly the result of more extended less entangled chains, which also lead to more facile crystallization, as well as faster flow.

Increased amounts of c-PCL were obtained from their U-ICs and were sufficient to permit melt-spinning of single filament fibers. These c-PCL fibers were tested mechanically and thermally before and after drawing, and their results were compared to fibers melt-spun from asr-PCL. Table 2 shows clearly that both before and after drawing the fibers obtained from c-PCL are superior in performance to the asr-PCL fibers. Figure 11 shows the strong correlation between the moduli of the undrawn and drawn c-PCL and asr-PCL fibers and their birefringence, which serves as a measure of the chain orientation in each fiber.

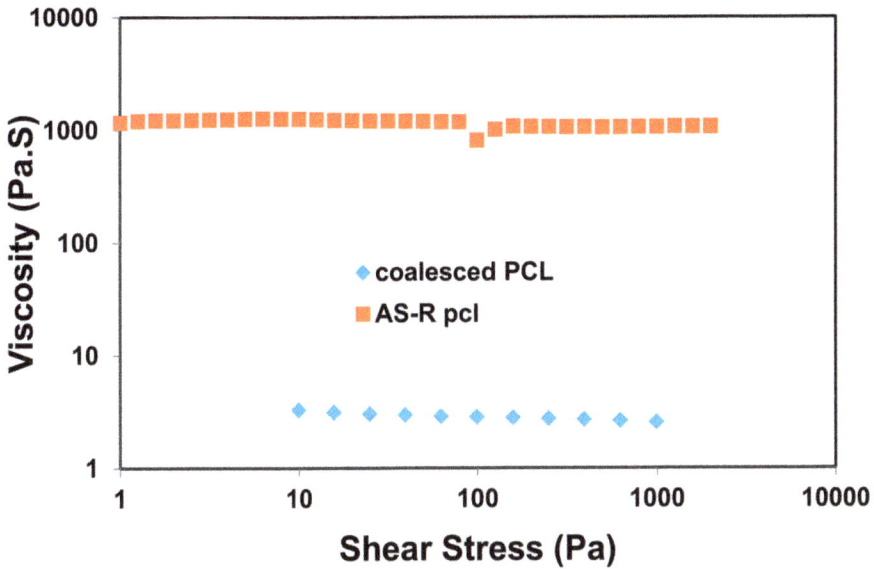

Figure 10. Comparison of frequency sweep rheology of as-received and coalesced PCL melts at $T = 100\ °C$. Unpublished research from Prof. Tonelli's research group.

Table 2. % Crystallinity and mechanical properties (mean ± standard error) of asr and c-PCL Fibers. Reprinted (adapted) with permission from Reference [73] Gurarslan, A.; Caydamli, Y.; Shen, J.; Tse, S.; Yetukuri, M.; Tonelli, A.E. *Biomacromolecules* **2015**, *16*, 890–893. Copyright 2015 American Chemical Society.

Physical properties	asr-PCL fiber	c-PCL fiber	Drawn asr-PCL fiber	Drawn c-PCL fiber
modulus (MPa)	41 ± 6	271 ± 20	465 ± 12	770 ± 32
elongation at break (mm)	197 ± 19	110 ± 8	32 ± 2	14 ± 2
% crystallinity	40.6	50.1	50.8	53.3

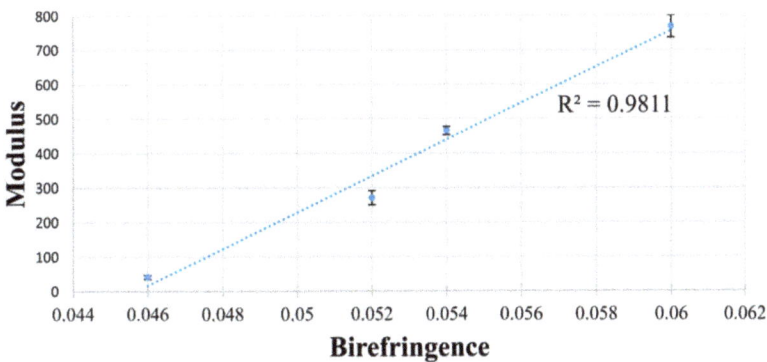

Figure 11. Correlation between modulus and birefringence of the four PCL fiber samples in Table 1. Error bars represent standard error. Reprinted (adapted) with permission from Gurarslan, A.; Caydamli, Y.; Shen, J.; Tse, S.; Yetukuri, M.; Tonelli, A.E. *Biomacromolecules* **2015**, *16*, 890–893 (Reference [73]). Copyright 2015 American Chemical Society.

The improved crystallizability of c-polymers recommends their use as self-nucleants for asr-samples of the same polymer, as demonstrated for c-PCL in Figure 12. The DSC cooling scans of c-PCL and an asr-PCL sample to which 2.5 wt % of c-PCL has been added (nuc-PCL) are shown there. The self-nucleated PCL sample clearly exhibits an enhanced crystallizability and a finer more uniform morphology both produced by the higher temperature and narrower range of crystallization of the added c-PCL self-nucleant.

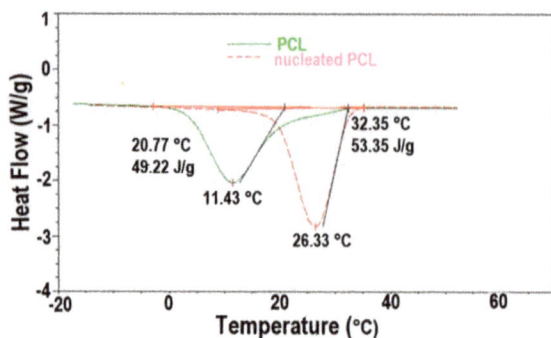

Figure 12. DSC −20 °C/min cooling scans of molten asr-PCL with and without 2.5 wt % c-PCL. Figure adapted with permission from Reference [64], Copyright 2011 Elsevier Ltd.

In Table 3, the densities and CO_2 permeabilities of melt-pressed asr- and nuc-poly(ethylene terephthalate) (PET) films are presented [69]. The nuc-PET film was obtained by melt-pressing a physical mixture of 95 wt % asr-PET and 5 wt % c-PET. DSC observations of both largely amorphous melt-quenched PET films indicated similar crystallinities of ~10%. Clearly the nuc-PET film is denser than the asr-PET film and is markedly less permeable to CO_2 even after both films were quenched from their melts into ice water. Both observations are consistent with the suggested higher orientation and more extended conformations of PET chains in the self-nucleated film, which likely increase their ordering and packing in the predominant amorphous domains.

Table 3. Densities of and CO_2 (0.2 MPa) permeabilities in PET films. Table adapted with permission from Reference [69], Copyright 2013 Wiley Periodicals, Inc.

PET samples	Sample density at 25 °C (g·cm^{-3})	Permeability (P × 10^{14}) (cm^3·s^{-1}·Pa^{-1})
asr-PET	1.368	1.64
nuc-PET	1.386	0.57

Another advantageous feature of using c-polymers as self nucleants is illustrated in Figure 13, where it can be seen that asr-PCL nucleated with c-PCL can itself be repeatedly used to nucleate additional asr-PCL. This means that very little c-polymer need be produced in order to serve as a self-nucleant for improving the morphologies and the resultant properties of semi-crystalline polymer materials. Additionally, using c-polymers as self-nucleants means the polymer materials they nucleate do not contain any foreign material, making them "Stealth" nucleants that are readily recyclable, and appropriate for medical applications (see Figure 14 for example).

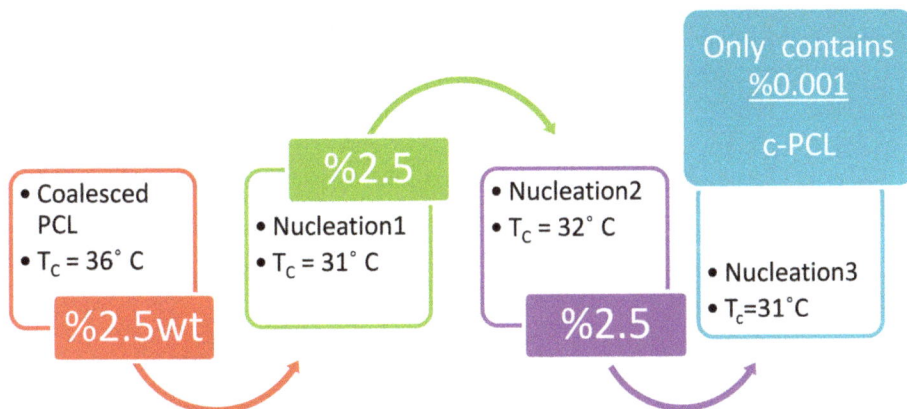

Figure 13. Sequential "Stealth" nucleation of molten asr-PCL (T_c = 11 °C) with c- and subsequently nuc-PCLs observed during −20 °C/min DSC cooling scans. Adapted with permission from Gurarslan, A.; Tonelli, A.E. ACS 2012, Spring 2012 National Meeting, Anaheim, CA (Reference [68]). Copyright 2012 American Chemical Society.

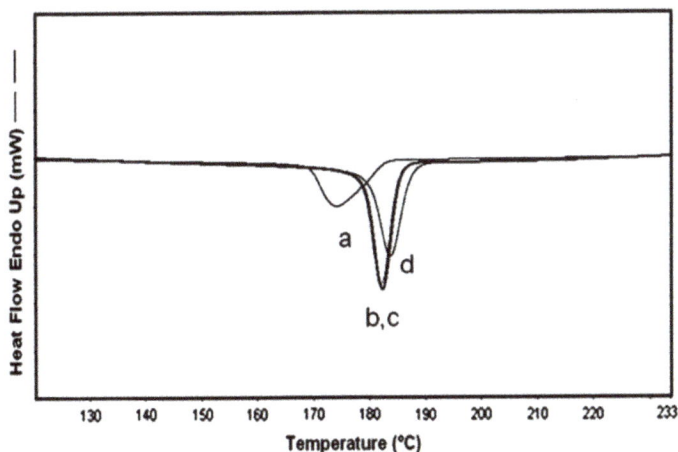

Figure 14. DSC cooling scans for (**a**) neat N-6 sample (M_W = 600,000 g/mol), with 2 wt % of (**b**) N-6 coalesced from a stoichiometric N6-α-CD-IC, (**c**) N-6 coalesced from a 3:1 (n-s)-N6-α-CD-IC, and (**d**) N-6 with 2 wt % talc. Adapted with permission from Reference [57] Mohan, A.; Joyner, X.; Kotek, R.; Tonelli, A.E. *Macromolecules* **2009**, *42*, 8983–8991. Copyright 2009 American Chemical Society.

3. Thermal Stability of the Chain Organization in Coalesced Polymer Samples

From DSC and density observations it is clear that c-PVAc is organized differently and behaves distinctly from asr-PVAc. Therefore, we conducted annealing studies to estimate the time taken for c-PVAc to revert back from its presumed extended un-entangled chain coalesced morphology to the randomly-coiling entangled chain morphology of asr-PVAc (see Figure 1). The T_gs observed for asr-PVAc before and after annealing at 70 °C for 14 days were 29.1 and 29.4 °C, respectively, i.e., essentially the same. The T_gs observed for c-PVAc obtained from its γ-CD after annealing at 70 °C for different times are presented in Table 4. They remain around 41 °C, so c-PVAc chains apparently remain largely extended and un-entangled even after annealing above their T_g, without apparently returning to entangled random coils as in the asr-PVAc. It should also be noted that single concentration (0.2 g/dL) viscometer flow times for asr- and c-PVAcs measured in dioxane before and after annealing were both ~400 s, suggesting that long-time annealing above T_g (at 70 °C) had not caused any PVAc degradation.

Table 4. T_gs of c-PVAc from its γ-CD-IC annealed above T_g at 70 °C for different times. Adapted with permission from Reference [72]. Copyright Joijode et al. North Carolina State University 2014.

Annealing time (days)	T_g (°C)
0	41.5
2	41.7
8	41.5
14	41.2

Even more intriguing are the tentative observations seen in Figure 15 [72], where DSC heating scans of a 50:50 physical mixture of asr-PVAc:c-PVAc recorded before and after long-time annealing above T_g are presented. While the initial heating scan reveals distinct T_gs for phase separated asr- and c-PVAcs, after annealing for 14 days, apparently the 50/50 asr-PVAC/c-PVAC mixture is now a well-mixed blend, because a single T_g between those of asr-PVAc (29 °C) and c-PVAc (41 °C), but much closer to that of c-PVAc, is observed around 38 °C. After four weeks of annealing above T_g at 70 °C, the 50/50 asr-PVAc/c-PVAc blend has a T_g (DSC not shown) and density (Table 5) nearly identical to that of neat c-PVAc (from its γ-CD-IC).

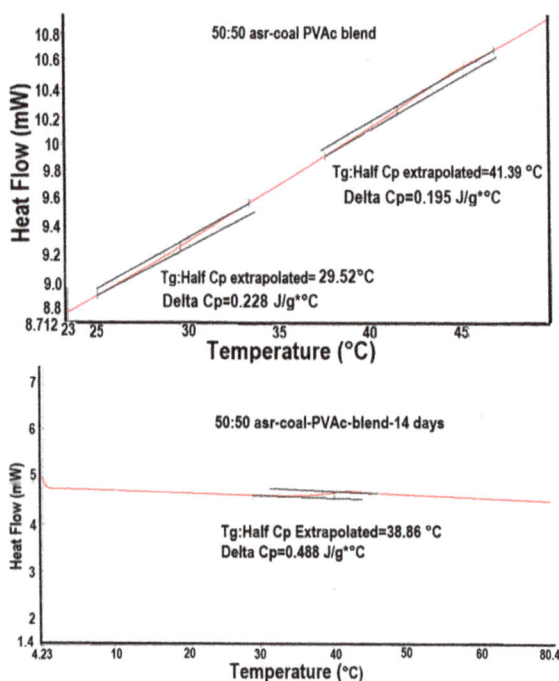

Figure 15. DSC scans of 50/50 physical mixture of asr-PVAc/c-PVAc: before (**top**); and after (**bottom**) annealing for 14 days above T_g at 70 °C. Adapted with permission from Reference [72]. Copyright Joijode et al. North Carolina State University 2014.

Table 5. Densities of asr-PVAc, c-PVAc (γ-CD), and their 50/50 blend after annealing at 70 °C for 4 weeks [#]. Adapted with permission from Reference [72]. Copyright Joijode et al. North Carolina State University 2014.

Sample *	Density (g/cm³) (above T_g) at 58 °C
asr-PVAc	1.044
c-PVAc	1.077
50/50 = asr/c-PVAc blend	1.076

* T_g of asr-PVAc film is ~29 °C and for the c-PVAc (γ-CD) and 50/50 blend films T_gs are both ~41 °C. [#] The densities of asr- and c-PVAc samples were measured as described in [64] by floatation using water and aq NaBr (21 wt %) (densities of 1.0 and 1.184 g/cm³, respectively, lower and higher than that of PVAc). Into a known volume of water, vol (H_2O), containing a magnetic stirring bar, were placed small pieces of both PVAc films pressed at 70 °C, which sank to the bottom. The NaBr/H_2O solution was slowly added from a burette, under stirring, until each PVAc film in turn rose from the bottom and was suspended in the aq solution, and the volume of added NaBr/H_2O, vol(NaBr/H_2O), was noted. The densities of asr- and c-PVAc films were then obtained at both below and above their glass-transition temperatures $\rho = \frac{[Vol(water) \times \rho \, (water) + vol \left(\frac{NaBr}{water} \right) \times \rho \left(\frac{NaBr}{water} \right)]}{Vol \, (water) + vol \left(\frac{NaBr}{water} \right)}$.

The observation of a single T_g for the 50/50 asr-PVAc/c-PVAc (γ-CD) physical mixture annealed a long time above their T_gs into a well-mixed blend is not surprising. However, the observation that the single T_g observed for their apparently well-mixed blend is close to that of c-PVAc, rather than asr-PVAc, is surprising. In other words, the blend is likely not characterized by completely entangled randomly-coiling PVAc chains, but rather by an organization of PVAc chains more similar to that of the initially coalesced neat c-PVAc (γ-CD) (see Figure 1).

Similarly, when semi-crystalline asr- and c-PCL (obtained from its α-CD-IC) samples were held in the melt for 1 week, 2 weeks, and 1 month at 90 °C, and subsequently observed by DSC, the following melt crystallization temperatures were observed at a −20 °C/min cooling rate: 35.6 (0), 33.4 (1 week), 32.6 (2 weeks), and 31.1 °C (1 month) for c-PCL and 11.4 (0) and 15 °C (2 weeks) for melt-annealed asr-PCL. In addition, viscometer flow times of 1 g/dL chloroform solutions of c-PCL before and after long term melt annealing were closely similar, indicating little if any thermal degradation had occurred. PCL coalesced from its U-IC before and after melt-annealing for 2–4 weeks were similarly crystallized at 33.3 (0), 30 (2 weeks), and 31.1 C (4 weeks) when cooled at −20 °C/min.

These results prompt the question: "What are the conformations and organization of the polymer chains in their c-samples, and why are they so resistant to long-time high temperature annealing?"

Rastogi et al. [75] recently observed that slow and carefully controlled melting of ultra-high molecular weight, metallocene catalyzed, single-chain crystalline polyethylenes (UHMW-PEs), as illustrated in the drawing at the left in Figure 16, can produce heterogeneous melts, with more and less entangled regions. They observed that it took more than a day of heating at 180 °C to produce a fully entangled melt of randomly-coiling PE chains. On the other hand, melts of those polymers initially coalesced from their CD-IC crystals, such as those reported here for c-PVAc, with presumably extended and largely un-entangled chains, apparently take much longer to return to their randomly-coiling entangled melt state.

Quoting Rastogi et al. [75], "crystals composed of single chains are feasible, where the chains are fully separated from each other. If such separation can be maintained in the melt a new melt state can be formed. Here we show that through slow and carefully controlled melting such polymer crystals form a heterogeneous melt with more entangled regions, where the chains are mixed, and less entangled ones, composed of individually separated chains. Chain reptation, required for the homogenization of the entanglement distribution, is found to be considerably hindered. The long-lived heterogeneous melt shows decreased melt viscosity and provides enhanced drawability on crystallization."

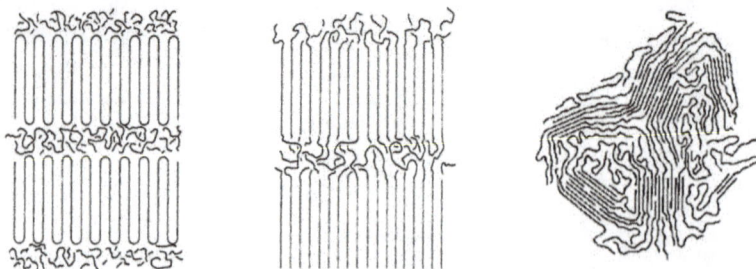

Figure 16. Idealized comparison of: regularly-folded single-chain lamellar crystals (**left**); irregularly-folded multi-chain "switchboard" lamellar crystals (**center**); and "fringed-micelle" semi-crystalline morphology (**right**). Adapted with permission from Reference [76]. Gurarslan, A.; Joijode, A.S.; Tonelli, A.E. *J. Polym. Sci. Part B Polym. Phys.* **2012**, *50*, 813–820. Copyright Wiley 2012.

In contrast, the usual morphology of semi-crystalline polymers with intimately mixed chains, as illustrated by the central drawing in Figure 16, is rapidly converted upon heating into a homogeneous entangled melt of randomly-coiling chains. Chain reptation is required to produce homogeneous melts from both semi-crystalline morphologies. However, in the case of the slow heating of metallocene-catalyzed UHMW-PE nascent single-chain crystals, the initial rapid entanglements formed between portions of PE chains melting and detaching from the ends of different single-crystals serve to retard the further entanglement of the central portions of the PE chains, leading to a heterogeneous melt whose ultimate homogenization is thus greatly retarded [77].

Because the coalescence process begins with a polycrystalline powder sample of the polymer-IC, with individual crystalline grains that are randomly oriented. Upon careful removal of the CD or U hosts, the volume of each resultant coalesced polymer region is significantly reduced (see Figure 17). Consequently, we envision the chains in a c-polymer sample to be possibly organized [77] as indicated in Figure 18.

Figure 17. Schematic representation of a channel structure polymer-CD-IC. Adapted with permission from Reference [57] Mohan, A.; Joyner, X.; Kotek, R.; Tonelli, A.E. *Macromolecules* **2009**, *42*, 8983–8991. Copyright 2009 American Chemical Society.

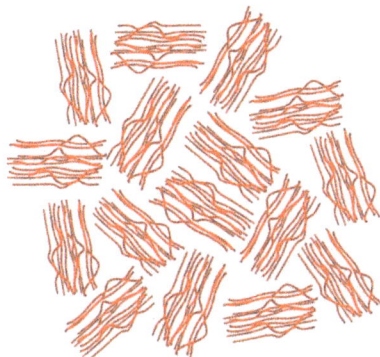

Figure 18. 2-D representation of a polymer sample coalesced from its CD-IC. Unpublished work from Prof. Tonelli's research group.

More recently, McLeish [78] has suggested "In the case of these heterogeneous melts, however, chains leaving and entering the original 'cells' (single-crystal interiors) must result in strong elastic deformations of the entangled material of the cell walls (detached chain-ends). This will result in a free-energy barrier to chain motion ... single-chain (non-cooperative) reptation may be strongly suppressed by elastic deformation of the partitioning regions (initially detached and entangled chain-ends, see Figure 19." Assuming the morphology of polymers initially coalesced from their CD-IC crystals resembles that of the right-hand drawing in Figure 19, but without tie-chains connecting the randomly arranged regions of extended and un-entangled chains, can we (or you) suggest why its transformation to a randomly-coiling entangled melt takes so long?

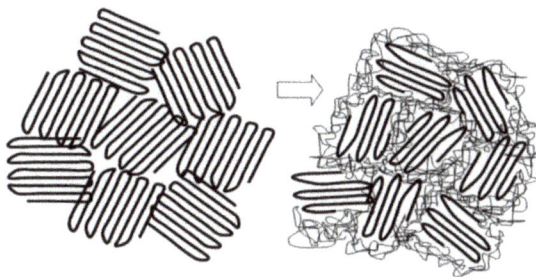

Figure 19. McLeish model for the heterogeneous melt structure produced by the slow heating low-temperature melting of the outer portions of the original single-chain UHMW-PE crystals (**left**) to form entangled normal melt regions. These partition the remaining crystal cores into "cells" (**right**). On further melting, the inner regions of these cells contain un-entangled melt. Figure adapted with permission from Reference [78]. Copyright 2007 The Royal Society of Chemistry Ltd.

It seems likely that the chain-ends emerging from each individually coalesced sample region, resulting from the individual polymer-CD-IC crystals from which they were coalesced, should be able to rapidly entangle. Once this occurs, however, it would appear difficult to entangle the central portions of the essentially parallel and extended coalesced chains, as in the case of the heterogeneous melts produced from nascent UHMW-PE single-chain crystalline samples. Then, why does it take so much longer for the melt of an initially coalesced polymer to become fully randomly-coiled and entangled, even though their chains are more than an order of magnitude shorter than those of the UHMW-PEs?

The reason may have its basis in the topological distinction(s) between their initial semi-crystalline and/or the presumed as-coalesced morphologies, as indicated in the right and left hand drawings, respectively, in Figures 18 and 19. As the nascent single-chain UHME-PE crystals melt by detachment from the ends of the single crystals and their chain-ends entangle, the un-entangled central portion of each PE chain remains more mobile, though their reptative motion is restricted by the entanglement of their previously detached chain-ends [75,78]. In addition to a reptation hindered by entanglement of their chain-ends, the un-entangled central chain portions in the coalesced polymer melt are initially more extended and closely packed with a density higher than that of its homogeneous melt. This likely reduces their mobility and may lead to a significant enhancement of the elastic strain placed on the already entangled chain-ends, thereby further retarding their full entanglement. The distinctions between the inter-chain packing densities and mobilities of the initially un-entangled chain portions of coalesced polymers and those of the un-entangled and mobile central portions of UHMW-PE chains individually detached from their single-chain crystals may account for the increased sluggishness of the transition to a fully entangled randomly-coiled melt exhibited by polymers coalesced from their CD-ICs.

4. Polarized Optical Examination of Solid and Molten asr- and c-PCL Films

Films of asr-PCL and c-PCL coalesced from its U-IC were examined with a Nikon Eclipse 50i POL polarized optical microscope (POM) (Nikon, Tokyo, Japan) as they were heated from room temperature to 90 °C well above their T_ms (~60 °C) and then cooled to 30 °C at a rate of 10 °C/min using a Mettler FP82HT hot stage (Mettler Toledo, Greifensee, Switzerland) controlled by a Mettler FP90 central processor.

Observations of both the asr- and c-PCL melts above T_m to 90 °C revealed no differences. No birefringence indicative of polymer chain ordering or anisotropic chain packing were observed. On cooling from their melts at 10 °C/min, the crystalline spherulites formed in the cooled c-PCL film were smaller and uniform in size compared to those in the asr-PCL film which were heterogeneous and larger in size.

Because the POM observations yielded no distinctions between the chain organizations in asr- and c-PCL, we will continue our search for an experimental probe that will be able to make this distinction. The distinct behaviors of these PCL melts must logically have a structural basis.

5. Summary and Conclusions

We have presented an up to date summary of our observations of bulk polymer samples that have been reorganized by the formation of and coalescence from their channel structure ICs formed with CD and U hosts. Amorphous polymers coalesced from their CD-ICs (and U-ICs) evidence significantly higher glass-transition temperatures, T_gs. Coalesced semi-crystalline polymers show higher melting and crystallization temperatures, and sometimes different crystalline polymorphs. Samples coalesced from ICs containing two or more guest homopolymers or guest block copolymers exhibit intimate mixing between their common homopolymer chains or between the blocks of the previously included copolymer. Most striking of the unique behaviors evidenced by coalesced polymer samples is the retention of their distinct behaviors even after long-term high temperature melt annealing.

Acknowledgments: We are indebted to the many previous graduate students, Post-Docs., and collaborators listed in References [2–74] who contributed to and made possible the research summarized here.

Author Contributions: All authors are former students in the Tonelli group and have contributed to the research as cited in the appropriate references.

Conflicts of Interest: The authors declare no conflict of interest.

References

1. Harada, A.; Kamachi, M. Complex formation between poly (ethylene glycol) and α-cyclodextrin. *Macromolecules* **1990**, *23*, 2821. [CrossRef]
2. Tonelli, A.E. Molecular processing of polymers with cyclodextrins. *Adv. Polym. Sci.* **2009**, *222*, 55–77.
3. Huang, L.; Tonelli, A. On the dynamics of dense polymer systems. *J. Macromol. Sci. Rev. Macromol. Chem. Phys.* **1998**, *38*, 781. [CrossRef]
4. Huang, L.; Vasanthan, N.; Tonelli, A.E. Polymer-polymer composites fabricated by the in situ release and coalescence of polymer chains from their inclusion compounds with urea into a carrier polymer phase. *J. Appl. Polym. Sci.* **1997**, *64*, 281–287. [CrossRef]
5. Huang, L.; Allen, E.G.; Tonelli, A.E. *Recent Research Developments in Macromolecular Research*; Pandalai, S.G., Ed.; Research Sign-Post: Trivandrum, India, 1997; Volume 2, p. 175.
6. Huang, L.; Allen, E.; Tonelli, A.E. Study of the Inclusion Compounds Formed between α-Cyclodextrin and High Molecular Weight Poly (ethylene oxide) and Poly (∈-caprolactone). *Polymer* **1998**, *39*, 4857–4865. [CrossRef]
7. Huang, L.; Tonelli, A.E. Inclusion Compounds as a Means to Fabricate Controlled Release Materials. In *Intelligent Materials for Controlled Release*; Dinh, S.M., DeNuzzio, J.D., Comfort, A.R., Eds.; ACS Symposium Series No. 728; American Chemical Society: Washington, DC, USA, 1999; Chapter 10.
8. Rusa, C.C.; Tonelli, A.E. Polymer/polymer inclusion compounds as a novel approach to obtaining a PLLA/PCL intimately compatible blend. *Macromolecules* **2000**, *33*, 5321–5324. [CrossRef]
9. Huang, L.; Gerber, M.; Taylor, H.; Lu, J.; Tapaszi, E.; Wutkowski, M.; Hill, M.; Nunalee, F.N.; Harvey, A.; Rusa, C.C.; et al. Creation of Polymer Films with Novel Structures by Processing with Inclusion Compounds. In *Film Formation in Coatings: Mechanisms, Properties, and Morphology*; Povder, T., Urban, M.W., Eds.; ACS Symposium Series No. 790; American Chemical Society: Washington, DC, USA, 2001; Chapter 14.
10. Lu, J.; Mirau, P.A.; Rusa, C.C.; Tonelli, A.E. Cyclodextrin: From Basic Research to Market. In Proceedings of the 10th International Cyclodextrin Symposium (CD-2000), Ann Arbor, MI, USA, 21–24 May 2000; Szejtli, J., Ed.; Mira Digital Publishing: Saint Louis, MO, USA, 2001.
11. Rusa, C.C.; Lu, J.; Huang, L.; Tonelli, A.E. Cyclodextrin: From Basic Research to Market. In Proceedings of the 10th International Cyclodextrin Symposium (CD-2000), Ann Arbor, MI, USA, 21–24 May 2000; Szejtli, J., Ed.; Mira Digital Publishing: Saint Louis, MO, USA, 2001.
12. Rusa, C.C.; Luca, C.; Tonelli, A.E. Polymer−Cyclodextrin Inclusion Compounds: Toward New Aspects of Their Inclusion Mechanism. *Macromolecules* **2001**, *34*, 1318–1322. [CrossRef]
13. Wei, M.; Tonelli, A.E. Compatiblization of polymers via coalescence from their common cyclodextrin inclusion compounds. *Macromolecules* **2001**, *34*, 4061–4065. [CrossRef]
14. Shuai, X.; Porbeni, F.E.; Wei, M.; Shin, I.D.; Tonelli, A.E. Formation of and coalescence from the inclusion complex of a biodegradable block copolymer and α-cyclodextrin: A novel means to modify the phase structure of biodegradable block copolymers. *Macromolecules* **2001**, *34*, 7355–7361. [CrossRef]
15. Huang, L.; Gerber, M.; Taylor, H.; Lu, J.; Tapazsi, E.; Wutkowski, M.; Hill, M.; Lewis, C.; Harvey, A.; Wei, M.; et al. Creation of novel polymer materials by processing with inclusion compounds. *Macromol. Symp.* **2001**, *176*, 129–144. [CrossRef]
16. Shuai, X.; Wei, M.; Porbeni, F.E.; Bullions, T.A.; Tonelli, A.E. Formation of and Coalescence from the Inclusion Complex of a Biodegradable Block Copolymer and α-Cyclodextrin. 2: A Novel Way To Regulate the Biodegradation Behavior of Biodegradable Block Copolymers. *Biomacromolecules* **2002**, *3*, 201–207. [CrossRef] [PubMed]
17. Bullions, T.A.; Wei, M.; Porbeni, F.E.; Gerber, M.J.; Peet, J.; Balik, M.; White, J.L.; Tonelli, A.E. Reorganization of the structures, morphologies, and conformations of bulk polymers via coalescence from polymer–cyclodextrin inclusion compounds. *J. Polym. Sci. Part B Polym. Phys.* **2002**, *40*, 992–1012. [CrossRef]
18. Shuai, X.; Porbeni, F.E.; Wei, M.; Bullions, T.; Tonelli, A.E. Formation of Inclusion Complexes of Poly (3-hydroxybutyrate) s with Cyclodextrins. 1. Immobilization of Atactic Poly (R,S-3-hydroxybutyrate) and Miscibility Enhancement between Poly (R,S-3-hydroxybutyrate) and Poly (∈-caprolactone). *Macromolecules* **2002**, *35*, 3126–3132. [CrossRef]

19. Wei, M.; Davis, W.; Urban, B.; Song, Y.; Porbeni, F.E.; Wang, X.; White, J.L.; Balik, C.M.; Rusa, C.C.; Fox, J.; et al. Manipulation of Nylon-6 crystal structures with its α-Cyclodextrin inclusion complex. *Macromolecules* **2002**, *35*, 8039–8044. [CrossRef]

20. Rusa, C.C.; Bullions, T.A.; Fox, J.; Porbeni, F.E.; Wang, X.; Tonelli, A.E. Inclusion compound formation with a new columnar cyclodextrin host. *Langmuir* **2002**, *18*, 10016–10032. [CrossRef]

21. Shuai, X.; Porbeni, F.E.; Wei, M.; Bullions, T.; Tonelli, A.E. Inclusion complex formation between α,γ-cyclodextrins and a triblock copolymer and the cyclodextrin-type-dependent microphase structures of their coalesced samples. *Macromolecules* **2002**, *35*, 2401–2405. [CrossRef]

22. Wei, M.; Shuai, X.; Tonelli, A.E. Melting and crystallization behaviors of biodegradable polymers enzymatically coalesced from their cyclodextrin inclusion complexes. *Biomacromolecules* **2003**, *4*, 783–792. [CrossRef] [PubMed]

23. Bullions, T.A.; Edeki, E.M.; Porbeni, F.E.; Wei, M.; Shuai, X.; Rusa, C.C.; Tonelli, A.E. Intimate blend of poly (ethylene terephthalate) and poly (ethylene 2,6-naphthalate) via formation with and coalescence from their common inclusion compound with γ-cyclodextrin. *J. Polym. Sci. Part B Polym. Phys.* **2003**, *41*, 139–148. [CrossRef]

24. Abdala, A.A.; Tonelli, A.E.; Khan, S.A. Modulation of hydrophobic interactions in associative polymers using inclusion compounds and surfactants. *Macromolecules* **2003**, *36*, 7833–7841. [CrossRef]

25. Tonelli, A.E. The potential for improving medical textiles with cyclodextrin inclusion compounds. *J. Text. Appar. Technol. Manag.* **2003**, *3*, 12.

26. Tonelli, A.E. Reorganization of the structures, morphologies, and conformations of polymers by coalescence from their crystalline inclusion compounds formed with cyclodextrins. *Macromol. Symp.* **2003**, *203*, 71–88. [CrossRef]

27. Wei, M.; Bullions, T.A.; Rusa, C.C.; Wang, X.; Tonelli, A.E. Unique morphological and thermal behaviors of reorganized poly (ethylene terephthalates). *J. Polym. Sci. Part B Polym. Phys.* **2004**, *42*, 386–394. [CrossRef]

28. Rusa, C.C.; Uyar, T.; Rusa, M.; Hunt, M.A.; Wang, X.; Tonelli, A.E. An intimate polycarbonate/poly (methyl methacrylate)/poly (vinyl acetate) ternary blend via coalescence from their common inclusion compound with γ-cyclodextrin. *J. Polym. Sci. Part B Polym. Phys.* **2004**, *42*, 4182–4194. [CrossRef]

29. Abdala, A.A.; Wu, W.; Olesen, K.R.; Jenkins, R.D.; Tonelli, A.E.; Khan, S. Solution rheology of hydrophobically modified associative polymers: Effects of backbone composition and hydrophobe concentration. *J. Rheol.* **2004**, *48*, 979–994. [CrossRef]

30. Wei, M.; Shin, I.D.; Urban, B.; Tonelli, A.E. Partial miscibility in a nylon-6/nylon-66 blend coalesced from their common α-cyclodextrin inclusion complex. *J. Polym. Sci. Part B Polym. Phys.* **2004**, *42*, 1369–1378. [CrossRef]

31. Rusa, C.C.; Shuai, X.; Bullions, T.A.; Wei, M.; Porbeni, F.E.; Lu, J.; Huang, L.; Fox, J.; Tonelli, A.E. Controlling the behaviors of biodegradable/bioabsorbable polymers with cyclodextrins. *J. Polym. Environ.* **2004**, *12*, 157–163. [CrossRef]

32. Uyar, T.; Rusa, M.; Tonelli, A.E. Polymerization of Styrene in Cyclodextrin Channels: Can Confined Free-Radical Polymerization Yield Stereoregular Polystyrene? *Makromol. Rapid Commun.* **2004**, *25*, 1382–1386. [CrossRef]

33. Rusa, C.C.; Wei, M.; Bullions, T.A.; Rusa, M.; Gomez, M.A.; Porbeni, F.E.; Wang, X.; Shin, I.D.; Balik, C.M.; White, J.L.; et al. Controlling the polymorphic behaviors of semicrystalline polymers with cyclodextrins. *Cryst. Growth Des.* **2004**, *4*, 1431–1441. [CrossRef]

34. Rusa, C.C.; Wei, M.; Shuai, X.; Bullions, T.A.; Wang, X.; Rusa, M.; Uyar, T.; Tonelli, A.E. Molecular mixing of incompatible polymers through formation of and coalescence from their common crystalline cyclodextrin inclusion compounds. *J. Polym. Sci. Part B Polym. Phys.* **2004**, *42*, 4207–4224. [CrossRef]

35. Rusa, M.; Aboelfotoh, O.; Kolbas, R.M.; Tonelli, A.E. Conformable nanoscale polymers through formation of cyclodextrin inclusion compounds. *PMSE Prepr.* **2004**, *90*, 620.

36. Rusa, M.; Wang, X.; Tonelli, A.E. Fabrication of Inclusion Compounds with Solid Host γ-Cyclodextrins and Water-Soluble Guest Polymers: Inclusion of Poly (N-acylethylenimine) s in γ-Cyclodextrin Channels As Monitored by Solution 1H NMR. *Macromolecules* **2004**, *37*, 6898–6903. [CrossRef]

37. Rusa, C.C.; Wei, M.; Bullions, T.A.; Shuai, X.; Uyar, T.; Tonelli, A.E. Nanostructuring polymers with cyclodextrins. *Polym. Adv. Technol.* **2005**, *16*, 269–275. [CrossRef]

38. Rusa, C.C.; Rusa, M.; Gomez, M.; Shin, I.D.; Fox, J.D.; Tonelli, A.E. Nanostructuring High Molecular Weight Isotactic Polyolefins via Processing with γ-Cyclodextrin Inclusion Compounds. Formation and Characterization of Polyolefin-γ-Cyclodextrin Inclusion Compounds. *Macromolecules* **2004**, *37*, 7992–7999. [CrossRef]

39. Hernández, R.; Rusa, M.; Rusa, C.C.; López, D.; Mijangos, C.; Tonelli, A.E. Controlling PVA hydrogels with γ-cyclodextrin. *Macromolecules* **2004**, *37*, 9620–9625. [CrossRef]

40. Jia, X.; Wang, X.; Tonelli, A.E.; White, J.L. Two-dimensional spin-diffusion NMR reveals differential mixing in biodegradable polymer blends. *Macromolecules* **2005**, *38*, 2775–2780. [CrossRef]

41. Uyar, T.; Rusa, C.C.; Wang, X.; Rusa, M.; Hacaloglu, J.; Tonelli, A.E. Intimate blending of binary polymer systems from their common cyclodextrin inclusion compounds. *J. Polym. Sci. Part B Polym. Phys.* **2005**, *43*, 2578–2593. [CrossRef]

42. Rusa, C.C.; Bridges, C.; Ha, S.-W.; Tonelli, A.E. Conformational Changes Induced in Bombyx m ori Silk Fibroin by Cyclodextrin Inclusion Complexation. *Macromolecules* **2005**, *38*, 5640–5646. [CrossRef]

43. Uyar, T.; Rusa, C.C.; Hunt, M.A.; Aslan, E.; Hacaloglu, J.; Tonelli, A.E. Reorganization and improvement of bulk polymers by processing with their cyclodextrin inclusion compounds. *Polymer* **2005**, *46*, 4762–4775. [CrossRef]

44. Uyar, T.; Aslan, E.; Tonelli, A.E.; Hacaloglu, J. Thermal degradation of polycarbonate, poly (vinyl acetate) and their blends. *Polym. Degrad. Stab.* **2006**, *91*, 2960–2967. [CrossRef]

45. Uyar, T.; Hunt, M.A.; Gracz, H.S.; Tonelli, A.E. Crystalline cyclodextrin inclusion compounds formed with aromatic guests: Guest-dependent stoichiometries and hydration-sensitive crystal structures. *Cryst. Growth Des.* **2006**, *6*, 1113–1119. [CrossRef]

46. Rusa, C.C.; Rusa, M.; Peet, J.; Uyar, T.; Fox, J.; Hunt, M.A.; Wang, X.; Balik, C.M.; Tonelli, A.E. The nano-threading of polymers. *J. Incl. Phenom. Macrocycl. Chem.* **2006**, *55*, 185–192. [CrossRef]

47. Pang, K.; Schmidt, B.; Kotek, R.; Tonelli, A.E. Reorganization of the chain packing between poly (ethylene isophthalate) chains via coalescence from their inclusion compound formed with γ-cyclodextrin. *J. Appl. Polym. Sci.* **2006**, *102*, 6049–6053. [CrossRef]

48. Uyar, T.; Gracz, H.S.; Rusa, M.; Shin, I.D.; El-Shafei, A.; Tonelli, A.E. Polymerization of styrene in γ-cyclodextrin channels: Lightly rotaxanated polystyrenes with altered stereosequences. *Polymer* **2006**, *47*, 6948–6955. [CrossRef]

49. Uyar, T.; Oguz, G.; Tonelli, A.E.; Hacaloglu, J. Thermal degradation processes of poly(carbonate) and poly(methyl methacrylate) in blends coalesced either from their common inclusion compound formed with γ-cyclodextrin or precipitated from their common solution. *Polym. Degrad. Stab.* **2006**, *91*, 2471–2481. [CrossRef]

50. Tonelli, A.E. *Nanofibers and Nanotechnology in Textiles*; Brown, P., Stevens, K., Eds.; Woodhead Publ. Ltd.: Cambridge, UK, 2007.

51. Vedula, J.; Tonelli, A.E. Reorganization of poly (ethylene terephthalate) structures and conformations to alter properties. *J. Polym. Sci. Part B Polym. Phys.* **2007**, *45*, 735–746. [CrossRef]

52. Uyar, T.; Rusa, C.C.; Tonelli, A.E.; Hacaloğlu, J. Pyrolysis mass spectrometry analysis of polycarbonate/poly (methyl methacrylate)/poly (vinyl acetate) ternary blends. *Polym. Degrad. Stab.* **2007**, *92*, 32–43. [CrossRef]

53. Martínez, G.; Gómez, M.A.; Villar-Rodil, S.; Garrido, L.; Tonelli, A.E.; Balik, C.M. Formation of crystalline inclusion compounds of poly (vinyl chloride) of different stereoregularity with γ-cyclodextrin. *J. Polym. Sci. Part A Polym. Chem.* **2007**, *45*, 2503–2513. [CrossRef]

54. Tonelli, A.E. Cyclodextrins as a means to nanostructure and functionalize polymers. *J. Incl. Phenom. Macrocycl. Chem.* **2008**, *60*, 197–202. [CrossRef]

55. Tonelli, A.E. Nanostructuring and functionalizing polymers with cyclodextrins. *Polymer* **2008**, *49*, 1725–1736. [CrossRef]

56. Tonelli, A.E. Organizational stabilities of bulk neat and well-mixed, blended polymer samples coalesced from their crystalline inclusion compounds formed with cyclodextrins. *J. Polym. Sci. Part B Polym. Phys.* **2009**, *47*, 1543–1553. [CrossRef]

57. Mohan, A.; Joyner, X.; Kotek, R.; Tonelli, A.E. Constrained/directed crystallization of nylon-6. I. nonstoichiometric inclusion compounds formed with cyclodextrins. *Macromolecules* **2009**, *42*, 8983–8991. [CrossRef]

58. Busche, B.J.; Tonelli, A.E.; Balik, C.M. Compatibilization of polystyrene/poly (dimethyl siloxane) solutions with star polymers containing a γ-cyclodextrin core and polystyrene arms. *Polymer* **2010**, *51*, 454–462. [CrossRef]

59. Busche, B.J.; Tonelli, A.E.; Balik, C.M. Morphology of polystyrene/poly (dimethyl siloxane) blends compatibilized with star polymers containing a γ-cyclodextrin core and polystyrene arms. *Polymer* **2010**, *51*, 1465–1471. [CrossRef]

60. Busche, B.J.; Tonelli, A.E.; Balik, C.M. Properties of polystyrene/poly (dimethyl siloxane) blends partially compatibilized with star polymers containing a γ-cyclodextrin core and polystyrene arms. *Polymer* **2010**, *51*, 6013–6020. [CrossRef]

61. Mohan, A.; Gurarslan, A.; Joyner, X.; Child, R.; Tonelli, A.E. Melt-crystallized nylon-6 nucleated by the constrained chains of its non-stoichiometric cyclodextrin inclusion compounds and the nylon-6 coalesced from them. *Polymer* **2011**, *52*, 1055–1062. [CrossRef]

62. Williamson, B.R.; Tonelli, A.E. Constrained polymer chain behavior observed in their non-stoichiometric cyclodextrin inclusion complexes. *J. Incl. Phenom. Macrocycl. Chem.* **2012**, *72*, 71–78. [CrossRef]

63. Gurarslan, A.; Tonelli, A.E. Single-component polymer composites. *Macromolecules* **2011**, *44*, 3856–3861. [CrossRef]

64. Williamson, B.R.; Krishnaswany, R.; Tonelli, A.E. Physical properties of poly (ε-caprolactone) coalesced from its α-cyclodextrin inclusion compound. *Polymer* **2011**, *52*, 4517–4527. [CrossRef]

65. Peet, J.; Rusa, C.C.; Hunt, M.A.; Tonelli, A.E.; Balik, C.M. Solid-State Complexation of Poly (Ethylene Glycol) with α-Cyclodextrin. *Macromolecules* **2005**, *38*, 537–541. [CrossRef]

66. Rusa, C.C.; Fox, J.; Tonelli, A.E. Competitive Formation of Polymer–Cyclodextrin Inclusion Compounds. *Macromolecules* **2003**, *36*, 2742–2747. [CrossRef]

67. Gurarslan, A.; Shen, J.; Tonelli, A.E. Behavior of poly (ε-caprolactone) s (PCLs) coalesced from their stoichiometric urea inclusion compounds and their use as nucleants for crystallizing PCL melts: Dependence on PCL molecular weights. *Macromolecules* **2012**, *45*, 2835–2840. [CrossRef]

68. Gurarslan, A.; Tonelli, A.E. Self-reinforced PCL/PCL composites. In Proceedings of the ACS 2012 National Meeting, Anaheim, CA, USA, 27–31 March 2011.

69. Joijode, A.S.; Hawkins, K.; Tonelli, A.E. Improving Poly (ethylene terephthalate) Through Self-nucleation. *Macromol. Mater. Eng.* **2013**, *298*, 1190–1200. [CrossRef]

70. Joijode, A.S.; Antony, G.J.; Tonelli, A.E.J. Glass-transition temperatures of nanostructured amorphous bulk polymers and their blends. *Polym. Sci. Part B Polym. Phys.* **2013**, *51*, 1041–1050. [CrossRef]

71. Williamson, B.R. Processing Polymers with Cyclodextrins. Ph.D. Thesis, North Carolina State University, Raleigh, NC, USA, 2010; p. 1332.

72. Joijode, A.S. Nano-Structuring Polymers by Processing Them with Small Molecule Hosts. Ph.D. Thesis, North Carolina State University, Raleigh, NC, USA, 2014; Chapter 3.

73. Gurarslan, A.; Caydamli, Y.; Shen, J.; Tse, S.; Yetukuri, M.; Tonelli, A.E. Coalesced poly (ε-caprolactone) fibers are stronger. *Biomacromolecules* **2015**, *16*, 890–893. [CrossRef] [PubMed]

74. Tonelli, A.E. Restructuring polymers via nanoconfinement and subsequent release. *Beilstein J. Org. Chem.* **2012**, *8*, 1318. [CrossRef] [PubMed]

75. Rastogi, S.; Lippits, D.R.; Peters, G.W.M.; Graf, R.; Yao, Y.; Spiess, H.W. Heterogeneity in polymer melts from melting of polymer crystals. *Nat. Mater.* **2005**, *4*, 635–641. [CrossRef] [PubMed]

76. Gurarslan, A.; Joijode, A.S.; Tonelli, A.E. Polymers coalesced from their cyclodextrin inclusion complexes: What can they tell us about the morphology of melt-crystallized polymers? *J. Polym. Sci. Part B Polym. Phys.* **2012**, *50*, 813–823. [CrossRef]

77. Beers, D.E.; Ramirez, J.E. Vectran high-performance fibre. *J. Text. Inst.* **1990**, *81*, 561–574. [CrossRef]

78. McLeish, T.C.B. A theory for heterogeneous states of polymer melts produced by single chain crystal melting. *Soft Matter* **2007**, *3*, 83–87. [CrossRef]

polymers

MDPI

Review

Preparation, Characterization and Application of Polysaccharide-Based Metallic Nanoparticles: A Review

Cong Wang, Xudong Gao, Zhongqin Chen, Yue Chen and Haixia Chen *

Tianjin Key Laboratory for Modern Drug Delivery & High-Efficiency, School of Pharmaceutical Science and Technology, Tianjin University, Tianjin 300072, China; 1015213012@tju.edu.cn (C.W.); xdgao@tju.edu.cn (X.G.); chenzhongqin@tju.edu.cn (Z.C.); chenyue0126@tju.edu.cn (Y.C.)
* Correspondence: chenhx@tju.edu.cn; Tel.: +86-22-2740-1483

Received: 8 November 2017; Accepted: 5 December 2017; Published: 8 December 2017

Abstract: Polysaccharides are natural biopolymers that have been recognized to be the most promising hosts for the synthesis of metallic nanoparticles (MNPs) because of their outstanding biocompatible and biodegradable properties. Polysaccharides are diverse in size and molecular chains, making them suitable for the reduction and stabilization of MNPs. Considerable research has been directed toward investigating polysaccharide-based metallic nanoparticles (PMNPs) through host–guest strategy. In this review, approaches of preparation, including top-down and bottom-up approaches, are presented and compared. Different characterization techniques such as scanning electron microscopy, transmission electron microscopy, dynamic light scattering, UV-visible spectroscopy, Fourier-transform infrared spectroscopy, X-ray diffraction and small-angle X-ray scattering are discussed in detail. Besides, the applications of PMNPs in the field of wound healing, targeted delivery, biosensing, catalysis and agents with antimicrobial, antiviral and anticancer capabilities are specifically highlighted. The controversial toxicological effects of PMNPs are also discussed. This review can provide significant insights into the utilization of polysaccharides as the hosts to synthesize MPNs and facilitate their further development in synthesis approaches, characterization techniques as well as potential applications.

Keywords: biopolymers; polysaccharides; polysaccharide-based metallic nanoparticles; host–guest strategy; preparation and characterization; application; toxicity evaluation

---

1. Introduction

Scientific research on nanotechnology has received tremendous interest in this century due to its interdisciplinary applications in the field of catalysis, biomedicine, fuel cells, magnetic data storage and energy technology [1]. Nanoparticles (NPs), with the size range from 10.0 to 100.0 nm in diameter, have unique features including high surface area, quantum property as well as adsorption and releasing properties, exhibiting great potential in multifunctional applications [2]. Among all nanoparticles, metallic nanoparticles (MNPs) are especially attractive owing to their unique properties and diverse applications [3]. It has been accepted that the size, morphology, dispersibility and physicochemical properties of MNPs are strongly associated with their applications, which are affected by the synthesized approach [4]. Thus, the investigations of searching new hosts for controlling the properties of MNPs have been one of the main objectives in MNPs research [5]. Moreover, conventional synthesis techniques of MNPs are chemicals and energy consuming, causing various risks to the environment [6]. The awareness of developing an alternatively green synthesis approach is evolving into another objective in MNPs research [7]. To these aims, an ideal scheme is to emphasize these two objectives in parallel.

Biopolymers are naturally abundant and environment friendly polymer alternatives, which are widely used in medical, agricultural and environmental industries due to their especially renewable, sustainable and nontoxic properties when compared to petroleum-based polymers [8]. According to the Food and Agriculture Organization (FAO, QC, Canada), around 35 million tons of natural fibers are harvested each year, which are a fundamental resource that can be used to produce biopolymers [9,10]. In the last decades, various biopolymers, including polysaccharides, proteins and nucleic acid, obtained from animals, plants and microbes, have been employed to packaging materials, drug delivery and regenerative medicine [11,12]. As significant types of biopolymers, natural polysaccharides in particular have some excellent properties owing to their chemical and structural diverse [13,14]. The differences ranging in charge, chain lengths, monosaccharides sequences, and stereochemistry give the highest capacity for the development of advanced functionalized materials and biomedicines [15]. Especially, the hydrophilic groups of polysaccharides can form non-covalent bonds with tissue cells and thus serve in cell–cell recognition and adhesion [16,17]. Moreover, for multi-gene, multi-step devastating disease such as cancer, diabetes and cardiovascular diseases, polysaccharides have a tendency to be more selectively to more than one specific site, eliminating the disadvantages of one-target strategy and preventing the over or under dosing of traditional delivery system [18,19]. Owing to the very safe and stable biodegradation, and biocompatibility, polysaccharides are considered as the most promising hosts in the synthesis of polysaccharide based metallic nanoparticles (PMNPs) with guest metallic ion and MNPs [20,21]. Besides, the host polysaccharides can assemble a carrier with metallic ion and hydrophobic chemical drugs such as doxorubicin, levofloxacin, cefotaxime, ceftriaxone, and ciprofloxacin for targeted drug delivery [22–24]. Polysaccharides can also act as good reducing and stabilizing agents to regulate the physical properties of PMNPs during synthesis process [25]. Over the last decades, remarkable progress has been achieved in the study of PMNPs through host–guest strategy (Figure 1) [26,27]. Therefore, it is necessary to review the recent progresses in PMNPs area.

Although there are several relevant reviews about MNPs, reviews that introduce nanoparticles from the perspective of green synthesis of PMNPs through host–guest strategy are still limited [28]. In this review, a basic introduction of the utilization of natural polysaccharides as guest molecules, conjugated with the host molecules metal ions or MNPs to form PMNPs is given. First, we provide an in-depth introduction on the two common synthesis approaches and characterization methods. Then, the applications of PMNPs in different fields are summarized and discussed in detail. Finally, the potential toxic risks are demonstrated with both in vitro and in vivo evaluation, and future perspectives are presented. Patented PMNPs are not discussed in this review.

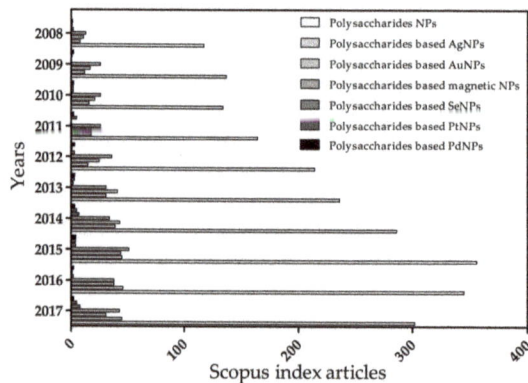

Figure 1. Scopus-indexed articles for polysaccharide based nanoparticles (NPs) and metallic nanoparticles (MNPs). (Archived until 6 November 2017).

2. Preparation of PMNPs

It is widely accepted that the size, structure and corresponding physical, chemical and biological properties of nanoparticles are largely dependent on the preparation method [29]. Therefore, the selection of synthesis approach is crucial in achieving the appropriate properties of nanoparticles [30]. Generally, nanoparticles are prepared through a variety of chemical and physical methods, which bring serious problems such as high energy consumption, use of large amount of toxic solvents, and generation of hazardous byproducts [31]. Moreover, nanoparticles synthesized from chemical approach are not suitable in biomedical applications owing to the presence of toxic capping agents [32]. Alternatively, developing green synthesis approaches that use mild reaction conditions and non-toxic reaction precursors can eliminate the drawbacks of conventional approaches [33].

Green synthesis of MNPs, using natural polysaccharides as stabilizing and reducing agents, are considered to be a promising area in nanotechnology [34]. Currently, two approaches are involved in the green synthesis of nanoparticles: top-down and bottom-up approaches (Figure 2) [35]. In top-down synthesis, the suitable starting materials are reduced in size using physical treatments such as mechanical milling, thermal ablation or chemical treatment such as chemical etching [36]. After preliminary treatment, the surface of the MNPs will be altered and the high temperature and pressure during the size reduction may induce the oxidation of the nanoparticles, which can affect their physical properties and surface chemistry [35]. Therefore, bottom-up synthesis (self-assembly) is the most frequently chosen approach in the preparation of PMNPs. In a typical bottom-up synthesis, a metallic precursor is either decomposed or reduced to zero-valent state to form the building blocks (smaller entities), followed by the nucleation and nanocrystals growth [37–39]. During this process, polysaccharides can act as the hosts to combine with guest metallic ions and MNPs through noncovalent bonding, and then the order of free energy is altered to realize the stabilization, morphological control and kinetic growth of the MNPs [40]. In addition, the stereogenic centers of polysaccharides will also benefit for the anchoring of the MNPs [1]. In contrast to the harsh reaction condition of top-down approach, the bottom-up synthesis process can occur in the bulk solution or in droplets, which is easy to regulate through controlling the process conditions [41]. Therefore, PMNPs, synthesized through bottom-up synthesis approach that involved in host–guest strategy, are relatively homogeneous compared to those synthesized through top-down approach.

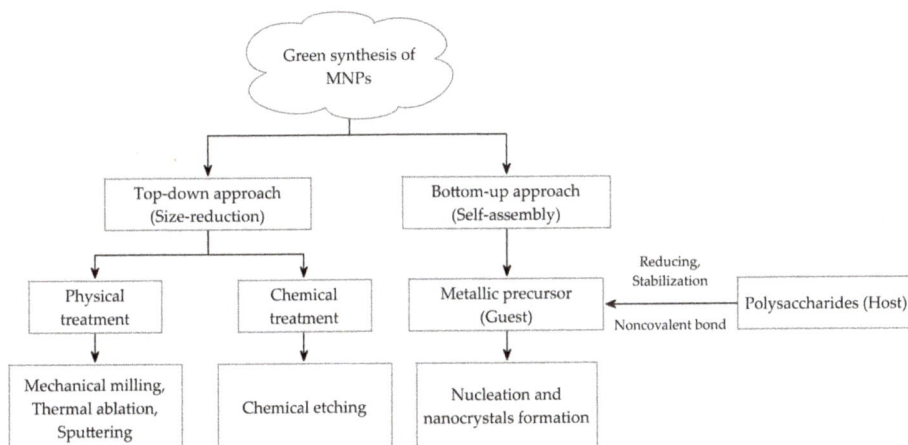

Figure 2. Overall approaches for the synthesis of PMNPs.

Although PMNPs with different properties were achieved by regulating the temperature, reaction time and different molar ratios of polysaccharides and metal ions in most studies, there are some attempts that try improving the efficiency in the preparation of PMNPs [42]. The application of microwave heating in the synthesis of sulfated chitosan coated AuNPs resulted in a lower Gibb's free energy and thus stimulated the activation of reaction [43,44]. Radiolytic reduction was also proven to be helpful in the nucleation, growth and aggregation of PMNPs during the synthesis process [45]. Owing to the different charges of metallic ions and polysaccharides in the solution, electrochemical synthetic techniques were considered to have great potential in the regulation of PMNPs synthesis. In addition, other techniques that could control the synthesis of PMNPs such as microemulsion and photoinduced reduction also need to be attempted in further study [46].

3. Characterization of PMNPs

Nanoparticles are of great scientific interests as they combine bulk materials and atomic or molecular structures together [35]. These unique properties are largely associated with their physical characteristics such as diameter, molar volume, morphology and dispersibility [31]. Usually, nanoparticles have a spherical structure [47]. As for sophisticated nanoparticles, core–shell structure and particle distribution are also attractive for specific applications [48]. Thus, many techniques have been explored to elucidate the structure and spectroscopic characters of PMNPs in the literature (Figure 3).

As the basic parameter of nanoparticles, size and morphology are firstly considered in many studies. Scanning electron microscopy (SEM) can give the information about the size, distribution and the shape of the nanoparticles [49]. However, the drying and contrasting process will alter the characteristics of nanoparticles, leading to imaging faults or artifacts [50]. Therefore, transmission electron microscopy (TEM) is introduced in the characterization of nanoparticles. It can be used to determine the particle size, dispersion and aggregation in aqueous environment with a high spatial resolution (<1.0 nm, Figure 3E) [51]. In addition, TEM can provide more details at the atomic scale such as crystal structure, which is more powerful and competitive than SEM [52]. However, many PMNPs are irregular in shape and thus hard to define. Moreover, they also tend to form large particles, and it is hard to attribute the reason to polydispersity or agglomeration [53]. In addition, it is difficult to measure the relevant size of some metallic colloids attached to drugs by TEM physiologically [49]. One technique that can solve this problem is dynamic light scattering (DLS), which can provide a measurement of particle size in solution [54]. In a DLS measurement, the nanoparticles will cause the fluctuations of the laser light intensity, which is recorded and used to determine the equivalent sphere hydrodynamic diameter of the particles [55]. DLS is also sensitive to flexible biological molecules, such as proteins and polysaccharides and thus suitable for the characterization of PMNPs [52]. Nevertheless, owing to the electron dispersion mechanism of DLS, it cannot provide information about the particle with heterogeneous size distributions [56]. Nanoparticles with similar sized distribution are also not well separated from each other in DLS measurement. Considering the diversity and ambiguity of nanoparticles, multiple techniques are commonly used to investigate nanoparticles size such as the combination of TEM and DLS [49].

UV-visible spectroscopy plays a significant role in the illustration of optical properties of PMNPs. It can monitor the quantitative formation and provide information for the size measurement of nanoparticles through different response to the electromagnetic waves, ranging from 190.0 nm to 700.0 nm [57]. The effects of concentration and pH on the stability and aggregation state of PMNPs in different time can also be recorded by UV-visible spectroscopy [35]. Fourier-transform infrared spectroscopy (FTIR) is basically applied to elucidate the functional groups from the spectrum. It can provide information about capping and stabilization of PMNPs, and therefore is utilized to demonstrate the conjugation between MNPs and polysaccharides [58]. X-ray diffraction (XRD) is the primary tool for the determination of the crystal property of PMNPs such as crystallite size and lattice strain [59].

Polymers **2017**, *9*, 689

Typically, XRD is useful for the characterization of the size and shape of crystalline PMNPs at the atomic scale. However, the requirement of single conformation and high atomic numbers of crystals limits the application of XRD [60]. In contrast to XRD, small-angle X-ray scattering (SAXS) can provide the information of the crystalline and amorphous PMNPs [51]. By analyzing the intensity of the X-ray, the size distribution, shape, orientation and structure of the nanoparticles can be well illustrated [61]. On the other hand, SAXS can give a holistic characteristic of the PMNPs, which is more effective than XRD. Nevertheless, the crystallite size of the PMNPs is not exactly the same as the particle size due to the polycrystalline aggregates of nanoparticles [62] and the lattice strain can provide detailed information about the distribution of lattice constant arising from crystal imperfections [63]. As for magnetic nanoparticles, a superconducting quantum interference device (SQUID) magnetometer is introduced to investigate their magnetization property, offering great insights into their application in specific field [64].

In addition to the above techniques, the number of novel methods for the characterization of PMNPs, such as scanning tunneling microscopy (STM), atomic-force microscopy (AFM), Raman scattering and electron spin resonance (ESR) spectroscopy, is growing rapidly, providing more evidence for their applications [35].

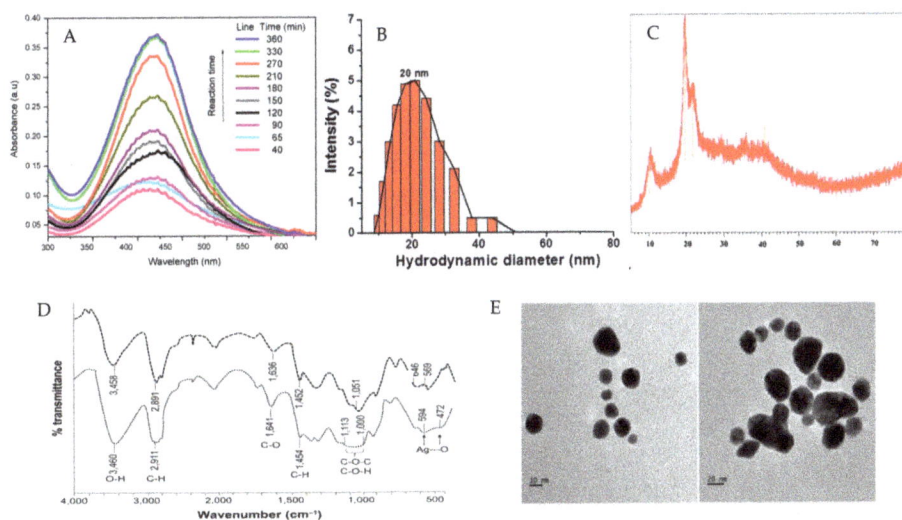

Figure 3. Overall scheme for the illustration of different methods to characterize PMNPs: (**A**) UV-vis spectra of AgNPs; (**B**) particle size distribution of AgNPs analyzed by DLS; (**C**) X-ray diffraction spectra of ZnSNPs; (**D**) FTIR spectra of AgNPs; and (**E**) TEM images of AuNPs. Reproduced with permission [23,43,65–67].

4. Application of PMNPs

Polysaccharide-based metallic nanoparticles have been extensively used in numerous technological fields owing to their remarkable physical, chemical and biological properties. Herein, the applications of PMNPs are introduced, specifically in wound healing, targeted delivery, biosensing, catalysis and agents with antimicrobial, antiviral and anticancer capabilities.

4.1. Antimicrobial and Antiviral Property of PMNPs

Microbial infections are responsible for most common clinical diseases worldwide, bringing big threats to human public health [68]. Currently, the antimicrobial agents in the market are quaternary ammonium salt, metal salt solutions and antibiotics [33]. Unfortunately, the poor effectiveness and overuse of these agents have led to growing drug resistance of pathogenic bacterial and fungi strains, especially multidrug resistance strains [69]. Infections caused by these strains are more difficult to cure and prevent. Therefore, it is urgent to find novel antimicrobial agents with low toxicity and high efficiency or alternative therapies to solve these problems. Among all the candidates used for the treatment of bacterial infections, MNPs, especially AgNPs, have drawn much attention due to their small size, large surface to volume ratio and tunable plasmon resonance characteristics [70,71]. Since the 1920s, AgNPs were officially approved by the FDA administration to be used in wound therapy as an antibacterial agent; the exploration of MNPs in antibacterial field has increased rapidly [72].

Nowadays, numerous kinds of PMNPs have been synthesized and demonstrated to have significant antimicrobial potential (Table 1). Various methods had been applied to evaluate the antimicrobial activity of PMNPs (Figure 4). Among the methods, agar well diffusion method is the most widely used one to assess the antimicrobial activities of PMNPs against different bacteria and fungi due to its easy, quick and intuitive properties (Figure 4A). Results showed that AgNPs stabilized by different polysaccharides have effective antimicrobial effects on both Gram-positive bacteria, such as *S. epidermidis*, *S. aureus*, *S. lutea*, *B. subtilis*, *L. fermentum*, *E. faecium*, *B. licheniformis*, *B. cereus*, *K. rhizophila*, *S. pyogenes*, *Actinomycetes*, *Staphylococcus*, *S. pneumoniae*, *L. monocytogenes*, and *E. faecalis*, and Gram-negative bacteria, such as *E. coli*, *P. aeruginosa*, *K. planticola*, *K. pneumoniae*, *V. parahaemolyticus*, *P. vulgaris*, *S. typhimurium*, *A. hydrophila*, and *V. cholerae*. In addition, they also exhibited an extensively antifungal activity against *C. albicans*, *F. oxysporum*, *A. niger*, *T. rubrum*, *C. krusei*, and *A. flavus*. It appeared that different kinds of strains had different sensitivities toward polysaccharide-based AgNPs. Gram-negative bacteria were more likely to be affected due to their membrane compositions and the negatively charged cell wall, which made it easier to attach the released Ag^+, which resulted in cell death [69]. Gram-positive bacteria were less susceptible to Ag^+ compared to Gram-negative bacteria [33]. Confocal laser scanning microscopy (CLSM) was another method that used to assess the antimicrobial activities of polysaccharide-based AgNPs by measuring the fluorescence intensity of cells (Figure 4B) [73]. Besides, light microscopy could show the reduced biofilm on the glass surface and SEM could observe the changes in the shape and appearance of bacteria cells, providing more straightforward visualization choices (Figure 4C,E) [44,74]. It has been reported that the binding of Ag^+ to oligonucleotides would cause changes in the fluorescence excitation and emission spectrum [72]. Thus, 3D fluorescence spectroscopy was introduced to investigate the interaction between Ag^+ and DNA of bacteria (Figure 4D). Furthermore, minimum inhibitory concentration (MIC) is widely applied to evaluate the antimicrobial activity of PMNPs against bacteria and fungi that cultured in liquid medium [23].

Regardless of the fact that most reported PMNPs with antimicrobial activity were AgNPs, there are other PMNPs that exhibited great antimicrobial potential. AuNPs stabilized by bacterial exopolysaccharides and sulfate chitosan exhibited excellent antimicrobial properties, as assessed by the approaches of MIC, agar plate and SEM [25,75,76]. Polysaccharide-based SeNPs showed effective inhibitory activity on the growth of both bacteria and fungi in the agar plate [39]. In addition, starch copped CuNPs were also demonstrated with antimicrobial capability [77]. Nevertheless, the mechanisms underlying the antimicrobial effects of PMNPs remained unclear. Other proposed mechanisms such as interaction with enzymes and DNA, free radical production should be given more concern [33].

Polymers **2017**, *9*, 689

Table 1. Summary of literature data regarding the antimicrobial property of PMNPs.

Resource	Polysaccharides	Metals	Diameter (nm)	Shape	Antimicrobial Strains	References
Lactobacillus plantarum	Exopolysaccharides	Au	10.0–20.0	Spherical/ellipsoidal	*E. coli, S. aureus, K. pneumoniae*	[25]
Pleurotus tuber-regium	Polysaccharides-protein complexes	Se	122.0	-	*Staphylococcus, T. rubrum*	[39]
-	Hydroxypropylcellulose	Ag	25.0–55.0	Spherical	*E. coli, B. subtilis, S. aureus, P. aeruginosa, S. epidermidis, A. niger, Actinomycetes*	[43]
-	6-O-chitosan sulfate	Au	15.0	Spherical	*E. coli*	[44]
Tamarind	Carboxymethyl polysaccharides	Ag	20.0–40.0	Spherical/polygonal	*E. coli, B. subtilis, S. typhimurium*	[54]
-	Agarose/dextran/gelatin	Fe_2O_3	10.0	Dumbbell shape	*S. aureus, A. hydrophila, S. pyogenes, P. aeruginosa*	[64]
-	Guar gum	Ag	16.0	Spherical	*B. subtilis*	[65]
-	Chitosan-g-poly(acrylamide)	ZnS	19.0–26.0	Triangular	*E. coli*	[66]
Astragalus membranaceus root	Crude polysaccharides	Ag	65.1	Spherical	*S. aureus, E. coli, S. epidermidis, P. aeruginosa*	[69]
-	Pullulan	Ag	2.0–30.0	Spherical/oval-shaped	*E. coli, K. pneumoniae, L. monocytogenes, P. aeruginosa, Aspergillus spp., Penicillum spp.*	[70]
-	Pectin	Ag	5.4–10.6	Spherical	*E. coli, S. epidermidis*	[73]
-	Chitosan	Ag/ZnO	10.0–65.0	Spherical/Uneven distribution	*E. coli, P. aeruginosa, L. fermentum, E. faecium, S. aureus, B. lichenjformis, B. subtilis, B. cereus, V. parahaemolyticus, P. vulgaris*	[23,74]
Bacillus megaterium	Exopolysaccharides	Au	5.0–20.0	Spherical	*E. coli, B. cereus, S. aureus, S. epidermidis, K. pneumoniae, S. typhi, P. arruginosa, V. cholerae, S. pneumoniae*	[75]
Seaweed *Chondracanthuschamissoi, LessoniaSpicata, Ulvasp*	Polysaccharides	Ag/Au	10.0/25.0	Spherical	*P. aeruginosa, S. typhimurium*	[76]
-	Starch	$Cu(NO_3)_2$	5.0–12.0	Spherical	*E. coli, S. aureus, Salmonella typhi*	[77]
Padina tetrastromatica	Fucoidan	Ag	17.0	Spherical	*B. subtilis, Bacillus sp. K. planticola, K. pneumoniae, S. nematodiphila, Streptococcus sp.*	[78]

Table 1. *Cont.*

Resource	Polysaccharides	Metals	Diameter (nm)	Shape	Antimicrobial Strains	References
-	β-glucan	Ag	15.0	-	*E. coli, Methylobacterium* spp., *Sphingomonas* spp.	[79]
Arthrobacter sp. B4	Exopolysaccharides	Ag	9.0–72.0	Face-centred-cubic	*P. aeruginosa, S. aureus, C. albicans, F. oxysporum*	[80]
Cordyceps sinensis (Berk.)	Exopolysaccharides	Ag	50.0	Spherical	*E. coli, S. aureus*	[81]
-	Xanthan gum/chitosan	Ag	5.0–20.0	Spherical	*E. coli, S. aureus*	[22,82,83]
-	Chitosan-carboxymethyl cellulose	Ag	5.0–20.0	Irregular shape	*E. coli, S. aureus, P. aeruginosa*	[84]
Bradyrhizobium japonicum 36	Exopolysaccharides	Ag	5.0–50.0	Rod/oval-shaped structures	*E. coli, S. aureus*	[85]
Klebsiella oxytoca	Exopolysaccharides	Ag	6.0–16.0	Spherical	*E. coli, K. rhizophila*	[86]
Lentinus squarrosulus (Mont.)	Hetero polysaccharides	Ag	1.3–4.5	Spherical	*E. coli*	[87]
Pleurotus florida	Glucan	Ag	1.3–2.5	Spherical	*K. pneumoniae*	[88]
Lactic acid bacterium	Exopolysaccharides	Ag	2.0–15.0	Spherical/triangular	*E. coli, K. pneumoniae, L. monocytogenes, P. aeruginosa*	[89]
-	Dextran/sucrose	Fe	5.8/7.3	Spherical	*E. coli, P. aeruginosa, E. faecalis, C. krusei*	[90]
-	Mesoporous starch	Ag	5.0–25.0	Spherical	*E. coli, S. aureus*	[91]
Anogeissus latifolia	Gum ghatti	Ag	5.5–5.9	Uneven shape	*E. coli, S. aureus, P. aeruginosa*	[92]
Marine macro algae (*U. faciata, P. capillacae, J. rubins, C. sinusa*)	Polysaccharides	Ag	7.0–20.0	Spherical	*E. coli, S. aureus*	[93]
Bacillus subtilis	Exopolysaccharides	Ag	1.1–6.7	Spherical	*S. aureus, P. aeruginosa*	[94]
Cochlospermum gossypium	Gum kondagogu	Ag	18.9–55.0	Spherical	*E. coli, S. aureus, P. aeruginosa*	[95]
Porphyra vietnamensis	Sulfated polysaccharides	Ag	10.0–16.0	Spherical	*E. coli, S. aureus*	[96]
Portulaca	Arabinogalactan	Ag	20.0–35.0	Spherical	*C. albicans, S. cerevisiae, A. niger, A. flavus*	[97]

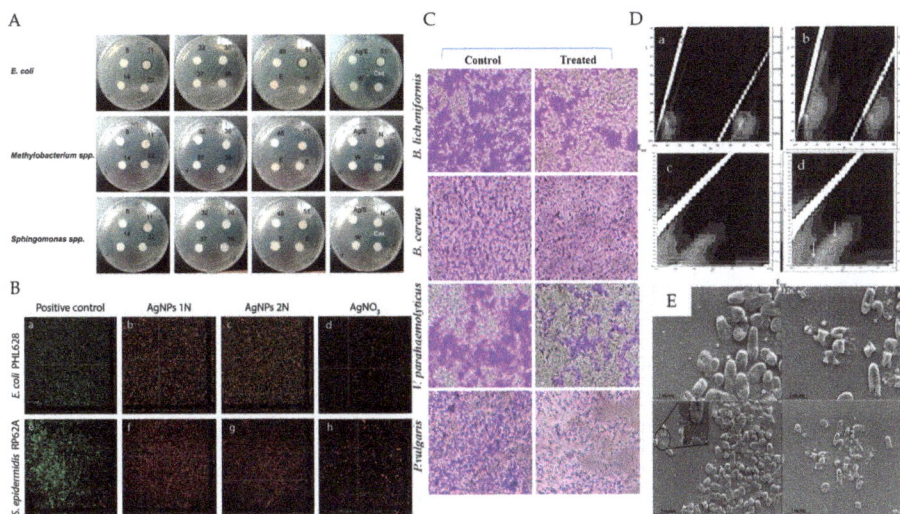

Figure 4. Overall scheme for the illustration of different methods to evaluate the antimicrobial activity of PMNPs: (**A**) antimicrobial tests on agar plate; (**B**) CLSM images of bacterial treated with p-AgNPs; (**C**) light microscopy images of bacterial treated with CS/Ag/ZnO nanocomposite; (**D**) fluorescent fingerprints of DNA of *E. coli* in the presence of exopolysaccharides based AgNPs; and (**E**) SEM images of bacterial treated with S-Chi@Au. Reproduced with permission [44,73,74,79,96].

In the last decades, viruses have become serious concerns due to their dramatic breakout worldwide, including the human immunodeficiency virus (HIV), influenza virus, Zika virus and severe acute respiratory syndrome virus (SARS). The infection of virus has brought unprecedented crisis for public health, leading to a staggering societal cost [98]. Unlike the pathogenic microorganisms, viruses are infectious agents that replicate only in living cells, which can be eliminated by antibodies produced by the immune system instead of antibiotics [99]. The viral diversity and rapid mutability properties make the diagnosis and prevention of viruses especially difficult in the clinic [100]. PMNPs have posed various unique advantages, which make them attractive in solving the virus problems. It had been reported that tailored glycan functionalized AuNPs (13.0 nm) could bind with the influenza viruses (vieH5N1, shaH1N1 and H7N9) and form an aggregation on the surface of virus, resulting in a discriminable plasmon band shift and color change subsequently [100,101]. The polysaccharide-capped AgNPs (15.0 nm) were demonstrated to absorb and quench dye-labeled single-stranded DNA, which could be utilized as an effective fluorescence sensing platform for human immunodeficiency virus (HIV) [102]. In addition, polysaccharide-coated AgNPs (10.0 nm, 25.0 nm) were also illustrated to inhibit the viral replication and reduce the progeny production of Tacaribe virus and Monkeypox virus [103,104]. Nevertheless, the antiviral mechanism of polysaccharide-based MNPs is still unclear [105]. It is critical to understand the precise interactions between nanoparticles and virus, especially their influences on the binding of virus to host receptors, in further research.

4.2. Anticancer Property of PMNPs

Cancer is a dominant factor of morbidity and mortality globally [106]. It has been estimated that about 90.5 million people are diagnosed with cancer, and the tendency is continually rising despite the preventive measures and therapeutic efforts in the past decades [107]. Recently, PMNPs, used in nanomedicine, provide an alternative opportunity in the prevention, diagnosis and treatment of cancer. Its sophisticated targeting strategies and multi-functional characters have remarkable advantages in the improvement of pharmacokinetics and pharmacodynamics profiles compared to conventional therapeutics [108]. Therefore, the application of PMNPs in cancer treatment as nanomedicine has emerged as a fruitful area in nanotechnology and extensive investigations have addressed this issue (Table 2).

3-(4,5-dimethylthiazol-2-yl)-2,5-diphenyltetrazolium bromide (MTT) assay is the most commonly used method to assess the effects of different toxicants on the viability of cells [109]. Pectin mediated AuPNs showed an obvious cytotoxicity on human breast adenocarcinoma cells MCF-7 and MDA-MB-231 with the concentration of 10 µg/mL via MTT reduction capability evaluation. Meanwhile, DNA of the cancer cells was also significantly fragmented indicating potential anticancer activity [110]. The noticeable alterations in the cellular shape and morphology were recognized as another hallmark of apoptosis of cancer cells [111]. It has been reported that chitosan based AgNPs (20.0 nm) reduced the viability of human lung adenocarcinoma cells A549/Lu, human hepatocellular carcinoma cells HepG2, human epidermic carcinoma cells KB and human breast carcinoma cells MCF-7 by reducing cell density and inducing the shrinkage and blebbing of the cells [23,112]. SeNPs have received additional attention in the prevention of cancers. Accumulated evidence illustrated that they could stimulate the immune system, modulate thioredoxin reductase activity, maintain cell redox balance and induce apoptosis of cancer cells [113]. D-glucose based SeNPs have been proven to exert anticancer activity against HepG2 cells at the cell cycle of S phase (Figure 5) [114]. In addition, absorption of polysaccharide-based SeNPs could also result in reactive oxygen species (ROS)-induced apoptosis of cancer cells through mitochondria mediated pathway [111].

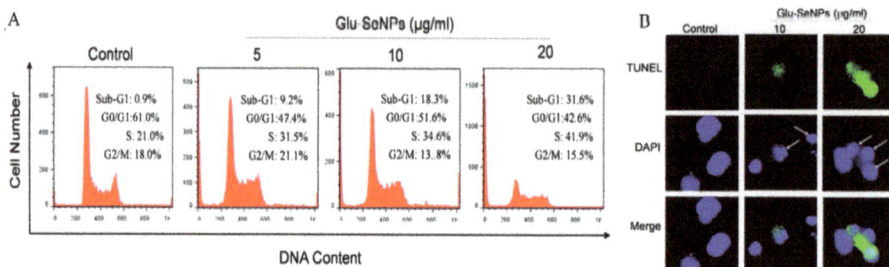

Figure 5. Apoptosis induced by Glu-SeNPs in cancer cells: (**A**) flow cytometric analysis of cancers cells; and (**B**) DNA fragmentation and nuclear condensation. Reproduced with permission [114].

Caspases-3 activation is another factor that involved in the execution phase of apoptosis [115]. Results showed that polysaccharide-based SeNPs and iron oxide NPs could significantly increase the activity of caspases-3 in a dose-dependent manner, resulting in the death of cancer cells [116,117]. Despite these results in the prevention of cancer cells, polysaccharide-based AuNPs were green synthesized to sufficiently eradicate tumors by means of heat and photothermal therapy [118,119]. In addition, the inhibition of *Tamarindus indica* polysaccharide-based AuNPs on tumor growth was also achieved by stimulating the proliferation of lymphocyte cells in vivo [111]. Therefore, PMNPs were good therapeutical agents to assist the cancer treatment.

Table 2. Summary of literature data regarding the anticancer property of PMNPs.

Resource	Polysaccharides	Metals	Diameter (nm)	Shape	Cancer types	References
Tamarindus indica	Galactoxyloglucan polysaccharides	Au	20.0	Spherical	Murine cancer cells (DLA, EAC)	[109]
Musa paradisiaca/Ganoderma lucidum	Pectin	Au	8.0	Spherical	Human breast adenocarcinoma cells (MCF-7/MDA-MB-231)	[110]
Tamarindus indica	Polysaccharides PST001	Au	15.0–20.0	Circular	Breast cancer cells (MCF7), Leukemia cells (K562)	[111]
-	Fucoidan-mimetic glycopolymer	Au	20.0–55.0	Spherical	Human colon cancer cells (HCT116)	[112]
Sargassum muticum	Aqueous extract	Fe₃O₄	-	-	HepG2, MCF-7, HeLa, Jurkat	[116]
Polyporus rhinocerus	Polysaccharide–protein complexes	Se	50.0	Spherical	Human lung adenocarcinoma cells (A549)	[117]
Halomonas maura	Sulfated exopolysaccharides	Au	70.0–107.0	Quasi-spherical	Breast cancer cells (MCF7) Glioma cells (GI-1)	[118]
-	Gum arabic	Au	0.9–2.3	Spherical	Human breast adenocarcinoma cells (MDA-MB-231)	[119]
Leuconostoc spp.	Dextran	Au	49.0	Spherical	Ehrlich ascites carcinoma (in vivo)	[120]
-	Chitosan	Ag	5.0–15.0	Spherical	A549, HepG2, Lu, KB, MCF-7	[23,121]
Chlorella vulgaris LARG-3	Polysaccharides	Pt	18.0–38.0	Quasi-spherical	Ovarian cancer A2780	[122]
Lentinus edodes	Lentinan	Se	28.0	Spherical	Human cervix carcinoma cells (HeLa)	[123]
-	Hyaluronic acid	Se	66.8	Spherical	Heps solid tumor (in vivo)	[124]
Undaria pinnatifida	Polysaccharides	Se	59.0	Spherical	Human melanoma cells (A375)	[125]
Spirulina	Polysaccharides	Se	20.0–50.0	Spherical	Human melanoma cells (A375)	[126]
Pleurotus tuber-regium	Polysaccharide–protein complexes	Se	44.0–220.0	Spherical	Human breast carcinoma (MCF-7)	[127]

4.3. Wound Healing Property of PMNPs

Wound healing is a complex physiological process that follows three overlapping phases including inflammation and tissue remodeling, the rate of which is affected by the type, size and depth of the wound, and especially the presence of bacterial infection [128]. Consequently, conditions that are unfavorable for the colonization of pathogenic bacteria and fungus or helpful for the host repair mechanisms are required to facilitate wound healing progress [129]. Traditional dressings (gauze and tulle) achieve their capability of healing wound by forming a barrier from the external microorganisms and maintain the dry environment of the wound instead of influencing the healing process directly [130]. PMNPs had been extensively proved with excellent antimicrobial activity, providing great potential in the healing of wounds.

Nowadays, several PMNPs had been reported to be suitable for wound management. Pectin (8.0 nm) and chitosan (40.5 nm) coated AgNPs exerted great antibacterial activity on *E. coli* (Gram-negative) and *S. epidermidis* (Gram-positive) both on planctonic and biofilm formation conditions despite the low free Ag$^+$ concentration [73,129]. It also exhibited cytocompatibility and capability in promoting fibroblasts proliferation through cytokine regulation [131]. Mannan sulfate AgNPs (20.0 nm) were reported to exhibit an enhanced cytocompatibility and promotion in the cellular uptake of murine macrophages, human skin fibroblasts and human keratinocytes. The excision and incision wound models in vivo also showed its repairs in wound area, wound contraction and epithelization period, suggesting promising potential for site-specific wound therapy [132]. In addition, xanthan based film incorporated with AgNPs could stimulate angiogenesis, which was the key factor of granulation tissue regeneration [133]. Nevertheless, more metallic nanoparticles preferred to use hydrogel as their reducing agents, which could be easily removed from the wound site, avoiding further trauma and facilitating re-epithelialization [134]. Usually, hydrogel mediated AgNPs revealed their wound healing ability through inhibition towards microorganisms [22,135,136]. Glucuronoxylan mediated AgNPs (9.3 nm) were proven to promote collagen content, which could stimulate the following re-epithelialization and granulation tissue formation (Figure 6) [137]. Recently, a novel bilayer composite was introduced in the wound dressing, which was more efficient than traditional single layer film. The upper layer, composed of chitosan based AgNPs, was designed to prevent bacterial penetration and ensured the permeability of oxygen to the wound site, while the lower layer, composed of cross-linked chitosan, was designed to improve cell proliferation activity [138]. This bilayer strategy had combined the advantages of both layers, which was an ideal strategy for the investigations of MNPs subsequently.

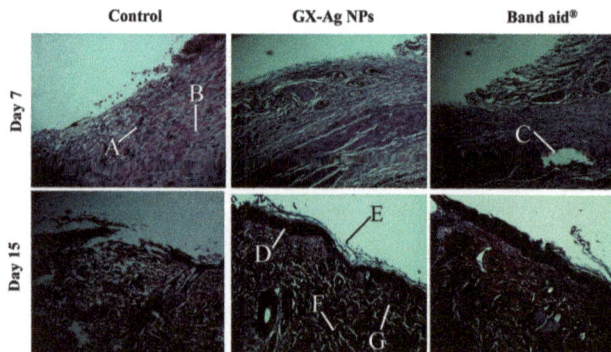

Figure 6. Histological examination in wound healing of glucuronoxylan-mediated AgNPs on Day 7, and 15 by Hematoxylin & eosin (H&E) staining: (**A**) neutrophils accumulation; (**B**) collagen deposition; (**C**) hair follicle; (**D**) epidermis; (**E**) stratum corneum; (**F**) dermis; and (**G**) fibroblasts. Reproduced with permission [137].

4.4. PMNPs in Targeted Delivery

Conventional chemotherapeutical agents have encountered various problems such as high toxicity, large volume of distribution, short lifetime in the body and low solubility, which led to a narrow therapeutical index [107]. Thus, it is critical to develop a directed therapy approaches and increase the efficiency of chemical agents. Nanoparticles represent unique physicochemical characters, which are small enough to traverse most biological membranes and avoid the uptake of the reticuloendothelial system [139,140]. These properties can be easily engineered with regard to the desired gene and drug delivery capacity and controlled release [141]. Over the years, PMNPs exhibited great potential in the targeted drug delivery applications and a summary of the current developments are shown in Table 3.

One common targeted delivery application of PMNPs was to transport the anticancer drugs to the specific sites of cancer cells. It can be observed that the metallic nanoparticles that correspond to anticancer drug delivery are AuNPs, which exhibited a pH dependent release behavior, resulted in a higher release efficiency at pH 5.7 than pH 7.4 [142,143]. Generally, the endosomal pH of cancer cells is acidic, while the endosomal pH of normal cells is neutral [144]. This property will ultimately improve the cytotoxic activity of drugs against cancer cells and reduce their toxicity on normal cells, which is are desirable characteristics for cancer targeted drug delivery [145]. Besides, polysaccharide-based AuNPs revealed desirable optical, electrical and chemical properties, which enabled to display high-amplitude photoacoustic signals in the cancer cells, suggesting promising potential in photoacoustic image-guided drug release and synergistic chemo-photoacoustic therapy [142,146]. Much evidence suggests that there are many receptors on the cell surface, which are responsible for the interaction between PMNPs and the microenvironment around the cells [147]. Based on this knowledge, hyaluronic acid supported AuNPs and *Gracilaria lemaneiformis* polysaccharide-based SeNPs achieved their anticancer effects by recognizing the receptors of CD44 and $\alpha_v\beta_3$ integrin in specific cancer cells, respectively [147,148], suggesting a new strategy for the design of targeted drug delivery. Another promising site-targeted type of MNPs was magnetic nanoparticles, which could be directed at the specific tissues by means of an external magnetic field and its significance in the anticancer drug delivery had been well discussed in the review [107].

Normally, biological therapeutical agents were susceptible to be hydrolyzed and digested in the gastrointestinal process and their low membrane permeability restricted the potential bioavailability in vivo [149]. Therefore, the possibilities of PMNPs as novel carriers for the delivery of biological agents were exploited. Recently, chitosan reduced AuNPs were demonstrated to enhance the transmucosal delivery of insulin and improve the pharmacodynamic activity [150]. PMNPs also exhibited a safe delivery of DNA in vivo, indicating great potential in gene therapy [151]. Nevertheless, the applications of PMNPs in the delivery of biological agents were still limited and more attempts should be given in this area.

Polymers **2017**, *9*, 689

Table 3. Summary of literature data regarding the targeted delivery property of PMNPs.

Resource	Polysaccharides	Metals	Diameter (nm)	Shape	Targeted delivery	References
Sphingomonas elodea	Gellan gum	Au	12.0–14.0	Spherical	Doxorubicin hydrochloride delivery	[24]
Lactobacillus plantarum	Exopolysaccharides	Au	20.0–30.0	Spherical/ellipsoidal	Levofloxacin, cefotaxime, ceftriaxone, ciprofloxacin delivery	[25]
-	Mannan sulfate	Ag	17.0–23.0	Spherical	Targeting in cellular uptake (J774A.1, TE 353.Sk and HaCaT cells)	[132]
Fucus vesiculosus	Fucoidan	Au	73.0–96.0	Spherical	Doxorubicin delivery	[142]
-	Chitosan-oligosaccharide	Au	58.8–64.8	Spherical	Paclitaxel delivery	[143]
-	β-cyclodextrin-hyaluronic acid	Au	2.2	Spherical	Doxorubicin hydrochloride, paclitaxel, topotecan hydrochloride, camptothecin, irinotecan hydrochloride delivery	[144]
-	Poly(acryl-amidoglycolic acid-co-vinylcaprolactam)-pectin	Ag	50.0–100.0	Spherical	5-fluorouracil delivery	[146]
-	Hyaluronic acid	Au	50.8–56.0	-	Binding with receptor CD44	[147]
Gracilaria lemaneiformis	Polysaccharides	Se	50.0	Near-spherical	$\alpha_v\beta_3$ integrin receptor mediated endocytosis	[148]
-	Chitosan	Au	10.0–50.0	-	Insulin delivery; bioadhesive and intestinal barrier bypass characters	[150]
Gynostemma pentaphyllum Makino	Folate-conjugated sulfated polysaccharides	Au	4.0–6.0	Spherical	Camptothecin delivery	[152]
Musa paradisiaca	Gal-Glc-[Gal-]GlcNAc	Au	1.7–1.9	Spherical	Polysaccharides of Targeting in *Streptococcus pneumoniae* type 14	[153]
-	β-cyclodextrin/chitosan	Fe	8.4–16.3	Spherical	Prodigiosin delivery	[154]
-	Gum saraya	Au	20.0–25.0	Spherical	Gemcitabine hydrochloride delivery	[155]
-	Dextran–lysozyme	Au	2.5–15.8	Spherical	Doxorubicin delivery	[156]
Saccharomyces cerevisiae	Mannan	Fe_3O_4	21.2–48.1	Ellipsoidal	Targeting in antigen-presenting cells/macrophage	[157,158]
-	Starch	Ag	11.5–19.3	Spherical	Targeting in mitochondrial membrane	[159]
-	k-carrageenan	Fe_3O_4	4.0	Spherical	Methotrexate	[160]

4.5. PMNPs for Biosensing

A biosensor is an analytical device capable of converting a biological event into a physico-chemical signal, which is highly specific and efficient in a low detection limit for the analysis [161]. Nowadays, various biosensors for the detection of chemicals, metal ions, proteins and gas had been reported. PMNPs, had been intensively applied in biosensing field owing to their outstanding optical, electronic and chemical properties [162]. A summary of the current developments in biosensing and probes are shown in Table 4.

Among all the biosensors, optical sensors are the most promising candidates as they are sensitive, flexible and convenient [163]. The basic principles for optic sensors are based on the determination of colorimetric absorbance, reflectance, luminescence, refractive index and light scattering changes [164]. On this basis, some PMNPs had been developed as biosensors and probes. It had been demonstrated that alginate and dextrin based AgNPs showed a concentration dependent changes in the absorbance of manganese (II) and copper (II), separately, suggesting a good application in the detection of metal ions [165,166]. Conventional methods of detection of melamine are complex and time-consuming. Chitosan stabilized AuNPs could lead to a color change in the presence of melamine with the limited concentration of 6×10^{-6} g/L, providing an alternative way for onsite detection of melamine [167]. Ammonia is a toxic pollutant that threatens the health of human [168]. In this regard, ammonia sensors that could detect the ambient ammonia concentrations before damage occurs are necessary. Polysaccharide-based AuNPs and AgNPs had excellent performances in the sensing of ammonia at room temperature with the detection limit of 1 ppb [169,170] and guar gum based AuNPs could present a wider detection range of ammonia from 0.1 parts-per-quadrillion (ppq) to 75,000 parts-per-million (ppm) due to the variations in electrical conductivity [171]. Another toxic pollutant ia hydrogen peroxide, which could also be successfully monitored by optic H_2O_2 sensor (polysaccharide-based AgNPs) in the concentration range of 0.001 to 10 mM [163]. Tin oxide (SnO_2), commonly known as semiconductor material, exhibited fast response, high sensitivity, low power consumption, mass-produced potency and wide operation working temperature range characteristics, which were especially suitable for the practical applications in gas sensors [172]. Therefore, polysaccharide-based Au-doped SnO_2 nanoparticles have been bio-green synthesized and their high sensitivity in the sensing of NO_2 and ethanol vapor make them possible agents for the monitoring of harmful gas [173,174]. In addition, MNPs coated with dextran provided possibilities in the measurement of amino acids and proteins such as cysteine and insulin with high sensitivity [175,176] and other applications in enzyme activity determination and large scale screening were also observed [177,178]. In some situations, fluorescent dyes could be introduced in the synthesis of PMNPs to improve their efficiency and this strategy could be an option for the future development of biosensors [179].

Table 4. Summary of literature data regarding the polysaccharide-based MNPs for biosensing.

Resource	Polysaccharides	Metals	Diameter (nm)	Shape	Biosensing applications	References
Ceratonia siliqua	Locust bean gum	Au-SnO$_2$/Ag	16.0–28.0	Spherical	Ethanol vapor sensing/hydrogen peroxide sensing	[163,164,174]
-	Alginate	Ag	10.0–20.0	Spherical	Detection of manganese (II) ions	[165]
-	Dextrin	Ag	15.0–28.0	Spherical	Detection of copper (II) ions	[166]
-	Chitosan	Ag/Au	7.3–8.8	Spherical	Detection of aromatic *ortho*-trihydroxy phenols/hydrogen sulfide/melamine	[167,180,181]
-	Guar gum	Au/Pd/Ag	6.0–10.0	Spherical	Sensor for the detection of ammonia level/electrocatalytic hydrazine	[169,171,182]
Cyamopsis tetragonoloba	Polysaccharides	Au	6.5	Spherical	Sensor for the detection of ammonia	[170]
C. arietinum L.	Water extracts	Au-SnO$_2$	25.0	Spherical	Sensor for the detection of NO$_2$	[173]
Leuconostoc mesenteroides T3	Dextran	Ag/Au	9.9–13.9	Spherical	Sensor for the detection of cysteine/insulin	[175,176]
-	Cellobiose	Au	10.7–33.5	-	Measurement of cellobiase activity	[177]
-	Hyaluronic acid	Au	14.0–19.0	Spherical	Hyaluronidase inhibitor screening	[178]
Bagasse	Xylan	Ag	20.0–35.0	Spherical	Detection of Hg^{2+}	[183]
-	β-cyclodextrin-dextran-*g*-stearic acid	Fe$_3$O$_4$	59.0–149.0	Micelles	Magnetic resonance imaging for monitoring cancer cells	[184]

4.6. PMNPs in Catalytic Application

Catalysis is an important process that associated with the chemical transformations, which is crucial for the development of modern industry [185]. As the key factors in catalysis, catalysts participate in the chemical reaction in a specific path without itself being consumed and they decide the rates of the reaction [1]. Ideally, catalysts can convert a large quantity of reactants, while consume less materials under mild reactive conditions [186]. However, traditional catalysts usually produce unwanted byproducts which have significant impacts on the environment [187]. In this regard, homogeneous catalysts that exhibited high selectivity had received much more attention. Among all the nanocatalysts, PMNPs are especially attractive due to their high surface area to volume ratios and high surface energy, which allowed their catalytic sites to be accessible [188]. Since catalytic reactions required transition metals, polysaccharides could provide a suitable functional support for dispersing the noble MNPs as hosts and make the size and shape of catalysts more controllable [1]. Thus, various studies had highlighted the possibilities of PMNPs in catalytic application (Table 5) and the reaction mechanisms were shown in Figure 7.

4-nitrophenol (4-NP) is a phenolic compound that widely exists in the wastewater [189]. It can cause serious damages to the central nervous system, kidney and liver in both animals and human beings [190]. Thus, various methods had been developed to reduce 4-NP to 4-aminophenol (4-AP), which was a common intermediate in the manufacture of antipyretic and analgesic drugs that friendly to the environment [191]. Commonly, the reduction of 4-NP to 4-AP was achieved in the presence of appropriate reducing agent named $NaBH_4$ and the reaction could be monitored by UV-visible spectroscopy [192]. Thus, this model evaluation system was used to determine the catalytic activity of PMNPs. Nowadays, AgNPs and AuNPs stabilized by different kinds of polysaccharides were the most effective nano catalysts that reported to involve the conversion of 4-NP. *Cordyceps sinensis* exopolysaccharide-based AgNPs (5.0 nm) showed a good capability in the reduction of 4-NP with the activity factor of 15.75 $s^{-1} \cdot g^{-1}$ [193]. AgNPs supported on xanthan gum (5.0–20.0 nm) could serve as a good catalyst for the reduction, and the large specific surface area of nanoparticles is favorable for the elevation of the efficiency [83]. *Portulaca arabinogalactan* stabilized AgNPs could facilitate electron transmission from BH_4^- to 4-NP thereby stimulating the reaction [97]. AuNPs stabilized by Locust bean gum, glucomannan and katira gum also revealed excellent catalytic ability in the reduction of 4-NP to 4-AP with their first-order rate constants were 14.46 $\times$ 10^{-2} min^{-1}, 6.03 $\times$ 10^{-3} s^{-1} and 2.67 $\times$ 10^{-2} min^{-1}, respectively [174,194,195]. Moreover, biometallic Ag-Au nanoparticles capped by hydroxyethyl starch-*g*-poly (11.1 nm) were synthesized and exhibited great recyclability (98–93%) after 4 cycles, providing an enhanced efficiency in the 4-NP reduction reaction [196].

Suzuki–Miyaura reaction, involving the C–C bond formation, is widely used for the synthesis of many carbon molecules, which is one of the most significant cross-coupling reaction [197]. The PdNPs catalysts are really unmatched choices for this reaction [198]. Recently, xylan-type hemicelluloses supported terpyridine-PdNPs showed high catalytic activity and stability for Suzuki–Miyaura reaction between arylboronic acid and aryl halide under aerobic condition, with a yield of 98%. It also could be recovered conveniently and rescued at least six times without significant changes in their catalytic activity [199]. PdNPs (3.0 nm) could form an alginate matrix with Cu alginate gels and resulted in an improvement in the activity and recyclability of Suzuki–Miyaura reaction [200]. Besides, starch derived PdNPs (1.5–4.5 nm) were tested in the microwave-assisted Heck, Suzuki and Sonogashira C–C coupling reactions and excellent catalytic performances were observed, confirming the catalytic potential of the Pd-supported catalysts [133].

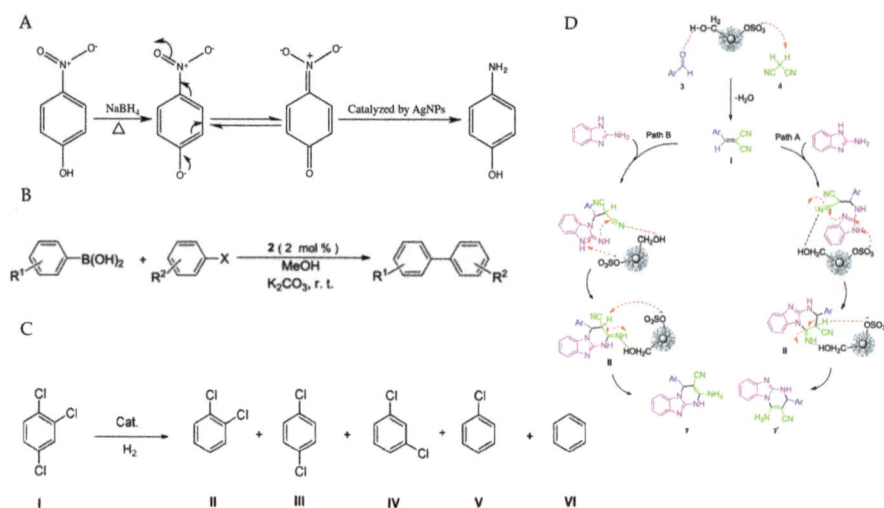

Figure 7. Overall scheme for the illustration of different catalytic reactions by PMNPs: (**A**) 4-NP reduction catalyzed by arabinogalactan stabilized AgNPs; (**B**) Suzuki coupling reaction of aryl halide with arylboronic acid catalyzed by xylan-type hemicelluloses supported terpyridine-PdNPs; (**C**) hydrogenation of 1,2,4-trichlorobenzene catalyzed by *Klebsiella oxytoca* BAS-10 exopolysaccharides supported PdFeNPs; and (**D**) Synthesis of imidazopyrimidine derivatives catalyzed by Irish moss/Fe$_3$O$_4$ nanoparticles. Reproduced with permission [97,185,199,201].

Klebsiella oxytoca BAS-10 exopolysaccharides supported PdNPs and PdFeNPs were investigated for catalyzing the hydrogenation of *trans-cinnamaldehyde* and 1,2,4-trichlorobenzene, respectively. Results showed that these two PMNPs could stimulate the hydrogenation under mild reaction conditions and their catalytic activities were maintained after several recycle experiments [201,202]. PMNPs also revealed remarkable catalytic ability in the degradation of dyes [203,204]. In addition, esterification of palm fatty acid distillate, benzylation of *o*-xylene, toluene and ethylene glycol oxidation as well as synthesis of imidazopyrimidine derivatives could also be catalyzed by different PMNPs [185,205,208]. Nevertheless, there still exist some problems in the isolation and recovery of these tiny nano catalysts from the reaction mixture through conventional methods due to their nano size and solvation properties [185]. The improvement of the recovery efficiency is obviously a significant objective in further research.

Table 5. Summary of literature data regarding the catalytic property of PMNPs.

Resource	Polysaccharides	Metals	Diameter (nm)	Shape	Reaction types	Reference
-	Xanthan	Ag	5.0–40.0	Spherical	4-NP reduction	[83]
Portulaca	Arabinogalactan	Ag	20.0–30.0	Spherical	4-NP reduction	[97]
-	Dextrin	Ag/Au	8.0–28.0	Spherical	4-NP reduction	[166]
Ceratonia siliqua	Locust bean gum	Au	-	Spherical	4-NP reduction	[174]
Chondrus crispus	Irish moss	Fe$_3$O$_4$	-	Homogeneous	Imidazopyrimidine derivatives synthesis	[185]
-	Alginate	Bi	5.0–8.0	Porous	4-NP reduction	[192]
Cordyceps sinensis	Exopolysaccharides	Ag	5.0	Spherical	4-NP reduction	[193]
-	Glucomannan	Au	12.0–31.0	Spherical	4-NP reduction	[194]
Cochlospermum religiosum	Katira gum	Au	6.9	Spherical	4-NP reduction	[195]
-	Starch-g-poly	Ag-Au	11.1	-	4-NP reduction	[196]
-	Xylan-type hemicellulose	Terpyridine-Pd	10.0–20.0	Particle	Suzuki-Miyaura reaction	[199]
-	Alginate	Pd-Cu	>10	Fibrils network	Suzuki-Miyaura reaction	[200]
Klebsiella oxytoca BAS-10	Exopolysaccharides	Fe/Fe-Pd	1.0–1.5	Cluster	Hydrodechlorination reaction	[201]
Klebsiella oxytoca BAS-10	Exopolysaccharides	Pd	30.0–550.0	Jagged undefined structures	Aqueous biphasic hydrogenation	[202]
-	Cellulose nanofibrils	Ag	25.2–18.0	Porous	Rhodamine B degradation	[203]
Corn	Crosslinked carboxymethyl starch/cellulose	ZnO/Zn	20.0–100.0	Spherical	Photodegradation of dyes	[204]
Algae	Algin	Al	4.0–5.0	Rough with wrinkled surface	Esterification reaction	[205]
-	Chitosan	ZnO	9.0	-	Benzylation of o-xylene	[206]
-	Sodium alginate	Cu-Mn	10.0–20.0	Spherical	Toluene oxidation	[207]
-	Dextrin	Au	8.4–12.0	-	Liquid phase oxidation of ethylene glycol	[208]
-	Chitin	Ag	5.5–15.2	Mesoporous, fibrous	p-NP reduction	[209]
-	Salep	Pd (II)	-	Rough	Suzuki coupling reaction	[210]
-	Alginate	Au	20.0–40.0	Centered cubic crystal lattice	Decoloration of Azo-Dyes	[211]
Bupleurum falcatum	Water extract	Au	8.2–12.8	Spherical	4-NP reduction	[212]
Acetobacter xylinum NCIM2526	Levan	Ag/Au	5.0–12.0	Spherical	4-NP reduction	[213]
-	Chitosan/corn starch/sodium alginate	ZnO	8.3–11.3	Hexagonal phase with Wurtzite structure	Photocatalytic reaction	[214]
Pleurotus florida	Glucan	Au	19.0–27.2	Spherical	4-NP reduction	[215]
-	Starch	Pd	1.5–4.5	Spherical	Heck reaction, Suzuki reaction, Sonogashira reaction	[216]

5. Toxicity of PMNPs

There is no doubt that polysaccharide-based metallic nanoparticles have made significant progress in many areas. Similar to most new technologies, the majority investigations in PMNPs have been focused on their potential applications and limited information is available on their toxicity to the cell, animals and the environment. Currently, some excellent review articles have provided significant insights into the toxicological significance and proposed mechanisms [217,218]. In fact, the risks associated with exposure to nanoparticles, the possible entry ways and metabolic mechanism have not been clarified until now, and several reports have illustrated that the nanoparticles deposit were responsible for toxic damages in different organs [219–221]. Although polysaccharides are attractive due to their low-toxic property and apparent lack of side effects [222], it is still necessary to evaluate the potential risks and toxicity of PMNPs in details.

The major entry portals of PMNPs are respiratory system, oral ingestion, and skin absorption [204]. After being administered, the nanoparticles can be transported via blood circulation into different organs. It had been reported that the colloidal gold nanoparticles (30.0 nm) could be quickly observed in rat platelets after intratracheal instillation and Tc-labelled carbon particles (99.0 nm) could get into the blood circulation in 1 min [223,224]. Hemolysis, commonly known as the rupturing of red blood cells, can be induced by the oxidative-stress caused by nanoparticles, causing lysis and death of cells. Thus, the toxic effects of PMNPs were evaluated with reference to hemolysis percentage at first. Gum karaya-based AuPNs (15.0–20.0 nm) and xanthan-based AuPNs (20.0–25.0 nm) were found to display lower hemolysis rates, even at test concentrations over 200 µg/mL [155,225]. However, AgNPs (1.3–4.2 nm), use polysaccharides from edible mushroom *L. squarrosulus* (Mont.) Singer as the reducing and stabilizing agent, were found to be compatible with human red blood cells at a dose of less than 5 µg/mL [87]. Starch-based AgNPs (5.8 nm) could also induce significant hemolysis and exhibited a concentration-dependent in hemagglutination than AuNPs and PtNPs [226]. Therefore, the hemolysis risk assessment of PMNPs was needed before utilization. Other toxic effects of PMNPs were evaluated with the reference of cytotoxicity in normal cells. Accordingly, Gellan gum (10.0–15.0 nm) and Fucan (210.4 nm) coated AgNPs had a slight toxic effect towards to the mouse embryonic fibroblasts (NIH3T3) cells, human renal (HEK 293) cells and murine fibroblasts (3T3) cells [227,228]. Porphyran (14.0 nm) and pectin (7.0–13.0 nm) based AuNPs had no significant changes in normal monkey kidney cells viability, even up to 100 µM concentration [229,230]. In addition, chitosan coated CuNPs (260.0 nm) could obviously increase the viability of human alveolar epithelial (A549) cells relative to the exposure of CuNPs. Inflammatory risks induced by chitosan coated CuNPs would be raised if administered via the lung [231]. Normally, macrophages will participate in the clearance of nanoparticles that have passed the mucociliary barrier, which will lead to the activation of pro-inflammatiory mediators, causing both acute and chronic inflammation subsequently [232]. The investigations of inflammation in vitro, driven by PMNPs had showed that chitosan coated Ag/ZnO nanoparticles (100.0–150.0 nm) exerted no significant cytotoxic effects on murine macrophages RAW264.7 cells at the concentrations of 50 µg/mI and the cell morphology of cells was also not affected [74]. Additionally, PMNPs exhibited lower cytotoxicity than the metal itself, and different fluid exposure processes showed a significant effect on the viability of macrophage cells [233,234]. Several pieces of evidence also suggested that PMNPs had no neurotoxic effects and phyto-toxicity potential [235,236]. The sub-acute oral toxicity assessment also demonstrated the limited influences of PMNPs on the hematological and biochemical indexes of rats at the dose of 1500 ppm for 28 days [229,237]. Similar results were observed in the zebra fish toxicity study in vivo [238]. In addition, skin permeation study also showed that PMNPs could be detected in the receptor compartment in intact skin [227]. Although previous report had clarified that MNPs with the diameter smaller than 10.0 nm were able to penetrate the skin and reach the deepest layers of the stratum corneum, it did not permeate the skin [239].

Recent progresses in the toxicity investigation of PMNPs had put forward both the safety and risks at the same time. Nevertheless, it was still essential to consider the possible threaten that brought

Polymers **2017**, *9*, 689

by MNPs in a long term exposure. Hence, more acute and sub-acute toxicity evaluation of PMNPs in vivo should be performed to avoid the potential hazards in the future.

6. Conclusions and Perspectives

As the research in MNPs increases, the awareness towards seeking renewable multifunctional hosts for the preparation of MNPs has increased in both academia and society. Using natural polysaccharides exhibited enormous potential in the green synthesis of MNPs owing to their non-toxic, biocompatible and biodegradable advantages. Through host–guest strategy, polysaccharides can act as the reducing and stabilizing agents of metallic ions and MNPs, providing an alternative way of solving the problems in conventional physical and chemical synthesis approaches. The present review has focused on the recent advances in the preparation, characterization and application of PMNPs. For the synthesis of MNPs employing natural polysaccharides through bottom-up synthesis approach can not only consume few chemicals and energy, but also showed a good control of the size and morphology property of MNPs. Despite the extensive applications of various techniques in the characterization of PMNPs, the combination of multiple techniques is considered to be more suitable for illustrating their properties because of their diverse and ambiguous properties. The illustration of the characteristics of the PMNPs will provide more insights into the synthesis mechanism, which will be beneficial to the targeted synthesis of the nanoparticles.

Although the development of green synthesis of PMNPs has achieved advanced progress, some problems still cannot be ignored. Due to significant variations in geography and time, the same species harvested in different seasons may lead to structural differences of the polysaccharides. Different synthesis methods of PMNPs also bring great challenges to their development and application. Thus, a commercially viable, eco-friendly and easy route for the synthesis of PMNPs is urgently needed. Another alternative attempt is to use synthetic polymers instead of the natural polysaccharides to synthesize PMNPs. The convenient management and adjustment of the synthetic process makes them to be a suitable host for the synthesis of homogeneous PMNPs.

Primarily PMNPs and their application in specific fields including wound healing, targeted delivery, biosensing, catalysis and agents with antimicrobial, antiviral and anticancer capabilities are well demonstrated in detail. The applications of the PMNPs are largely associated with the characteristics of the metallic ions. Therefore, the guest metallic ions that exhibited good property in the field of optics, images, diagnosis and nanomedicine shall be considered in the synthesis of PMNPs. On the other hand, the controversial toxicological evaluations of PMNPs in recent years are introduced both in vitro and in vivo. Long-term toxicity investigations, such as for acute and sub-acute toxicity evaluation of PMNPs, are still required to elucidate their potential risks. Despite the remarkable advances, the recovery efficiency of PMNPs remains low. Further investigations that introduce physical methods such as the recovery of magnetic PMNPs through an external magnetic field to improve the recovery efficiency are urgently needed. In addition, different methods that involve host–guest strategy in the synthesis of PMNPs are also encouraged.

Acknowledgments: This work was supported by the grant from the National Natural Science Foundation of China (NSFC 31371879) and National High Technology Research and Development Program ("863" Program) of China (Grant No. SS2013AA100207).

Author Contributions: Haixia Chen conceived and designed the manuscript; Cong Wang wrote the paper; Cong Wang, Xudong Gao and Zhongqin Chen contributed literature survey; and Xudong Gao, Zhongqin Chen and Yue Chen critically reviewed the manuscript.

Conflicts of Interest: The authors declare no conflict of interest.

Abbreviations

The following abbreviations are used in this manuscript: Metallic nanoparticles, MNPs; Polysaccharides based metallic nanoparticles, PMNPs; Confocal laser scanning microscopy, CLSM; Scan electron microscopy, SEM; Transmission electron microscopy, TEM; X-ray diffraction, XRD; small-angle X-ray scattering, SAXS; Scanning tunneling microscopy, STM; Dynamic light scattering, DLS; Atomic-force microscopy, AFM; Electron spin resonance, ESR; Minimum inhibitory concentration, MIC; Fourier transform infrared spectroscopy, FTIR; Reactive oxygen species, ROS; Human immunodeficiency virus, HIV; Severe acute respiratory syndrome, SARS; 3-(4,5-dimethylthiazol-2-yl)-2,5-diphenyltetrazolium bromide, MTT; Hematoxylin & eosin, H&E; *Escherichia coli, E. coli; Staphylococcus epidermidis, S. epidermidis; Staphylococcus aureus, S. aureus; Pseudomonas aeruginosa, P. aeruginosa; Sarcina lutea, S. lutea; Bacillus subtilis, B. subtilis; Klebsiella planticola, K. planticola; Klebsiella pneumoniae, K. pneumoniae; Candida albicans, C. albicans; Fusarium oxysporum, F. oxysporum; Lactobacillus fermentum, L. fermentum; Enterococcus faecium, E. faecium; Bacillus licheniformis, B. licheniformis; Bacillus cereus, B. cereus; Vibrio parahaemolyticus, V. parahaemolyticus; Proteus vulgaris, P. vulgaris; Salmonella typhimurium, S. typhimurium; Kocuria rhizophila, K. rhizophila; Aeromonas hydrophila, A. hydrophila; Streptococcus pyogenes, S. pyogenes; Aspergillus niger, A. niger; Trichophyton rubrum, T. rubrum; Vibrio cholerae, V. cholerae; Streptococcus pneumoniae, S. pneumoniae; Listeria monocytogenes, L. monocytogenes; Enterococcus faecalis, E. faecalis; Candida krusei, C. krusei; Saccharomyces cerevisiae, S. cerevisiae; Aspergillus flavus, A. flavus.*

References

1. Datta, K.K.R.; Reddy, B.V.S.; Zboril, R. Polysaccharides as functional scaffolds for noble metal nanoparticles and their catalytic applications. In *Encyclopedia of Nanoscience and Nanotechnology*; American Scientific Publishers: Valencia, CA, USA, 2016; pp. 1–20. ISBN 1588831590.

2. Huang, H.; Yuan, Q.; Yang, X. Preparation and characterization of metal-chitosan nanocomposites. *Colloids Surf. B Biointerfaces* **2004**, *39*, 31–37.

3. Huang, X.; Jain, P.K.; El-Sayed, I.H.; El-Sayed, M.A. Plasmonic photothermal therapy (PPTT) using gold nanoparticles. *Lasers Med. Sci.* **2008**, *23*, 217–228.

4. Sengupta, S.; Eavarone, D.; Capila, I.; Zhao, G.; Watson, N.; Kiziltepe, T.; Sasisekharan, R. Temporal targeting of tumour cells and neovasculature with a nanoscale delivery system. *Nature* **2005**, *436*, 568–572.

5. Neville, F.; Pchelintsev, N.A.; Broderick, M.J.F.; Gibson, T.; Millner, P.A. Novel one-pot synthesis and characterization of bioactive thiol-silicate nanoparticles for biocatalytic and biosensor applications. *Nanotechnology* **2009**, *20*, 55612.

6. Ahmed, S.; Ahmad, M.; Swami, B.L.; Ikram, S. Green synthesis of silver nanoparticles using *Azadirachta indica* aqueous leaf extract. *J. Radiat. Res. Appl. Sci.* **2016**, *9*, 1–7.

7. Medina-Ramirez, I.; Bashir, S.; Luo, Z.; Liu, J.L. Green synthesis and characterization of polymer-stabilized silver nanoparticles. *Colloids Surf. B Biointerfaces* **2009**, *73*, 185–191.

8. Zhao, C.; Li, J.; He, B.; Zhao, L. Fabrication of hydrophobic biocomposite by combining cellulosic fibers with polyhydroxyalkanoate. *Cellulose* **2017**, *24*, 2265–2274.

9. Siqueira, G.; Bras, J.; Dufresne, A. Cellulosic bionanocomposites: A review of preparation, properties and applications. *Polymers* **2010**, *2*, 728–765.

10. Pauly, M.; Keegstra, K. Cell-wall carbohydrates and their modification as a resource for biofuels. *Plant J.* **2008**, *54*, 559–568.

11. Emna, C.; Fatma, G.; Satinder, K.B. Biopolymers Synthesis and Application. In *Biotransformation of Waste Biomass into High Value Biochemicals*; Springer: New York, NY, USA, 2014; Chapter 17; pp. 415–443, ISBN 9781461480051.

12. Pasqui, D.; De Cagna, M.; Barbucci, R. Polysaccharide-based hydrogels: The key role of water in affecting mechanical properties. *Polymers* **2012**, *4*, 1517–1534.

13. Liu, Z.; Jiao, Y.; Wang, Y.; Zhou, C.; Zhang, Z. Polysaccharides-based nanoparticles as drug delivery systems. *Adv. Drug Deliv. Rev.* **2008**, *60*, 1650–1662.

14. Yang, J.; Han, S.; Zheng, H.; Dong, H.; Liu, J. Preparation and application of micro/nanoparticles based on natural polysaccharides. *Carbohydr. Polym.* **2015**, *123*, 53–66.

15. Lee, J.W.; Park, J.H.; Robinson, J.R. Bioadhesive-based dosage forms: The next generation. *J. Pharm. Sci.* **2000**, *89*, 850–866.

16. Debele, T.A.; Mekuria, S.L.; Tsai, H.C. Polysaccharide based nanogels in the drug delivery system: Application as the carrier of pharmaceutical agents. *Mater. Sci. Eng. C* **2016**, *68*, 964–981.

17. Wang, J.; Chen, H.; Wang, Y.; Xing, L. Synthesis and characterization of a new *Inonotus obliquus* polysaccharide-iron(III) complex. *Int. J. Biol. Macromol.* **2015**, *75*, 210–217.

18. Li, W.; Yuan, G.; Pan, Y.; Wang, C.; Chen, H. Network Pharmacology Studies on the Bioactive Compounds and Action Mechanisms of Natural Products for the Treatment of Diabetes Mellitus: A Review. *Front. Pharmacol.* **2017**, *8*, 1–10.

19. Wang, C.; Chen, Z.; Pan, Y.; Gao, X.; Chen, H. Anti-diabetic effects of *Inonotus obliquus* polysaccharides-chromium(III) complex in type 2 diabetic mice and its sub-acute toxicity evaluation in normal mice. *Food Chem. Toxicol.* **2017**, *108*, 498–509.

20. Rao, K.M.; Kumar, A.; Haider, A.; Han, S.S. Polysaccharides based antibacterial polyelectrolyte hydrogels with silver nanoparticles. *Mater. Lett.* **2016**, *184*, 189–192.

21. Kim, C.; Tonga, G.Y.; Yan, B.; Kim, C.S.; Kim, S.T.; Park, M.-H.; Zhu, Z.; Duncan, B.; Creran, B.; Rotello, V.M. Regulating exocytosis of nanoparticles via host–guest chemistry. *Org. Biomol. Chem.* **2015**, *13*, 2474–2479.

22. Dhar, S.; Maheswara Reddy, E.; Shiras, A.; Pokharkar, V.; Prasad, B.L. Natural gum reduced/stabilized gold nanoparticles for drug delivery formulations. *Chemistry* **2008**, *14*, 10244–10250.

23. Tran, H.V.; Tran, L.D.; Ba, C.T.; Vu, H.D.; Nguyen, T.N.; Pham, D.G.; Nguyen, P.X. Synthesis, characterization, antibacterial and antiproliferative activities of monodisperse chitosan- based silver nanoparticles. *Colloids Surf. A Physicochem. Eng. Asp.* **2010**, *360*, 32–40.

24. Narayanan, G.; Aguda, R.; Hartman, M.; Chung, C.C.; Boy, R.; Gupta, B.S.; Tonelli, A.E. Fabrication and Characterization of Poly(ε-caprolactone)/α-Cyclodextrin Pseudorotaxane Nanofibers. *Biomacromolecules* **2016**, *17*, 271–279.

25. Moreno-Trejo, M.B.; Sánchez-Domínguez, M. Mesquite gum as a novel reducing and stabilizing agent for modified tollens synthesis of highly concentrated Ag nanoparticles. *Materials* **2016**, *9*, 817.

26. Chen, W.H.; Lei, Q.; Luo, G.F.; Jia, H.Z.; Hong, S.; Liu, Y.X.; Cheng, Y.J.; Zhang, X.Z. Rational Design of Multifunctional Gold Nanoparticles via Host-Guest Interaction for Cancer-Targeted Therapy. *ACS Appl. Mater. Interfaces* **2015**, *7*, 17171–17180.

27. Noël, S.; Léger, B.; Ponchel, A.; Philippot, K.; Denicourt-Nowicki, A.; Roucoux, A.; Monflier, E. Cyclodextrin-based systems for the stabilization of metallic(0) nanoparticles and their versatile applications in catalysis. *Catal. Today* **2014**, *235*, 20–32.

28. Yao, X.; Zhu, Q.; Li, C.; Yuan, K.; Che, R.; Zhang, P.; Yang, C.; Lu, W.; Wu, W.; Jiang, X. Carbamoylmannose enhances the tumor targeting ability of supramolecular nanoparticles formed through host–guest complexation of a pair of homopolymers. *J. Mater. Chem. B* **2017**, *5*, 834–848.

29. Nair, L.S.; Laurencin, C.T. Silver nanoparticles: Synthesis and therapeutic applications. *J. Biomed. Nanotechnol.* **2007**, *3*, 301–316.

30. Sharma, D.; Kanchi, S.; Bisetty, K. Biogenic synthesis of nanoparticles: A review. *Arab. J. Chem.* **2015**. [CrossRef]

31. Ahmed, S.; Ahmad, M.; Swami, B.L.; Ikram, S. A review on plants extract mediated synthesis of silver nanoparticles for antimicrobial applications: A green expertise. *J. Adv. Res.* **2016**, *7*, 17–28.

32. Kaliaraj, G.S.; Subramaniyan, B.; Manivasagan, P. Green Synthesis of Metal Nanoparticles Using Seaweed Polysaccharides. In *Seaweed Polysaccharides*; Elsevier: Amsterdam, The Netherlands, 2017; Chapter 7; pp. 101–109, ISBN 9780128098165.

33. Shukla, A.K.; Iravani, S. Metallic nanoparticles: Green synthesis and spectroscopic characterization. *Environ. Chem. Lett.* **2017**, *15*, 223–231.

34. Singh, P.; Kim, Y.J.; Zhang, D.; Yang, D.C. Biological Synthesis of Nanoparticles from Plants and Microorganisms. *Trends Biotechnol.* **2016**, *34*, 588–599.

35. Thakkar, K.N.; Mhatre, S.S.; Parikh, R.Y. Biological synthesis of metallic nanoparticles. *Nanomed. Nanotechnol. Biol. Med.* **2010**, *6*, 257–262.

36. Mittal, A.K.; Chisti, Y.; Banerjee, U.C. Synthesis of metallic nanoparticles using plant extracts. *Biotechnol. Adv.* **2013**, *31*, 346–356.

37. Yip, J.; Liu, L.; Wong, K.H.; Leung, P.H.M.; Yuen, C.W.M.; Cheung, M.C. Investigation of antifungal and antibacterial effects of fabric padded with highly stable selenium nanoparticles. *J. Appl. Polym. Sci.* **2014**, *131*, 8886–8893.

38. Li, H.; Yang, Y.W. Gold nanoparticles functionalized with supramolecular macrocycles. *Chin. Chem. Lett.* **2013**, *24*, 545–552.

39. Cram, D.J.; Cram, J.M. Host-Guest Chemistry. *Science* **1974**, *183*, 803–809.
40. Niu, Z.; Li, Y. Removal and utilization of capping agents in nanocatalysis. *Chem. Mater.* **2014**, *26*, 72–83.
41. Chan, H.K.; Kwok, P.C.L. Production methods for nanodrug particles using the bottom-up approach. *Adv. Drug Deliv. Rev.* **2011**, *63*, 406–416.
42. Raghunandan, D.; Basavaraja, S.; Mahesh, B.; Balaji, S.; Manjunath, S.Y.; Venkataraman, A. Biosynthesis of stable polyshaped gold nanoparticles from microwave-exposed aqueous extracellular anti-malignant guava (*Psidium guajava*) leaf extract. *Nanobiotechnology* **2009**, *5*, 34–41.
43. Hussain, M.A.; Shah, A.; Jantan, I.; Shah, M.R.; Tahir, M.N.; Ahmad, R.; Bukhari, S.N.A. Hydroxypropylcellulose as a novel green reservoir for the synthesis, stabilization, and storage of silver nanoparticles. *Int. J. Nanomed.* **2015**, *10*, 2079–2088.
44. Ehmann, H.M.A.; Breitwieser, D.; Winter, S.; Gspan, C.; Koraimann, G.; Maver, U.; Sega, M.; Köstler, S.; Stana-Kleinschek, K.; Spirk, S.; et al. Gold nanoparticles in the engineering of antibacterial and anticoagulant surfaces. *Carbohydr. Polym.* **2015**, *117*, 34–42.
45. Abedini, A.; Daud, A.; Abdul Hamid, M.; Kamil Othman, N.; Saion, E. A review on radiation-induced nucleation and growth of colloidal metallic nanoparticles. *Nanoscale Res. Lett.* **2013**, *8*, 474.
46. Iravani, S.; Korbekandi, H.; Mirmohammadi, S.V.; Zolfaghari, B. Synthesis of silver nanoparticles: Chemical, physical and biological methods. *Res. Pharm. Sci.* **2014**, *9*, 385–406.
47. Sokolov, S.V.; Batchelor-Mcauley, C.; Tschulik, K.; Fletcher, S.; Compton, R.G. Are Nanoparticles Spherical or Quasi-Spherical? *Chemistry* **2015**, *21*, 10741–10746.
48. Lechner, M.; Mächtle, W. Characterization of nanoparticles. *Macromol. Symp.* **1999**, *7*, 1–7.
49. Hall, J.B.; Dobrovolskaia, M.A.; Patri, A.K.; McNeil, S.E. Characterization of nanoparticles for therapeutics. *Nanomedicine* **2007**, *2*, 789–803.
50. Bootz, A.; Vogel, V.; Schubert, D.; Kreuter, J. Comparison of scanning electron microscopy, dynamic light scattering and analytical ultracentrifugation for the sizing of poly(butyl cyanoacrylate) nanoparticles. *Eur. J. Pharm. Biopharm.* **2004**, *57*, 369–375.
51. Lin, P.C.; Lin, S.; Wang, P.C.; Sridhar, R. Techniques for physicochemical characterization of nanomaterials. *Biotechnol. Adv.* **2014**, *32*, 711–726.
52. Dobrovolskaia, M.A.; Patri, A.K.; Zheng, J.; Clogston, J.D.; Ayub, N.; Aggarwal, P.; Neun, B.W.; Hall, J.B.; McNeil, S.E. Interaction of colloidal gold nanoparticles with human blood: Effects on particle size and analysis of plasma protein binding profiles. *Nanomed. Nanotechnol. Biol. Med.* **2009**, *5*, 106–117.
53. Powers, K.W.; Palazuelos, M.; Moudgil, B.M.; Roberts, S.M. Characterization of the size, shape, and state of dispersion of nanoparticles for toxicological studies. *Nanotoxicology* **2007**, *1*, 42–51.
54. Sanyasi, S.; Majhi, R.K.; Kumar, S.; Mishra, M.; Ghosh, A.; Suar, M.; Satyam, P.V.; Mohapatra, H.; Goswami, C.; Goswami, L. Polysaccharide-capped silver Nanoparticles inhibit biofilm formation and eliminate multi-drug-resistant bacteria by disrupting bacterial cytoskeleton with reduced cytotoxicity towards mammalian cells. *Sci. Rep.* **2016**, *6*, 24929.
55. Berne, B.J.; Pecora, R. *Dynamic Light Scattering: With Applications to Chemistry, Biology, and Physics*, 1st ed.; Dover Publications: New York, NY, USA, 2000; pp. 3–24, ISBN 0486411559.
56. Ito, T.; Sun, L.; Bevan, M.A.; Crooks, R.M. Comparison of nanoparticle size and electrophoretic mobility measurements using a carbon-nanotube-based coulter counter, dynamic light scattering, transmission electron microscopy, and phase analysis light scattering. *Langmuir* **2004**, *20*, 6940–6945.
57. Upstone, S. Ultraviolet/visible light absorption spectrophotometry in clinical chemistry. *Encycl. Anal. Chem.* **2000**, 1699–1714. [CrossRef]
58. Liu, X.M.; Sheng, G.P.; Luo, H.W.; Zhang, F.; Yuan, S.J.; Xu, J.; Zeng, R.J.; Wu, J.G.; Yu, H.Q. Contribution of extracellular polymeric substances (EPS) to the sludge aggregation. *Environ. Sci. Technol.* **2010**, *44*, 4355–4360.
59. Khorsand Zak, A.; Abd. Majid, W.H.; Abrishami, M.E.; Yousefi, R. X-ray analysis of ZnO nanoparticles by Williamson-Hall and size-strain plot methods. *Solid State Sci.* **2011**, *13*, 251–256.
60. Sapsford, K.E.; Tyner, K.M.; Dair, B.J.; Deschamps, J.R.; Medintz, I.L. Analyzing nanomaterial bioconjugates: A review of current and emerging purification and characterization techniques. *Anal. Chem.* **2011**, *83*, 4453–4488.
61. Caminade, A.M.; Laurent, R.; Majoral, J.P. Characterization of dendrimers. *Adv. Drug Deliv. Rev.* **2005**, *57*, 2130–2146.
62. Stanjek, H.; Häusler, W. Basics of X-ray diffraction. *Hyperfine Interact.* **2004**, *154*, 107–119.

63. Bindu, P.; Thomas, S. Estimation of lattice strain in ZnO nanoparticles: X-ray peak profile analysis. *J. Theor. Appl. Phys.* **2014**, *8*, 123–134.

64. Tadic, M.; Panjan, M.; Damnjanovic, V.; Milosevic, I. Magnetic properties of hematite (α-Fe$_2$O$_3$) nanoparticles prepared by hydrothermal synthesis method. *Appl. Surf. Sci.* **2014**, *320*, 183–187.

65. Das, T.; Yeasmin, S.; Khatua, S.; Acharya, K.; Bandyopadhyay, A. Influence of a blend of guar gum and poly(vinyl alcohol) on long term stability, and antibacterial and antioxidant efficacies of silver nanoparticles. *RSC Adv.* **2015**, *5*, 54059–54069.

66. Gupta, D.; Singh, D.; Kothiyal, N.C.; Saini, A.K.; Singh, V.P.; Pathania, D. Synthesis of chitosan-*g*-poly(acrylamide)/ZnS nanocomposite for controlled drug delivery and antimicrobial activity. *Int. J. Biol. Macromol.* **2015**, *74*, 547–557.

67. Vidya, S.M.; Mutalik, S.; Bhat, K.U.; Huilgol, P.; Avadhani, K. Preparation of gold nanoparticles by novel bacterial exopolysaccharide for antibiotic delivery. *Life Sci.* **2016**, *153*, 171–179.

68. El-Rafie, M.H.; Mohamed, A.A.; Shaheen, T.I.; Hebeish, A. Antimicrobial effect of silver nanoparticles produced by fungal process on cotton fabrics. *Carbohydr. Polym.* **2010**, *80*, 779–782.

69. Ma, Y.; Liu, C.; Qu, D.; Chen, Y.; Huang, M.; Liu, Y. Antibacterial evaluation of sliver nanoparticles synthesized by polysaccharides from *Astragalus* membranaceus roots. *Biomed. Pharmacother.* **2017**, *89*, 351–357.

70. Kanmani, P.; Lim, S.T. Synthesis and characterization of pullulan-mediated silver nanoparticles and its antimicrobial activities. *Carbohydr. Polym.* **2013**, *97*, 421–428.

71. Chopra, I. The increasing use of silver-based products as antimicrobial agents: A useful development or a cause for concern? *J. Antimicrob. Chemother.* **2007**, *59*, 587–590.

72. Gwinn, E.G.; O'Neill, P.; Guerrero, A.J.; Bouwmeester, D.; Fygenson, D.K. Sequence-dependent fluorescence of DNA-hosted silver nanoclusters. *Adv. Mater.* **2008**, *20*, 279–283.

73. Pallavicini, P.; Arciola, C.R.; Bertoglio, F.; Curtosi, S.; Dacarro, G.; D'Agostino, A.; Ferrari, F.; Merli, D.; Milanese, C.; Rossi, S.; et al. Silver nanoparticles synthesized and coated with pectin: An ideal compromise for anti-bacterial and anti-biofilm action combined with wound-healing properties. *J. Colloid Interface Sci.* **2017**, *498*, 271–281.

74. Thaya, R.; Malaikozhundan, B.; Vijayakumar, S.; Sivakamavalli, J.; Jeyasekar, R.; Shanthi, S.; Vaseeharan, B.; Ramasamy, P.; Sonawane, A. Chitosan coated Ag/ZnO nanocomposite and their antibiofilm, antifungal and cytotoxic effects on murine macrophages. *Microb. Pathog.* **2016**, *100*, 124–132.

75. Sathiyanarayanan, G.; Vignesh, V.; Saibaba, G.; Vinothkanna, A.; Dineshkumar, K.; Viswanathan, M.B.; Selvin, J. Synthesis of carbohydrate polymer encrusted gold nanoparticles using bacterial exopolysaccharide: A novel and greener approach. *RSC Adv.* **2014**, *4*, 22817–22827.

76. Geraldo, D.A.; Needhan, P.; Chandia, N.; Arratia-Pérez, R.; Mora, G.C.; Villagra, N. Green synthesis of polysaccharides-based gold and silver nanoparticles and their promissory biological activity. *Biointerface Res. Appl. Chem.* **2016**, *6*, 1263–1271.

77. Valodkar, M.; Rathore, P.S.; Jadeja, R.N.; Thounaojam, M.; Devkar, R.V.; Thakore, S. Cytotoxicity evaluation and antimicrobial studies of starch capped water soluble copper nanoparticles. *J. Hazard. Mater.* **2012**, *201–202*, 244–249.

78. Rajeshkumar, S. Phytochemical constituents of fucoidan (*Padina tetrastromatica*) and its assisted AgNPs for enhanced antibacterial activity. *IET Nanobiotechnol.* **2016**, *11*, 292–299.

79. Goyal, G.; Hwang, J.; Aviral, J.; Seo, Y.; Jo, Y.; Son, J.; Choi, J. Green synthesis of silver nanoparticles using β-glucan, and their incorporation into doxorubicin-loaded water-in-oil nanoemulsions for antitumor and antibacterial applications. *J. Ind. Eng. Chem.* **2017**, *47*, 179–186.

80. Yumei, L.; Yamei, L.; Qiang, L.; Jie, B. Rapid Biosynthesis of Silver Nanoparticles Based on Flocculation and Reduction of an Exopolysaccharide from *Arthrobacter* sp. B4: Its Antimicrobial Activity and Phytotoxicity. *J. Nanomater.* **2017**, *2017*, 9703614.

81. Chen, X.; Yan, J.-K.; Wu, J.-Y. Characterization and antibacterial activity of silver nanoparticles prepared with a fungal exopolysaccharide in water. *Food Hydrocoll.* **2015**, *53*, 69–74.

82. Emam, H.E.; Zahran, M.K. Ag^0 nanoparticles containing cotton fabric: Synthesis, characterization, color data and antibacterial action. *Int. J. Biol. Macromol.* **2015**, *75*, 106–114.

83. Xu, W.; Jin, W.; Lin, L.; Zhang, C.; Li, Z.; Li, Y.; Song, R.; Li, B. Green synthesis of xanthan conformation-based silver nanoparticles: Antibacterial and catalytic application. *Carbohydr. Polym.* **2014**, *101*, 961–967.

84. Ghasemzadeh, H.; Mahboubi, A.; Karimi, K.; Hassani, S. Full polysaccharide chitosan-CMC membrane and silver nanocomposite: Synthesis, characterization, and antibacterial behaviors. *Polym. Adv. Technol.* **2016**, *27*, 1204–1210.

85. Rasulov, B.; Rustamova, N.; Yili, A.; Zhao, H.Q.; Aisa, H.A. Synthesis of silver nanoparticles on the basis of low and high molar mass exopolysaccharides of Bradyrhizobium japonicum 36 and its antimicrobial activity against some pathogens. *Folia Microbiol.* **2016**, *61*, 283–293.

86. Baldi, F.; Daniele, S.; Gallo, M.; Paganelli, S.; Battistel, D.; Piccolo, O.; Faleri, C.; Puglia, A.M.; Gallo, G. Polysaccharide-based silver nanoparticles synthesized by Klebsiella oxytoca DSM 29614 cause DNA fragmentation in *E. coli* cells. *BioMetals* **2016**, *29*, 321–331.

87. Manna, D.K.; Mandal, A.K.; Sen, I.K.; Maji, P.K.; Chakraborti, S.; Chakraborty, R.; Islam, S.S. Antibacterial and DNA degradation potential of silver nanoparticles synthesized via green route. *Int. J. Biol. Macromol.* **2015**, *80*, 455–459.

88. Sen, I.K.; Mandal, A.K.; Chakraborti, S.; Dey, B.; Chakraborty, R.; Islam, S.S. Green synthesis of silver nanoparticles using glucan from mushroom and study of antibacterial activity. *Int. J. Biol. Macromol.* **2013**, *62*, 439–449.

89. Kanmani, P.; Lim, S.T. Synthesis and structural characterization of silver nanoparticles using bacterial exopolysaccharide and its antimicrobial activity against food and multidrug resistant pathogens. *Process Biochem.* **2013**, *48*, 1099–1106.

90. Iconaru, S.L.; Prodan, A.M.; Motelica-Heino, M.; Sizaret, S.; Predoi, D. Synthesis and characterization of polysaccharide-maghemite composite nanoparticles and their antibacterial properties. *Nanoscale Res. Lett.* **2012**, *7*, 576.

91. White, R.J.; Budarin, V.L.; Moir, J.W.B.; Clark, J.H. A sweet killer: Mesoporous polysaccharide confined silver nanoparticles for antibacterial applications. *Int. J. Mol. Sci.* **2011**, *12*, 5782–5796.

92. Kora, A.; Beedu, S.; Jayaraman, A. Size-controlled green synthesis of silver nanoparticles mediated by gum ghatti (*Anogeissus latifolia*) and its biological activity. *Org. Med. Chem. Lett.* **2012**, *2*, 17.

93. El-Rafie, H.M.; El-Rafie, M.H.; Zahran, M.K. Green synthesis of silver nanoparticles using polysaccharides extracted from marine macro algae. *Carbohydr. Polym.* **2013**, *96*, 403–410.

94. Selvakumar, R.; Aravindh, S.; Ashok, A.M.; Balachandran, Y.L. A facile synthesis of silver nanoparticle with SERS and antimicrobial activity using *Bacillus subtilis* exopolysaccharides. *J. Exp. Nanosci.* **2013**, *8080*, 1–13.

95. Kora, A.J.; Sashidhar, R.B.; Arunachalam, J. Gum kondagogu (*Cochlospermum gossypium*): A template for the green synthesis and stabilization of silver nanoparticles with antibacterial application. *Carbohydr. Polym.* **2010**, *82*, 670–679.

96. Venkatpurwar, V.; Pokharkar, V. Green synthesis of silver nanoparticles using marine polysaccharide: Study of in-vitro antibacterial activity. *Mater. Lett.* **2011**, *65*, 999–1002.

97. Anuradha, K.; Bangal, P.; Madhavendra, S.S. Macromolecular arabinogalactan polysaccharide mediated synthesis of silver nanoparticles, characterization and evaluation. *Macromol. Res.* **2016**, *24*, 152–162.

98. Gostin, L.O.; Hodge, J.G. Zika virus and global health security. *Lancet Infect. Dis.* **2016**, *16*, 1099–1100.

99. Koonin, E.V.; Senkevich, T.G.; Dolja, V.V. The ancient Virus World and evolution of cells. *Biol. Direct* **2006**, *1*, 29.

100. Zheng, L.; Wei, J.; Lv, X.; Bi, Y.; Wu, P.; Zhang, Z.; Wang, P.; Liu, R.; Jiang, J.; Cong, H.; et al. Detection and differentiation of influenza viruses with glycan-functionalized gold nanoparticles. *Biosens. Bioelectron.* **2017**, *91*, 46–52.

101. Wei, J.; Zheng, L.; Lv, X.; Bi, Y.; Chen, W.; Zhang, W.; Shi, Y.; Zhao, L.; Sun, X.; Wang, F.; et al. Analysis of Influenza Virus Receptor Specificity Using Glycan-Functionalized Gold Nanoparticles. *ACS Nano* **2014**, *8*, 4600–4607.

102. Yan, J.-K.; Ma, H.-L.; Cai, P.-F.; Wu, J.-Y. Highly selective and sensitive nucleic acid detection based on polysaccharide-functionalized silver nanoparticles. *Spectrochim. Acta Part A Mol. Biomol. Spectrosc.* **2015**, *134*, 17–21.

103. Speshock, J.L.; Murdock, R.C.; Braydich-Stolle, L.K.; Schrand, A.M.; Hussain, S.M. Interaction of silver nanoparticles with Tacaribe virus. *J. Nanobiotechnol.* **2010**, *8*, 19.

104. Rogers, J.V.; Parkinson, C.V.; Choi, Y.W.; Speshock, J.L.; Hussain, S.M. A preliminary assessment of silver nanoparticle inhibition of monkeypox virus plaque formation. *Nanoscale Res. Lett.* **2008**, *3*, 129–133.

105. Chen, Y.-S.; Hung, Y.-C.; Lin, W.-H.; Huang, G.S. Assessment of gold nanoparticles as a size-dependent vaccine carrier for enhancing the antibody response against synthetic foot-and-mouth disease virus peptide. *Nanotechnology* **2010**, *21*, 195101.

106. Zheng, Y.; Wang, W.; Li, Y. Antitumor and immunomodulatory activity of polysaccharide isolated from *Trametes orientalis*. *Carbohydr. Polym.* **2015**, *131*, 248–254.

107. Tietze, R.; Zaloga, J.; Unterweger, H.; Lyer, S.; Friedrich, R.P.; Janko, C.; Pöttler, M.; Dürr, S.; Alexiou, C. Magnetic nanoparticle-based drug delivery for cancer therapy. *Biochem. Biophys. Res. Commun.* **2015**, *468*, 463–470.

108. Wicki, A.; Witzigmann, D.; Balasubramanian, V.; Huwyler, J. Nanomedicine in Cancer Therapy: Challenges, Opportunities, and Clinical Applications. *J. Control. Release* **2014**, *200*, 138–157.

109. Joseph, M.M.; Aravind, S.R.; George, S.K.; Pillai, K.R.; Mini, S.; Sreelekha, T.T. Antitumor activity of galactoxyloglucan-gold nanoparticles against murine ascites and solid carcinoma. *Colloids Surf. B Biointerfaces* **2014**, *116*, 219–227.

110. Suganya, K.S.U.; Govindaraju, K.; Kumar, V.G.; Karthick, V.; Parthasarathy, K. Pectin mediated gold nanoparticles induces apoptosis in mammary adenocarcinoma cell lines. *Int. J. Biol. Macromol.* **2016**, *93*, 1030–1040.

111. Joseph, M.M.; Aravind, S.R.; Varghese, S.; Mini, S.; Sreelekha, T.T. PST-Gold nanoparticle as an effective anticancer agent with immunomodulatory properties. *Colloids Surf. B Biointerfaces* **2013**, *104*, 32–39.

112. Tengdelius, M.; Gurav, D.; Konradsson, P.; Påhlsson, P.; Griffith, M.; Oommen, O.P. Synthesis and anticancer properties of fucoidan-mimetic glycopolymer coated gold nanoparticles. *Chem. Commun.* **2015**, *2*, 8532–8535.

113. Chen, T.; Wong, Y.S. Selenocystine induces apoptosis of A375 human melanoma cells by activating ROS-mediated mitochondrial pathway and p53 phosphorylation. *Cell. Mol. Life Sci.* **2008**, *65*, 2763–2775.

114. Nie, T.; Wu, H.; Wong, K.-H.; Chen, T. Facile synthesis of highly uniform selenium nanoparticles using glucose as the reductant and surface decorator to induce cancer cell apoptosis. *J. Mater. Chem. B* **2016**, *4*, 2351–2358.

115. Budihardjo, I.; Oliver, H.; Lutter, M.; Luo, X.; Wang, X. Biochemical pathways of caspase activation during apoptosis. *Annu. Rev. Cell Dev. Biol.* **1999**, *15*, 269–290.

116. Namvar, F.; Rahman, H.S.; Mohamad, R.; Baharara, J.; Mahdavi, M.; Amini, E.; Chartrand, M.S.; Yeap, S.K. Cytotoxic effect of magnetic iron oxide nanoparticles synthesized via seaweed aqueous extract. *Int. J. Nanomed.* **2014**, *9*, 2479–2488.

117. Wu, H.; Zhu, H.; Li, X.; Liu, Z.; Zheng, W.; Chen, T.; Yu, B.; Wong, K.H. Induction of apoptosis and cell cycle arrest in A549 human lung adenocarcinoma cells by surface-capping selenium nanoparticles: An effect enhanced by polysaccharide-protein complexes from *Polyporus rhinocerus*. *J. Agric. Food Chem.* **2013**, *61*, 9859–9866.

118. Raveendran, S.; Chauhan, N.; Palaninathan, V.; Nagaoka, Y.; Yoshida, Y.; Maekawa, T.; Kumar, D.S. Extremophilic polysaccharide for biosynthesis and passivation of gold nanoparticles and photothermal ablation of cancer cells. *Part. Part. Syst. Charact.* **2015**, *32*, 54–64.

119. Liu, C.P.; Lin, F.S.; Chien, C.T.; Tseng, S.Y.; Luo, C.W.; Chen, C.H.; Chen, J.K.; Tseng, F.G.; Hwu, Y.; Lo, L.W.; et al. In-situ formation and assembly of gold nanoparticles by gum Arabic as efficient photothermal agent for killing cancer cells. *Macromol. Biosci.* **2013**, *13*, 1314–1320.

120. Medhat, D.; Hussein, J.; El-Naggar, M.E.; Attia, M.F.; Anwar, M.; Latif, Y.A.; Booles, H.F.; Morsy, S.; Farrag, A.R.; Khalil, W.K.B.; et al. Effect of Au-dextran NPs as anti-tumor agent against EAC and solid tumor in mice by biochemical evaluations and histopathological investigations. *Biomed. Pharmacother.* **2017**, *91*, 1006–1016.

121. Arjunan, N.; Kumari, H.L.J.; Singaravelu, C.M.; Kandasamy, R.; Kandasamy, J. Physicochemical investigations of biogenic chitosan-silver nanocomposite as antimicrobial and anticancer agent. *Int. J. Biol. Macromol.* **2016**, *92*, 77–87.

122. Estrela-Llopis, V.R.; Chevichalova, A.V.; Trigubova, N.A.; Ryzhuk, E.V. Heterocoagulation of polysaccharide-coated platinum nanoparticles with ovarian-cancer cells. *Colloid J.* **2014**, *76*, 609–621.

123. Jia, X.; Liu, Q.; Zou, S.; Xu, X.; Zhang, L. Construction of selenium nanoparticles/β-glucan composites for enhancement of the antitumor activity. *Carbohydr. Polym.* **2015**, *117*, 434–442.

124. Ren, Y.; Zhao, T.; Mao, G.; Zhang, M.; Li, F.; Zou, Y.; Yang, L.; Wu, X. Antitumor activity of hyaluronic acid-selenium nanoparticles in Heps tumor mice models. *Int. J. Biol. Macromol.* **2013**, *57*, 57–62.

125. Chen, T.; Wong, Y.S.; Zheng, W.; Bai, Y.; Huang, L. Selenium nanoparticles fabricated in *Undaria pinnatifida* polysaccharide solutions induce mitochondria-mediated apoptosis in A375 human melanoma cells. *Colloids Surf. B Biointerfaces* **2008**, *67*, 26–31.

126. Yang, F.; Tang, Q.; Zhong, X.; Bai, Y.; Chen, T.; Zhang, Y.; Li, Y.; Zheng, W. Surface decoration by *Spirulina* polysaccharide enhances the cellular uptake and anticancer efficacy of selenium nanoparticles. *Int. J. Nanomed.* **2012**, *7*, 835–844.

127. Wu, H.; Li, X.; Liu, W.; Chen, T.; Li, Y.; Zheng, W.; Man, C.W.-Y.; Wong, M.-K.; Wong, K.-H. Surface decoration of selenium nanoparticles by mushroom polysaccharides–protein complexes to achieve enhanced cellular uptake and antiproliferative activity. *J. Mater. Chem.* **2012**, *22*, 9602–9610.

128. Martin, C.; Low, W.L.; Amin, M.C.I.M.; Radecka, I.; Raj, P.; Kenward, K. Current trends in the development of wound dressings, biomaterials and devices. *Pharm. Pat. Anal.* **2013**, *2*, 341–359.

129. El-Feky, G.S.; Sharaf, S.S.; El Shafei, A.; Hegazy, A.A. Using chitosan nanoparticles as drug carriers for the development of a silver sulfadiazine wound dressing. *Carbohydr. Polym.* **2017**, *158*, 11–19.

130. Kamoun, E.A.; Chen, X.; Mohy Eldin, M.S.; Kenawy, E.R.S. Crosslinked poly(vinyl alcohol) hydrogels for wound dressing applications: A review of remarkably blended polymers. *Arab. J. Chem.* **2015**, *8*, 1–14.

131. Keleştemur, S.; Kilic, E.; Uslu, Ü.; Cumbul, A.; Ugur, M.; Akman, S.; Culha, M. Wound healing properties of modified silver nanoparticles and their distribution in mouse organs after topical application. *Nano Biomed. Eng.* **2012**, *4*, 170–176.

132. Mugade, M.; Patole, M.; Pokharkar, V. Bioengineered mannan sulphate capped silver nanoparticles for accelerated and targeted wound healing: Physicochemical and biological investigations. *Biomed. Pharmacother.* **2017**, *91*, 95–110.

133. Huang, J.; Ren, J.; Chen, G.; Deng, Y.; Wang, G.; Wu, X. Evaluation of the Xanthan-Based Film Incorporated with Silver Nanoparticles for Potential Application in the Nonhealing Infectious Wound. *J. Nanomater.* **2017**, *2017*, 6802397.

134. Singla, R.; Soni, S.; Patial, V.; Kulurkar, P.M.; Kumari, A.; Mahesh, S.; Padwad, Y.S.; Yadav, S.K. In vivo diabetic wound healing potential of nanobiocomposites containing bamboo cellulose nanocrystals impregnated with silver nanoparticles. *Int. J. Biol. Macromol.* **2017**, *105*, 45–55. [CrossRef]

135. Haseeb, M.T.; Hussain, M.A.; Abbas, K.; Youssif, B.G.M.; Bashir, S.; Yuk, S.H.; Bukhari, S.N.A. Linseed hydrogel-mediated green synthesis of silver nanoparticles for antimicrobial and wound-dressing applications. *Int. J. Nanomed.* **2017**, *12*, 2845–2855.

136. Gupta, A.; Low, W.L.; Radecka, I.; Britland, S.T.; Mohd Amin, M.C.I.; Martin, C. Characterisation and in vitro antimicrobial activity of biosynthetic silver-loaded bacterial cellulose hydrogels. *J. Microencapsul.* **2016**, *33*, 725–734.

137. Muhammad, G.; Hussain, M.A.; Amin, M.; Hussain, S.Z.; Hussain, I.; Abbas Bukhari, S.N.; Naeem-ul-Hassan, M. Glucuronoxylan-mediated silver nanoparticles: Green synthesis, antimicrobial and wound healing applications. *RSC Adv.* **2017**, *7*, 42900–42908.

138. Ding, L.; Shan, X.; Zhao, X.; Zha, H.; Chen, X.; Wang, J.; Cai, C.; Wang, X.; Li, G.; Hao, J.; et al. Spongy bilayer dressing composed of chitosan–Ag nanoparticles and chitosan–Bletilla striata polysaccharide for wound healing applications. *Carbohydr. Polym.* **2017**, *157*, 1538–1547.

139. Mulens, V.; Morales, M.D.P.; Barber, D.F.; Barber, D.F. Development of Magnetic Nanoparticles for Cancer Gene Therapy: A Comprehensive Review. *ISRN Nanomater.* **2013**, *2013*, 1–14.

140. Venkatesan, J.; Anil, S.; Kim, S.-K.; Shim, M. Seaweed Polysaccharide-Based Nanoparticles: Preparation and Applications for Drug Delivery. *Polymers* **2016**, *8*, 30.

141. Maiyo, F.; Singh, M. Selenium nanoparticles: Potential in cancer gene and drug delivery. *Nanomedicine* **2017**, *12*, 1075–1089. [CrossRef]

142. Manivasagan, P.; Bharathiraja, S.; Bui, N.Q.; Jang, B.; Oh, Y.O.; Lim, I.G.; Oh, J. Doxorubicin-loaded fucoidan capped gold nanoparticles for drug delivery and photoacoustic imaging. *Int. J. Biol. Macromol.* **2016**, *91*, 578–588.

143. Manivasagan, P.; Bharathiraja, S.; Bui, N.Q.; Lim, I.G.; Oh, J. Paclitaxel-loaded chitosan oligosaccharide-stabilized gold nanoparticles as novel agents for drug delivery and photoacoustic imaging of cancer cells. *Int. J. Pharm.* **2016**, *511*, 367–379.

144. Li, N.; Chen, Y.; Zhang, Y.-M.; Yang, Y.; Su, Y.; Chen, J.-T.; Liu, Y. Polysaccharide-Gold Nanocluster Supramolecular Conjugates as a Versatile Platform for the Targeted Delivery of Anticancer Drugs. *Sci. Rep.* **2015**, *4*, 4164.

145. Aryal, S.; Grailer, J.J.; Pilla, S.; Steeber, D.A.; Gong, S. Doxorubicin conjugated gold nanoparticles as water-soluble and pH-responsive anticancer drug nanocarriers. *J. Mater. Chem.* **2009**, *19*, 7879–7884.

146. Reddy, P.R.S.; Eswaramma, S.; Rao, K.S.V.K.; Lee, Y.I. Dual responsive pectin hydrogels and their silver nanocomposites: Swelling studies, controlled drug delivery and antimicrobial applications. *Bull. Korean Chem. Soc.* **2014**, *35*, 2391–2399.

147. Rau, L.R.; Tsao, S.W.; Liaw, J.W.; Tsai, S.W. Selective Targeting and Restrictive Damage for Nonspecific Cells by Pulsed Laser-Activated Hyaluronan-Gold Nanoparticles. *Biomacromolecules* **2016**, *17*, 2514–2521.

148. Jiang, W.; Fu, Y.; Yang, F.; Yang, Y.; Liu, T.; Zheng, W.; Zeng, L.; Chen, T. Gracilaria lemaneiformis polysaccharide as integrin-targeting surface decorator of selenium nanoparticles to achieve enhanced anticancer efficacy. *ACS Appl. Mater. Interfaces* **2014**, *6*, 13738–13748.

149. Silva-Cunha, A.; Chéron, M.; Grossiord, J.L.; Puisieux, F.; Seiller, M. W/O/W multiple emulsions of insulin containing a protease inhibitor and an absorption enhancer: Biological activity after oral administration to normal and diabetic rats. *Int. J. Pharm.* **1998**, *169*, 33–44.

150. Bhumkar, D.R.; Joshi, H.M.; Sastry, M.; Pokharkar, V.B. Chitosan reduced gold nanoparticles as novel carriers for transmucosal delivery of insulin. *Pharm. Res.* **2007**, *24*, 1415–1426.

151. Kievit, F.M.; Veiseh, O.; Bhattarai, N.; Fang, C.; Gunn, J.W.; Lee, D.; Ellenbogen, R.G.; Olson, J.M.; Zhang, M. PEI-PEG-chitosan-copolymer-coated iron oxide nanoparticles for safe gene delivery: Synthesis, complexation, and transfection. *Adv. Funct. Mater.* **2009**, *19*, 2244–2251.

152. Chen, T.; Xu, S.; Zhao, T.; Zhu, L.; Wei, D.; Li, Y.; Zhang, H.; Zhao, C. Gold nanocluster-conjugated amphiphilic block copolymer for tumor-targeted drug delivery. *ACS Appl. Mater. Interfaces* **2012**, *4*, 5766–5774.

153. Safari, D.; Marradi, M.; Chiodo, F.; Th Dekker, H.A.; Shan, Y.; Adamo, R.; Oscarson, S.; Rijkers, G.T.; Lahmann, M.; Kamerling, J.P.; et al. Gold nanoparticles as carriers for a synthetic Streptococcus pneumoniae type 14 conjugate vaccine. *Nanomedicine* **2012**, *7*, 651–662.

154. Rastegari, B.; Karbalaei-Heidari, H.R.; Zeinali, S.; Sheardown, H. The enzyme-sensitive release of prodigiosin grafted β-cyclodextrin and chitosan magnetic nanoparticles as an anticancer drug delivery system: Synthesis, characterization and cytotoxicity studies. *Colloids Surf. B Biointerfaces* **2017**, *158*, 589–601.

155. Pooja, D.; Panyaram, S.; Kulhari, H.; Reddy, B.; Rachamalla, S.S.; Sistla, R. Natural polysaccharide functionalized gold nanoparticles as biocompatible drug delivery carrier. *Int. J. Biol. Macromol.* **2015**, *80*, 48–56.

156. Cai, H.; Yao, P. In situ preparation of gold nanoparticle-loaded lysozyme–dextran nanogels and applications for cell imaging and drug delivery. *Nanoscale* **2013**, *5*, 2892–2900.

157. Vu-Quang, H.; Yoo, M.K.; Jeong, H.J.; Lee, H.J.; Muthiah, M.; Rhee, J.H.; Lee, J.H.; Cho, C.S.; Jeong, Y.Y.; Park, I.K. Targeted delivery of mannan-coated superparamagnetic iron oxide nanoparticles to antigen-presenting cells for magnetic resonance-based diagnosis of metastatic lymph nodes in vivo. *Acta Biomater.* **2011**, *7*, 3935–3945.

158. Yoo, M.K.; Park, I.Y.K.; Kim, I.Y.; Kwon, J.S.; Jeong, H.J.; Jeong, Y.Y.; Cho, C.S. Superparamagnetic iron oxide nanoparticles coated with mannan for macrophage targeting. *J. Nanosci. Nanotechnol.* **2008**, *8*, 5196–5202.

159. Chichova, M.; Shkodrova, M.; Vasileva, P.; Kirilova, K.; Doncheva-Stoimenova, D. Influence of silver nanoparticles on the activity of rat liver mitochondrial ATPase. *J. Nanopart. Res.* **2014**, *16*, 2243.

160. Mahdavinia, G.R.; Mosallanezhad, A.; Soleymani, M.; Sabzi, M. Magnetic- and pH-responsive κ-carrageenan/chitosan complexes for controlled release of methotrexate anticancer drug. *Int. J. Biol. Macromol.* **2017**, *97*, 209–217.

161. Assa, F.; Jafarizadeh-Malmiri, H.; Ajamein, H.; Anarjan, N.; Vaghari, H.; Sayyar, Z.; Berenjian, A. A biotechnological perspective on the application of iron oxide nanoparticles. *Nano Res.* **2016**, *9*, 2203–2225.

162. Taton, T.A. Scanometric DNA Array Detection with Nanoparticle Probes. *Science* **2000**, *289*, 1757–1760.

163. Tagad, C.K.; Kim, H.U.; Aiyer, R.C.; More, P.; Kim, T.; Moh, S.H.; Kulkarni, A.; Sabharwal, S.G. A sensitive hydrogen peroxide optical sensor based on polysaccharide stabilized silver nanoparticles. *RSC Adv.* **2013**, *3*, 22940–22943.

164. Tagad, C.K.; Dugasani, S.R.; Aiyer, R.; Park, S.; Kulkarni, A.; Sabharwal, S. Green synthesis of silver nanoparticles and their application for the development of optical fiber based hydrogen peroxide sensor. *Sens. Actuators B Chem.* **2013**, *183*, 144–149.

165. Narayanan, K.B.; Han, S.S. Colorimetric detection of manganese (II) ions using alginate-stabilized silver nanoparticles. *Res. Chem. Intermed.* **2017**, *43*, 5665–5674.

166. Bankura, K.; Rana, D.; Mollick, M.M.R.; Pattanayak, S.; Bhowmick, B.; Saha, N.R.; Roy, I.; Midya, T.; Barman, G.; Chattopadhyay, D. Dextrin-mediated synthesis of Ag NPs for colorimetric assays of Cu^{2+} ion and Au NPs for catalytic activity. *Int. J. Biol. Macromol.* **2015**, *80*, 309–316.

167. Guan, H.; Yu, J.; Chi, D. Label-free colorimetric sensing of melamine based on chitosan-stabilized gold nanoparticles probes. *Food Control* **2013**, *32*, 35–41.

168. Narasimhan, L.R.; Goodman, W.; Patel, C.K.N. Correlation of breath ammonia with blood urea nitrogen and creatinine during hemodialysis. *Proc. Natl. Acad. Sci. USA* **2001**, *98*, 4617–4621.

169. Pandey, S.; Goswami, G.K.; Nanda, K.K. Green synthesis of biopolymer-silver nanoparticle nanocomposite: An optical sensor for ammonia detection. *Int. J. Biol. Macromol.* **2012**, *51*, 583–589.

170. Pandey, S.; Goswami, G.K.; Nanda, K.K. Green synthesis of polysaccharide/gold nanoparticle nanocomposite: An efficient ammonia sensor. *Carbohydr. Polym.* **2013**, *94*, 229–234.

171. Pandey, S.; Nanda, K.K. Au Nanocomposite Based Chemiresistive Ammonia Sensor for Health Monitoring. *ACS Sens.* **2016**, *1*, 55–62.

172. Dai, Z.; Xu, L.; Duan, G.; Li, T.; Zhang, H.; Li, Y.; Wang, Y.; Wang, Y.; Cai, W. Fast-Response, Sensitivitive and Low-Powered Chemosensors by Fusing Nanostructured Porous Thin Film and IDEs-Microheater Chip. *Sci. Rep.* **2013**, *3*, 1669.

173. Gattu, K.P.; Kashale, A.A.; Ghule, K.; Ingole, V.H.; Sharma, R.; Deshpande, N.G.; Ghule, A.V. NO_2 sensing studies of bio-green synthesized Au-doped SnO_2. *J. Mater. Sci. Mater. Electron.* **2017**, *28*, 13209–13216.

174. Tagad, C.K.; Rajdeo, K.S.; Kulkarni, A.; More, P.; Aiyer, R.C.; Sabharwal, S.; Hu, J.; Cai, W.; Cai, W.; Calderer, J. Green synthesis of polysaccharide stabilized gold nanoparticles: Chemo catalytic and room temperature operable vapor sensing application. *RSC Adv.* **2014**, *4*, 24014–24019.

175. Davidović, S.; Lazić, V.; Vukoje, I.; Papan, J.; Anhrenkiel, S.P.; Dimitrijević, S.; Nedeljković, J.M. Dextran coated silver nanoparticles—Chemical sensor for selective cysteine detection. *Colloids Surf. B Biointerfaces* **2017**, *160*. [CrossRef]

176. Lee, K.C.; Chiang, H.L.; Chiu, W.R.; Chen, Y.C. Molecular recognition between insulin and dextran encapsulated gold nanoparticles. *J. Mol. Recognit.* **2016**, *29*, 528–535.

177. Lai, C.; Zeng, G.M.; Huang, D.L.; Zhao, M.H.; Wei, Z.; Huang, C.; Xu, P.; Li, N.J.; Zhang, C.; Chen, M.; et al. Synthesis of gold-cellobiose nanocomposites for colorimetric measurement of cellobiase activity. *Spectrochim. Acta Part A Mol. Biomol. Spectrosc.* **2014**, *132*, 369–374.

178. Shen, M.Y.; Chao, C.F.; Wu, Y.J.; Wu, Y.H.; Huang, C.P.; Li, Y.K. A design for fast and effective screening of hyaluronidase inhibitor using gold nanoparticles. *Sens. Actuators B Chem.* **2013**, *181*, 605–610.

179. Li, Q.; Sun, A.; Si, Y.; Chen, M.; Wu, L. One-Pot Synthesis of Polysaccharide-Diphenylalanine Ensemble with Gold Nanoparticles and Dye for Highly Efficient Detection of Glutathione. *Chem. Mater.* **2017**, *29*, 6758–6765.

180. Chen, Z.; Zhang, X.; Cao, H.; Huang, Y. Chitosan-capped silver nanoparticles as a highly selective colorimetric probe for visual detection of aromatic ortho-trihydroxy phenols. *Analyst* **2013**, *138*, 2343–2349.

181. Sergeev, A.A.; Mironenko, A.Y.; Nazirov, A.E.; Leonov, A.A.; Voznesenskii, S.S. Nanocomposite Polymer Structures for Optical Sensors of Hydrogen Sulfide. *Tech. Phys.* **2017**, *62*, 1277–1280.

182. Rastogi, P.K.; Ganesan, V.; Krishnamoorthi, S. Palladium nanoparticles decorated gaur gum based hybrid material for electrocatalytic hydrazine determination. *Electrochim. Acta* **2014**, *125*, 593–600.

183. Luo, Y.; Shen, S.; Luo, J.; Wang, X.; Sun, R. Green synthesis of silver nanoparticles in xylan solution via Tollens reaction and their detection for Hg^{2+}. *Nanoscale* **2015**, *7*, 690–700.

184. Su, H.; Liu, Y.; Wang, D.; Wu, C.; Xia, C.; Gong, Q.; Song, B.; Ai, H. Amphiphilic starlike dextran wrapped superparamagnetic iron oxide nanoparticle clsuters as effective magnetic resonance imaging probes. *Biomaterials* **2013**, *34*, 1193–1203.

185. Hemmati, B.; Javanshir, S.; Dolatkhah, Z. Hybrid magnetic Irish moss/Fe_3O_4 as a nano-biocatalyst for synthesis of imidazopyrimidine derivatives. *RSC Adv.* **2016**, *6*, 50431–50436.

186. Lee, J.S.; Saka, S. Biodiesel production by heterogeneous catalysts and supercritical technologies. *Bioresour. Technol.* **2010**, *101*, 7191–7200.

187. Grunes, J.; Zhu, J.; Somorjai, G.A. Catalysis and nanoscience. *Chem. Commun.* **2003**, *18*, 2257–2260.

188. Králik, M.; Biffis, A. Catalysis by metal nanoparticles supported on functional organic polymers. *J. Mol. Catal. A Chem.* **2001**, *177*, 113–138.

189. Chang, Y.C.; Chen, D.H. Catalytic reduction of 4-nitrophenol by magnetically recoverable Au nanocatalyst. *J. Hazard. Mater.* **2009**, *165*, 664–669.

190. Aditya, T.; Pal, A.; Pal, T. Nitroarene reduction: A trusted model reaction to test nanoparticle catalysts. *Chem. Commun.* **2015**, *51*, 9410–9431.

191. Rode, C.V.; Vaidya, M.J.; Chaudhari, R.V. Synthesis of p-Aminophenol by Catalytic Hydrogenation of Nitrobenzene. *Org. Process Res. Dev.* **1999**, *3*, 465–470.

192. Zhou, J.; Gao, J.; Xu, X.; Hong, W.; Song, Y.; Xue, R.; Zhao, H.; Liu, Y.; Qiu, H. Synthesis of porous Bi@Cs networks by a one-step hydrothermal method and their superior catalytic activity for the reduction of 4-nitrophenol. *J. Alloys Compd.* **2017**, *709*, 206–212.

193. Zheng, Z.; Huang, Q.; Guan, H.; Liu, S. In situ synthesis of silver nanoparticles dispersed or wrapped by a *Cordyceps sinensis* exopolysaccharide in water and their catalytic activity. *RSC Adv.* **2015**, *5*, 69790–69799.

194. Gao, Z.; Su, R.; Huang, R.; Qi, W.; He, Z. Glucomannan-mediated facile synthesis of gold nanoparticles for catalytic reduction of 4-nitrophenol. *Nanoscale Res. Lett.* **2014**, *9*, 404.

195. Maity, S.; Sen, I.K.; Islam, S.S. Green synthesis of gold nanoparticles using gum polysaccharide of *Cochlospermum religiosum* (katira gum) and study of catalytic activity. *Phys. E Low-Dimens. Syst. Nanostruct.* **2012**, *45*, 130–134.

196. Tripathy, T.; Kolya, H.; Jana, S.; Senapati, M. Green synthesis of Ag-Au bimetallic nanocomposites using a biodegradable synthetic graft copolymer; hydroxyethyl starch-*g*-poly(acrylamide-*co*-acrylic acid) and evaluation of their catalytic activities. *Eur. Polym. J.* **2017**, *87*, 113–123.

197. Yang, Y.; Buchwald, S.L. Ligand-controlled palladium-catalyzed regiodivergent suzuki-miyaura cross-coupling of allylboronates and aryl halides. *J. Am. Chem. Soc.* **2013**, *135*, 10642–10645.

198. Elazab, H.A.; Siamaki, A.R.; Moussa, S.; Gupton, B.F.; El-Shall, M.S. Highly efficient and magnetically recyclable graphene-supported Pd/Fe$_3$O$_4$ nanoparticle catalysts for Suzuki and Heck cross-coupling reactions. *Appl. Catal. A Gen.* **2015**, *491*, 58–69.

199. Chen, W.; Zhong, L.; Peng, X.; Lin, J.; Sun, R. Xylan-type hemicelluloses supported terpyridine-palladium(II) complex as an efficient and recyclable catalyst for Suzuki-Miyaura reaction. *Cellulose* **2014**, *21*, 125–137.

200. Chtchigrovsky, M.; Lin, Y.; Ouchaou, K.; Chaumontet, M.; Robitzer, M.; Quignard, F.; Taran, F. Dramatic effect of the gelling cation on the catalytic performances of alginate-supported palladium nanoparticles for the Suzuki-Miyaura reaction. *Chem. Mater.* **2012**, *24*, 1505–1510.

201. Arcon, I.; Paganelli, S.; Piccolo, O.; Gallo, M.; Vogel-Mikus, K.; Baldi, F. XAS analysis of iron and palladium bonded to a polysaccharide produced anaerobically by a strain of *Klebsiella oxytoca*. *J. Synchrotron Radiat.* **2015**, *22*, 1215–1226.

202. Paganelli, S.; Piccolo, O.; Baldi, F.; Tassini, R.; Gallo, M.; La Sorella, G. Aqueous biphasic hydrogenations catalyzed by new biogenerated Pd-polysaccharide species. *Appl. Catal. A Gen.* **2013**, *451*, 144–152.

203. Chook, S.W.; Chia, C.H.; Chan, C.H.; Chin, S.X.; Zakaria, S.; Sajab, M.S.; Huang, N.M. A porous aerogel nanocomposite of silver nanoparticles-functionalized cellulose nanofibrils for SERS detection and catalytic degradation of rhodamine B. *RSC Adv.* **2015**, *5*, 88915–88920.

204. Lin, S.T.; Thirumavalavan, M.; Jiang, T.Y.; Lee, J.F. Synthesis of ZnO/Zn nano photocatalyst using modified polysaccharides for photodegradation of dyes. *Carbohydr. Polym.* **2014**, *105*, 1–9.

205. Cheryl-Low, Y.L.; Theam, K.L.; Lee, H.V. Alginate-derived solid acid catalyst for esterification of low-cost palm fatty acid distillate. *Energy Convers. Manag.* **2015**, *106*, 932–940.

206. Sherly, K.B.; Rakesh, K. Synthesis and catalytic activity of polysaccharide templated nanocrystalline sulfated zirconia. *AIP Conf. Proc.* **2014**, *128*, 128–131.

207. Behar, S.; Gonzalez, P.; Agulhon, P.; Quignard, F.; Świerczyński, D. New synthesis of nanosized Cu-Mn spinels as efficient oxidation catalysts. *Catal. Today* **2012**, *189*, 35–41.

208. Porta, F.; Rossi, M. Gold nanostructured materials for the selective liquid phase catalytic oxidation. *J. Mol. Catal. A Chem.* **2003**, *204–205*, 553–559.

209. Wang, Y.; Kong, Q.; Ding, B.; Chen, Y.; Yan, X.; Wang, S.; Chen, F.; You, J.; Li, C. Bioinspired catecholic activation of marine chitin for immobilization of Ag nanoparticles as recyclable pollutant nanocatalysts. *J. Colloid Interface Sci.* **2017**, *505*, 220–229.

210. Pourjavadi, A.; Motamedi, A.; Marvdashti, Z.; Hosseini, S.H. Magnetic nanocomposite based on functionalized salep as a green support for immobilization of palladium nanoparticles: Reusable heterogeneous catalyst for Suzuki coupling reactions. *Catal. Commun.* **2017**, *97*, 27–31.

211. Li, Y.; Li, G.; Li, W.; Yang, F.; Liu, H. Greenly Synthesized Gold–Alginate Nanocomposites Catalyst for Reducing Decoloration of Azo-Dyes. *Nano* **2015**, *10*, 1550108.

212. Lee, Y.J.; Cha, S.-H.; Lee, K.J.; Kim, Y.S.; Cho, S.; Park, Y. Plant Extract (*Bupleurum falcatum*) as a Green Factory for Biofabrication of Gold Nanoparticles. *Nat. Prod. Commun.* **2015**, *10*, 1593–1596.

213. Ahmed, K.B.A.; Kalla, D.; Uppuluri, K.B.; Anbazhagan, V. Green synthesis of silver and gold nanoparticles employing levan, a biopolymer from *Acetobacter xylinum* NCIM 2526, as a reducing agent and capping agent. *Carbohydr. Polym.* **2014**, *112*, 539–545.

214. Thirumavalavan, M.; Yang, F.M.; Lee, J.F. Investigation of preparation conditions and photocatalytic efficiency of nano ZnO using different polysaccharides. *Environ. Sci. Pollut. Res.* **2013**, *20*, 5654–5664.

215. Sen, I.K.; Maity, K.; Islam, S.S. Green synthesis of gold nanoparticles using a glucan of an edible mushroom and study of catalytic activity. *Carbohydr. Polym.* **2013**, *91*, 518–528.

216. Budarin, V.L.; Clark, J.H.; Luque, R.; Macquarrie, D.J.; White, R.J. Palladium nanoparticles on polysaccharide-derived mesoporous materials and their catalytic performance in C–C coupling reactions. *Green Chem.* **2008**, *10*, 382–387.

217. Yah, C.S.; Iyuke, S.E.; Simate, G.S. A review of nanoparticles toxicity and their routes of exposures. *Iran. J. Pharm. Sci.* **2012**, *8*, 299–314.

218. Medina, C.; Santos-Martinez, M.J.; Radomski, A.; Corrigan, O.I.; Radomski, M.W. Nanoparticles: Pharmacological and toxicological significance. *Br. J. Pharmacol.* **2009**, *150*, 552–558.

219. Lam, C.-W.; James, J.T.; McCluskey, R.; Hunter, R.L. Pulmonary toxicity of single-wall carbon nanotubes in mice 7 and 90 days after intratracheal instillation. *Toxicol. Sci.* **2004**, *77*, 126–134.

220. Radomski, A.; Jurasz, P.; Alonso-Escolano, D.; Drews, M.; Morandi, M.; Malinski, T.; Radomski, M.W. Nanoparticle-induced platelet aggregation and vascular thrombosis. *Br. J. Pharmacol.* **2005**, *146*, 882–893.

221. Hussain, S.M.; Javorina, A.K.; Schrand, A.M.; Duhart, H.M.H.M.; Ali, S.F.; Schlager, J.J. The interaction of manganese nanoparticles with PC-12 cells induces dopamine depletion. *Toxicol. Sci.* **2006**, *92*, 456–463.

222. Xie, J.-H.; Jin, M.-L.; Morris, G.A.; Zha, X.-Q.; Chen, H.-Q.; Yi, Y.; Li, J.-E.; Wang, Z.-J.; Gao, J.; Nie, S.-P.; et al. Advances on Bioactive Polysaccharides from Medicinal Plants. *Crit. Rev. Food Sci. Nutr.* **2016**, *56*, S60–S84.

223. Berry, J.; Arnoux, B.; Stanislas, G.; Galle, P.; Chretien, J. A microanalytic study of particles transport across the alveoli: Role of blood platelets. *Biomedicine* **1977**, *27*, 354–357.

224. Nemmar, A.; Hoet, P.H.M.; Vanquickenborne, B.; Dinsdale, D.; Thomeer, M.; Hoylaerts, M.F.; Vanbilloen, H.; Mortelmans, L.; Nemery, B. Passage of inhaled particles into the blood circulation in humans. *Circulation* **2002**, *105*, 411–414.

225. Pooja, D.; Panyaram, S.; Kulhari, H.; Rachamalla, S.S.; Sistla, R. Xanthan gum stabilized gold nanoparticles: Characterization, biocompatibility, stability and cytotoxicity. *Carbohydr. Polym.* **2014**, *110*, 1–9.

226. Asharani, P.V.; Sethu, S.; Vadukumpully, S.; Zhong, S.; Lim, C.T.; Hande, M.P.; Valiyaveettil, S. Investigations on the structural damage in human erythrocytes exposed to silver, gold, and platinum nanoparticles. *Adv. Funct. Mater.* **2010**, *20*, 1233–1242.

227. Dhar, S.; Murawala, P.; Shiras, A.; Pokharkar, V.; Prasad, B.L.V. Gellan gum capped silver nanoparticle dispersions and hydrogels: Cytotoxicity and in vitro diffusion studies. *Nanoscale* **2012**, *4*, 563–567.

228. Amorim, M.O.R.; Gomes, D.L.; Dantas, L.A.; Viana, R.L.S.; Chiquetti, S.C.; Almeida-Lima, J.; Silva Costa, L.; Rocha, H.A.O. Fucan-coated silver nanoparticles synthesized by a green method induce human renal adenocarcinoma cell death. *Int. J. Biol. Macromol.* **2016**, *93*, 57–65.

229. Venkatpurwar, V.; Mali, V.; Bodhankar, S.; Pokharkar, V. In vitro cytotoxicity and *in vivo* sub-acute oral toxicity assessment of porphyran reduced gold nanoparticles. *Toxicol. Environ. Chem.* **2012**, *94*, 1357–1367.

230. Reena, K.; Balashanmugam, P.; Gajendiran, M.; Antony, S.A. Synthesis of Leucas Aspera Extract Loaded Gold-PLA-PEG-PLA Amphiphilic Copolymer Nanoconjugates: In Vitro Cytotoxicity and Anti-Inflammatory Activity Studies. *J. Nanosci. Nanotechnol.* **2016**, *16*, 4762–4770.

231. Worthington, K.L.S.; Adamcakova-Dodd, A.; Wongrakpanich, A.; Mudunkotuwa, I.A.; Mapuskar, K.A.; Joshi, V.B.; Allan Guymon, C.; Spitz, D.R.; Grassian, V.H.; Thorne, P.S.; et al. Chitosan coating of copper nanoparticles reduces *in vitro* toxicity and increases inflammation in the lung. *Nanotechnology* **2013**, *24*, 395101.

232. Li, J.J.; Muralikrishnan, S.; Ng, C.T.; Yung, L.Y.; Bay, B.H. Nanoparticle-induced pulmonary toxicity. *Exp. Biol. Med.* **2010**, *235*, 1025–1033.

233. Hwang, P.A.; Lin, X.Z.; Kuo, K.L.; Hsu, F.Y. Fabrication and cytotoxicity of fucoidan-cisplatin nanoparticles for macrophage and tumor cells. *Materials* **2017**, *10*, 291.

234. Braydich-Stolle, L.K.; Breitner, E.K.; Comfort, K.K.; Schlager, J.J.; Hussain, S.M. Dynamic characteristics of silver nanoparticles in physiological fluids: Toxicological implications. *Langmuir* **2014**, *30*, 15309–15316.

235. Borysov, A.; Krisanova, N.; Chunihin, O.; Ostapchenko, L.; Pozdnyakova, N.; Borisova, T. A comparative study of neurotoxic potential of synthesized polysaccharide-coated and native ferritin-based magnetic nanoparticles. *Croat. Med. J.* **2014**, *55*, 195–205.

236. Iram, F.; Iqbal, M.S.; Athar, M.M.; Saeed, M.Z.; Yasmeen, A.; Ahmad, R. Glucoxylan-mediated green synthesis of gold and silver nanoparticles and their phyto-toxicity study. *Carbohydr. Polym.* **2014**, *104*, 29–33.

237. Dhar, S.; Mali, V.; Bodhankar, S.; Shiras, A.; Prasad, B.L.V.; Pokharkar, V. Biocompatible gellan gum-reduced gold nanoparticles: Cellular uptake and subacute oral toxicity studies. *J. Appl. Toxicol.* **2011**, *31*, 411–420.

238. Devendiran, R.M.; Chinnaiyan, S. kumar; Yadav, N.K.; Moorthy, G.K.; Ramanathan, G.; Singaravelu, S.; Sivagnanam, U.T.; Perumal, P.T. Green synthesis of folic acid-conjugated gold nanoparticles with pectin as reducing/stabilizing agent for cancer theranostics. *RSC Adv.* **2016**, *6*, 29757–29768.

239. Baroli, B.; Ennas, M.G.; Loffredo, F.; Isola, M.; Pinna, R.; López-Quintela, M.A. Penetration of metallic nanoparticles in human full-thickness skin. *J. Investig. Dermatol.* **2007**, *127*, 1701–1712.

polymers

MDPI

Review

Preparation and Material Application of Amylose-Polymer Inclusion Complexes by Enzymatic Polymerization Approach

Saya Orio, Kazuya Yamamoto and Jun-ichi Kadokawa *

Department of Chemistry, Biotechnology, and Chemical Engineering, Graduate School of Science and Engineering, Kagoshima University, 1-21-40 Korimoto, Kagoshima 860-0065, Japan; k3951278@kadai.jp (S.O.); yamamoto@eng.kagoshima-u.ac.jp (K.Y.)
* Correspondence: kadokawa@eng.kagoshima-u.ac.jp; Tel.: +81-99-285-7743

Received: 23 November 2017; Accepted: 13 December 2017; Published: 18 December 2017

Abstract: This review presents our researches on the preparation and material application of inclusion complexes that comprises an amylose host and polymeric guests through phosphorylase-catalyzed enzymatic polymerization. Amylose is a well-known polysaccharide and forms inclusion complexes with various hydrophobic small molecules. Pure amylose is produced by enzymatic polymerization by using α-D-glucose 1-phosphate as a monomer and maltooligosaccharide as a primer catalyzed by phosphorylase. We determined that a propagating chain of amylose during enzymatic polymerization wraps around hydrophobic polymers present in the reaction system to form inclusion complexes. We termed this polymerization "vine-twining polymerization" because it is similar to the way vines of a plant grow around a rod. Hierarchical structured amylosic materials, such as hydrogels and films, were fabricated by inclusion complexation through vine-twining polymerization by using copolymers covalently grafted with hydrophobic guest polymers. The enzymatically produced amyloses induced complexation with the guest polymers in the intermolecular graft copolymers, which acted as cross-linking points to form supramolecular hydrogels. By including a film-formable main-chain in the graft copolymer, a supramolecular film was obtained through hydrogelation. Supramolecular polymeric materials were successfully fabricated through vine-twining polymerization by using primer-guest conjugates. The products of vine-twining polymerization form polymeric continuums of inclusion complexes, where the enzymatically produced amylose chains elongate from the conjugates included in the guest segments of the other conjugates.

Keywords: amylose host; enzymatic polymerization; hierarchical structured material; inclusion complex; vine-twining polymerization

---

1. Introduction

Biopolymers such as polysaccharides, proteins, and nucleic acids are common in Nature and play important in vivo roles [1–3]. The biological functions of polymers such as polysaccharides are achieved through both their primary chemical structures and controlled higher-order structure. Amylose is a natural linear polysaccharide with a left-handed helical conformation, which consists of glucose residues linked through α(1→4)-glycosidic linkages [3]. It is a main component of starch and functions as an energy storage molecule with the other component of starch, amylopectin. The seclusion of hydroxy groups in the glucose units to the outer side of the helix creates a hydrophobic cavity inside the helices. Therefore, amylose can act as a host to form host-guest inclusion complexes with hydrophobic guest molecules of low molecular weight through hydrophobic interactions (Figure 1) [4,5]. In addition to the traditional functions of these inclusion complexes, they can be manipulated to form higher-order materials with extended functionalities and properties suitable for specific practical applications. Polymeric guest molecules with high molecular weights are promising candidates for complexation with amylose, compared to low molecular weight guests, to achieve new functionalities. However, a limited number of studies have been reported on the complexation of amylose and polymeric guest molecules (Figure 1). Because weak hydrophobic interactions drive the incorporation of guest molecules into the amylose cavity, amylose does not have the ability to encapsulate large polymeric guests into its cavity. For the direct incorporation of polymeric guests, hydrophilic groups can be introduced at the polymer chain ends, which enhance the degree of complexation in aqueous media [6,7]. Additional methods to directly form amylose-polymer inclusion complexes include inclusion polymerization and guest-exchange approaches [8–10].

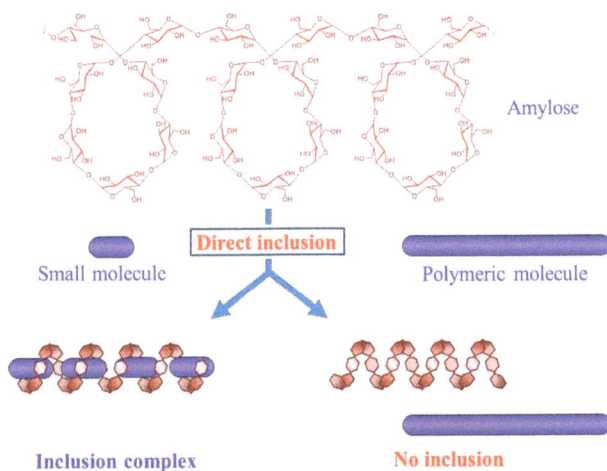

Figure 1. Amylose forms inclusion complex with relatively low molecular weight (small) hydrophobic molecule but largely, does not form it with polymeric molecule.

Recently, it has been accepted that the enzymatic approach is a powerful tool to precisely synthesize polysaccharides [11–16], and amylose with a well-defined structure can be synthesized by phosphorylase-catalyzed enzymatic polymerization of α-D-glucose 1-phosphate (G-1-P) and maltooligosaccharide as a monomer and primer, respectively (Figure 2) [17–20]. The polymerization is analogous to living polymerization because there are no significant termination or chain-transfer reactions. Accordingly, the molecular weight of the produced amylose can be controlled by monomer/primer feed ratios and typically result in narrow distributions [21]. By means of this enzymatic polymerization for the direct production of amylose, we developed an efficient method for the formation of inclusion complexes with synthetic polymers. The elongation of the short α(1→4)-glucan (maltooligosaccharide) to the longer α(1→4)-glucan (amylose) is considered to provide sufficient dynamic field for more facile complexation of polymeric guests compared to the direct complexation between the polymeric host (amylose) and guest [22–27]. The polymerization propagation is similar to the way that the vines of plants grow, twining around a rod (Figure 3). Accordingly, we proposed that this polymerization method for the production of amylose-polymer inclusion complexes should be called "vine-twining polymerization". Furthermore, the vine-twining approach has been applied to the dynamic preparation of supramolecular networks/higher-order materials [28]. This review summarizes the preparation and material application of amylose-polymer inclusion complexes fabricated by the vine-twining polymerization approach achieved in our research group.

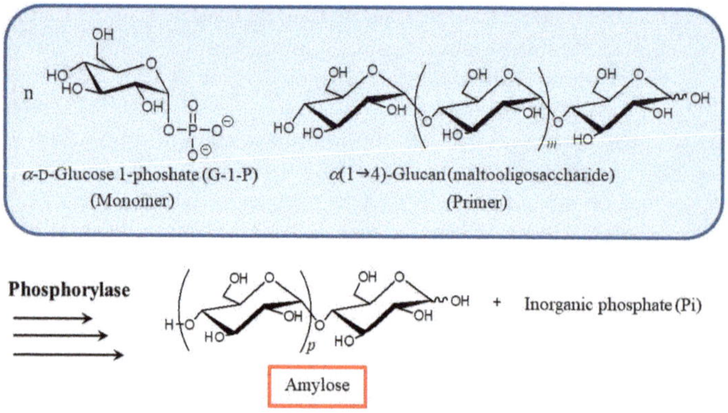

Figure 2. Phosphorylase-catalyzed enzymatic polymerization to produce amylose with well-defined structure.

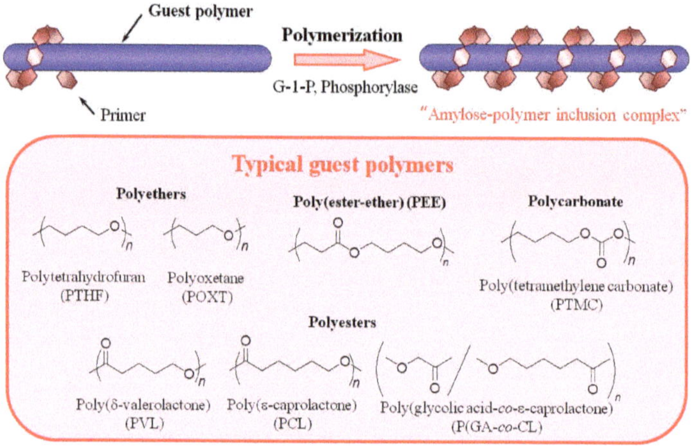

Figure 3. Image for vine-twining polymerization and typical guest polymers.

2. Preparation of Amylose-Polymer Inclusion Complexes by Enzymatic Polymerization Filed (Vine-Twining Polymerization)

In the following section, we discuss typical characteristics required by guest polymers to dynamically form inclusion complexes with amylose in vine-twining polymerization. As mentioned previously, hydrophobicity is required for inclusion complexation in the cavity of amylose. As vine-twining polymerization is conducted in an aqueous buffer solvent, the guest polymers must be able to be dispersed in aqueous media. Therefore, relatively polar groups, such as ethers and esters, should be present in the main-chain of the guest polymers. The guest polymer must also be slender without bulky substituents because of the cavity size of the amylose helix is not sufficiently large to encapsulate most bulky molecules. Based on the above features, hydrophobic synthetic polymers shown in Figure 3 have been found to act as guest polymers for the formation of inclusion complexes with amylose in vine-twining polymerization.

The first example of vine-twining polymerization was reported using polytetrahydrofuran (PTHF) as a hydrophobic guest polyether [29,30]. The structure of PTHF has been identified as suitable for guest polymers because it is generally hydrophobic, but includes relatively polar ether groups without any side groups. When the phosphorylase-catalyzed enzymatic polymerization of G-1-P from maltooligosaccharide (maltoheptaose, G$_7$) was performed in the presence of PTHF dispersed in aqueous buffer, the product was gradually precipitated from the reaction media. The subsequent characterization of the isolated product supported the proposed structure of an amylose-PTHF inclusion complex. Mixing amylose and PTHF in aqueous buffer did not result in the formation of an inclusion complex, strongly suggesting the inclusion occurs during or a result of enzymatic polymerization.

To investigate the effect of the structures of polyethers on the formation of inclusion complexes in vine-twining polymerization, the phosphorylase-catalyzed enzymatic polymerization of G-1-P was conducted using polyethers with different alkyl chain lengths including PTHF (4 methylenes), polyoxetane (POXT, 3 methylenes), and poly(ethylene glycol) (PEG, 2 methylenes) [30]. Consequently, the hydrophobic POXT formed an inclusion complex with amylose, whereas vine-twining polymerization with PEG did not induce inclusion complexation. This was likely due to the hydrophilic nature of PEG, resulting in much less hydrophobic interaction with the amylose cavity. These results highlight the importance of the hydrophobicity of guest polymers in forming inclusion complexes with amylose via vine-twining polymerization.

Hydrophobic polyesters, including poly(ε-caprolactone) (PCL), poly(δ-valerolactone) (PVL), and poly(glycolic acid-*co*-ε-caprolactone) (P(GA-*co*-CL)), have also been used as guest polymers in vine-twining polymerization to form inclusion complexes with amylose, as they contain relatively polar ester bonds in the main-chain [31–33]. However, when the homopolyester poly(glycolic acid), was used as a guest polymer it was not able to form an inclusion complex with amylose due to its high crystallinity and low dispersibility in aqueous media.

An inclusion complex was formed via vine-twining polymerization with a hydrophobic poly(ester-ether) (PEE, –CH$_2$CH$_2$C(C=O)OCH$_2$CH$_2$CH$_2$CH$_2$O–) composed of alternating ester and ether linkages [32]. A hydrophobic polycarbonate, poly(tetramethylene carbonate) (PTMC), with relatively polar carbonate bonds, also formed an inclusion complex with amylose via vine-twining polymerization [34]. On the other hand, a hydrophilic poly(ester-ether) (–CH$_2$CH$_2$C(=O)OCH$_2$CH$_2$O–) with a short methylene length, could not form an inclusion complex with amylose under the same conditions.

In addition to their inability to form inclusion complexes with hydrophilic polymers, it is difficult to produce inclusion complexes from polymers with strong hydrophobicity owing to aggregation in aqueous buffer. For example, the strongly hydrophobic polyoxepane (6 methylenes), did not form an inclusion complex with amylose via vine-twining polymerization.

Based on the aforementioned results regarding the formation of inclusion complexes through vine-twining polymerization, we have speculated that moderate hydrophobicity of the guest polymers is required for complexation with amylose. Indeed, amylose exhibits different complexation behaviors depending on subtle changes in the structures of the hydrophobic guest polymers. For example, amylose selectively included PTHF in a mixture of PTHF and POXT in vine-twining polymerization system, owing to the slight difference the hydrophobicity of the potential guest polymers (Figure 4) [35]. Also, in a mixture of PCL and PVL, amylose selectively formed an inclusion complex with PVL during vine-twining polymerization (Figure 4) [36].

Amylose selectively included a specific range of molecular weights of guest polymers in vine-twining polymerization. Synthetic polymers are generally mixtures of different molecular weight analogs, which possess different properties. For example, the molecular weight of PTHF polymers affect its hydrophobicity and water-solubility, where low molecular weight PTHF exhibits good water-solubility, whereas those with larger molecular weight are hydrophobic and insoluble in water. When several vine-twining polymerization systems were studied using PTHFs with different average molecular weights, the specific range of molecular weights of all PTHFs were suitably recognized by amylose to form inclusion complexes [24].

Besides the chemical structure and molecular weight, amylose also showed selectivity towards chirality in guest polymers in vine-twining polymerization. The selective inclusion of chiral molecules by amylose was achieved using chiral polyesters, poly(lactide)s (PLAs) as guest polymers with three stereoisomers, i.e., poly(L-lactide) (PLLA), poly(D-lactide) (PDLA), and racemic poly(DL-lactide) (PDLLA, Figure 5) [37]. When vine-twining polymerization was conducted using PLLA, an inclusion complex was formed, whereas the PDLA and PDLLA polymers did not achieve inclusion complexation.

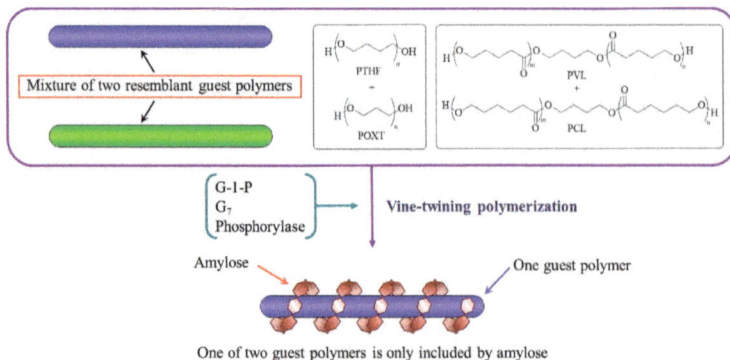

Figure 4. Amylose selectively includes one of two resembling polyethers or polyesters mixture; G-1-P = α-D-glucose 1-phosphate, G$_7$ = maltoheptaose, PTHF = polytetrahydrofuran, POXT = polyoxetane, PVL = poly(δ-valerolactone), PCL = poly(ε-caprolactone).

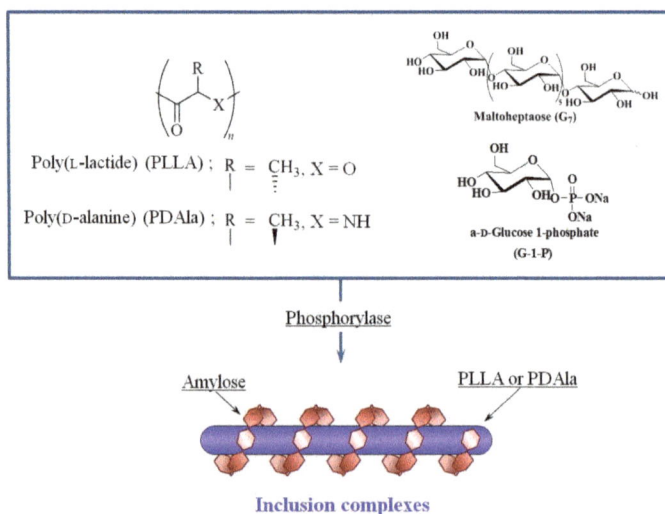

Figure 5. Stereoselective inclusion complexation by amylose in vine-twining polymerization using poly(L-lactide) (PLLA) and poly(D-alanine) (PDAla).

The selective inclusion based on chirality was also observed in vine-twining polymerization using chiral polyalanine (PAlas) stereoisomers as guest polymers (Figure 5) [38]. An inclusion complex was formed with poly(D-alanine) (PDAla), whereas inclusion complexes were not obtained with poly(L-alanine) (PLAla) or poly(DL-alanine) (PDLAla).

The stereoselective inclusion behavior of amylose toward PLLA and PDAla in vine-twining polymerization can be explained by the helical direction of the host and guest polymers. The left-handed helical conformation of the guest polymers PLLA and PDAla is the same direction as that of the host amylose, resulting in their efficient inclusion. In contrast, the opposite and irregular helical conformations of PDLA/PLAla and PDLLA/PDLAla, respectively, are not suitable for binding by the amylose helix.

3. Hierarchical Structured Materials from Amylose-Polymer Inclusion Complexes by Vine-Twining Polymerization

The vine-twining polymerization approach has been applied to the fabrication of hierarchical structured materials, such as gels and films, based on amylose-polymer inclusion complexes [28]. To construct such materials, supramolecular networks, which are hierarchically composed of inclusion complexes as crosslinking points, were designed as vine-twining polymerization products by using graft copolymers with hydrophobic graft chains. Significantly, the enzymatically produced amylose chains include the hydrophobic graft chains as guest polymers to produce inclusion complexes, which act as crosslinking points to hierarchically construct a supramolecular network structure in aqueous media, forming hydrogels (Figure 6). The hydrophobicity of the graft chains as guest polymers is necessary, while the graft copolymer should generally be water-soluble to successfully form hydrogels.

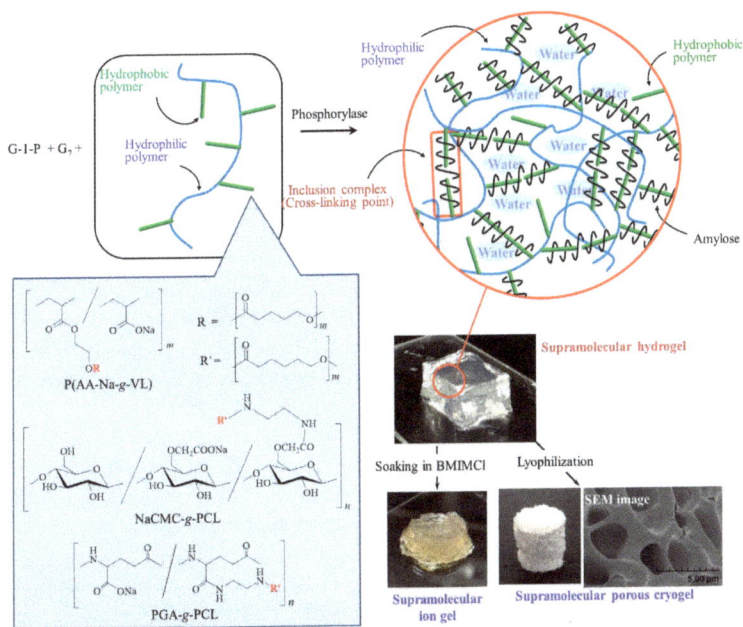

Figure 6. Preparation of hierarchical structured materials by vine-twining polymerization using graft copolymers having hydrophilic main chains and hydrophobic guest graft chains and conversion of supramolecular hydrogel into cryo- and ion gels; G-1-P = α-D-glucose 1-phosphate, G_7 = maltoheptaose, P(AA-Na-g-VL) = poly(acrylic acid sodium salt-*graft*-δ-valerolactone), NaCMC-g-PCL = carboxymethyl cellulose sodium salt-*graft*-poly(ε-caprolactone), PGA-g-PCL = poly(γ-glutamic acid)-*graft*-poly(ε-caprolactone).

The hierarchical formation of a hydrogel was achieved by the phosphorylase-catalyzed polymerization of G-1-P from G_7 in the presence of a water-soluble copolymer composed of hydrophobic PVL graft chains, poly(acrylic acid sodium salt-*graft*-δ-valerolactone) (P(AA-Na-g-VL)), by vine-twining polymerization (Figure 6) [39]. The enzymatic reaction mixture was completely converted into the hydrogel form. The enzymatically produced amylose included the PVL graft chains to form inclusion complexes as the polymerization progressed, which acted as cross-linking points for hydrogelation. Furthermore, the hydrogels were enzymatically disrupted and reproduced through combination of the β-amylase-catalyzed hydrolysis of amylose and the reformation of amylose by the phosphorylase-catalyzed polymerization.

A film was constructed through the hierarchical formation of a hydrogel by vine-twining polymerization using another graft copolymer, carboxymethyl cellulose sodium salt-*graft*-poly (ε-caprolactone) (NaCMC-*g*-PCL) (Figure 6). The reaction mixture was completely converted into the hydrogel by the vine-twining polymerization [40]. The film was formed by moisturizing the powdered sample prepared by lyophilization of the hydrogel.

The mechanical properties of the hydrogels obtained by vine-twining polymerization using PAA-Na-*g*-PVL and NaCMC-*g*-PCL were insufficient for further applications. To improve the mechanical properties of the hydrogels, poly(γ-glutamic acid) (PGA) was used as the main-chain of a graft copolymer (Figure 6) [41], because its shows better water retention and moisturizing properties. Indeed, vine-twining polymerization using poly(γ-glutamic acid)-*graft*-poly(ε-caprolactone) (PGA-*g*-PCL) resulted in a hydrogel with self-standing properties, indicating much better mechanical properties compared to the aforementioned hydrogels. The prepared hydrogel exhibited macroscopic interfacial healing behavior upon the phosphorylase-catalyzed enzymatic polymerization. The hydrogel formed initially from the vine-twining polymerization was cut into two pieces, and G-1-P and phosphorylase-containing sodium acetate buffer was dropped on the surface of the hydrogels. After the surfaces were placed in contact with one another, the materials were left standing for enzymatic polymerization. Consequently, the two hydrogel pieces were fused at the contacted area. Such behavior of the hydrogels on a macroscopic level was induced by the complexation of the enzymatically produced amyloses with the PCL graft chains at the interface. In addition, a porous cryogel and an ion gel were obtained by lyophilization and soaking of the hydrogel in an ionic liquid of 1-butyl-3-methylimidazolium chloride (BMIMCl) (Figure 6).

Supramolecular polymers composed of amylose-PTHF and amylose-PLLA inclusion complexes were dynamically formed by vine-twining polymerization using primer–guest conjugates, i.e., maltoheptaose-*block*-polytetrahydrofuran (G$_7$-*block*-PTHF) and maltoheptaose-*block*-poly(L-lactide) (G$_7$-*block*-PLLA, Figure 7a) [42,43]. In these systems, an enzymatically propagating amylose chain included a PTHF or PLLA segment of another conjugate, whereby consecutive inclusion led to the formation of linear supramolecular polymers.

Vine-twining polymerization using a branched maltoheptaose-(poly(L-lactide))$_2$ (G$_7$-PLLA$_2$) conjugate resulted in a hyperbranched supramolecular polymer (Figure 7b) [44]. The hyperbranched product formed an ion gel with BMIMCl, which was further converted into a hydrogel upon exchange of the dispersion media by soaking in water. Lyophilization of the resulting hydrogel produced a porous cryogel.

The relative chain orientation of amylose and PLLA in the supramolecular polymers was investigated using two G$_7$-*block*-PLLA conjugates, which were composed of a G$_7$ moiety interconnected to the carboxylate or hydroxy terminus of PLLA [45]. Enzymatic polymerization in the presence of the two PLLA conjugates formed amylose-PLLA supramolecular polymers by vine-twining polymerization. This suggested that, regardless of the chain orientation of PLLA, the amylose cavity recognized the PLLA segment for complexation. Conversely, the phosphorylase-catalyzed enzymatic polymerization in the presence of the two G$_7$-*block*-PDLA conjugates under similar conditions only resulted in the formation of amylose-PDLA diblock copolymers, which did not form an inclusion complex structure. These results indicated that chirality in PLAs affected the inclusion behavior of the amylose cavity, irrespective of the PLA chain orientation. The left-handed helices of both the amylose and PLLA induce inclusion complexation, whereas complexation was not significantly affected by the orientation of the methyl substituents in PLA, which oppositely change according to the relative chain orientation.

Polymers **2017**, *9*, 729

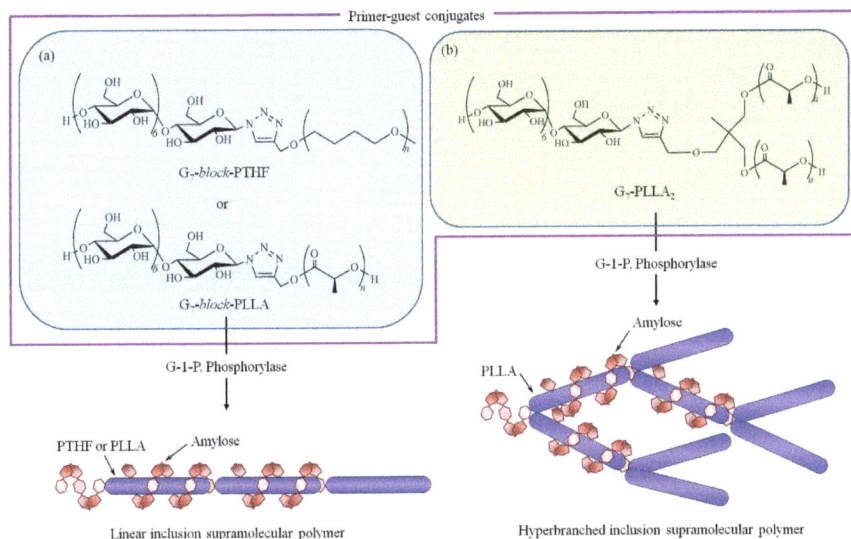

Figure 7. Formation of (**a**) linear and (**b**) hyperbranched supramolecular polymers by vine-twining polymerization using primer–guest conjugates; G-1-P = α-D-glucose 1-phosphate, G_7-*block*-PTHF = maltoheptaose-*block*-polytetrahydrofuran, G_7-*block*-PLLA = maltoheptaose-*block*-poly(L-lactide), G_7-PLLA$_2$ = branched maltoheptaose-(poly(L-lactide))$_2$.

4. Conclusions

In this review, we presented our studies on the precision preparation of amylose-polymer inclusion complexes through the phosphorylase-catalyzed enzymatic polymerization of G-1-P in the presence of synthetic hydrophobic polymers by vine-twining polymerization. The results of the vine-twining polymerization study suggested that amylose exhibited different inclusion behaviors depending on the specific interactions with the guest polymers according to subtle changes in their structures. Moreover, hierarchical structured materials could be dynamically fabricated by vine-twining polymerization using designed graft copolymers composed of hydrophilic main chains and hydrophobic guest graft chains. Vine-twining polymerization using guest-primer conjugates afforded the dynamic formation of the supramolecular polymers composed of a continuum of inclusion complexes. Because of the production of a structurally defined amylose host by phosphorylase catalysis, the precision preparation of controlled amylosic host-guest polymeric complexes was achieved. Therefore, the vine-twining polymerization method can be applied to the production of additional amylosic inclusion complexes with regularly controlled nanostructure, and will contribute to further developments of host-guest chemistry in the future.

Acknowledgments: The authors are indebted to the co-workers, whose names are found in references from their papers, for their enthusiastic collaborations. The authors also gratefully thank for financial supports from a Grant-in-Aid for Scientific Research from Ministry of Education, Culture, Sports, and Technology, Japan (No. 14550830, 17550118, 19550126, 24350062, and 17K06001) and Sekisui Foundation.

Author Contributions: The manuscript was written through contributions of all authors. All authors have given approval to the final version of the manuscript.

Conflicts of Interest: The authors declare no conflict of interest.

References

1.   Berg, J.M.; Tymoczko, J.L.; Stryer, L. *Biochemistry*, 7th ed.; W.H. Freeman: New York, NY, USA, 2012; pp. 1041, 1043, 1048, 1054.

2. Kasapis, S.; Norton, I.T.; Ubbink, J.B. *Modern Biopolymer Science: Bridging the Divide between Fundamental Treatise and Industrial Application*; Academic Press: San Diego, CA, USA, 2009.

3. Schuerch, C. Polysaccharides. In *Encyclopedia of Polymer Science and Engineering*, 2nd ed.; Mark, H.F., Bilkales, N., Overberger, C.G., Eds.; John Wiley & Sons: New York, NY, USA, 1986; Volume 13, pp. 87–162.

4. Sarko, A.; Zugenmaier, P. Crystal structures of amylose and its derivatives. In *Fiber Diffraction Methods*; ACS Symposium Series 141; French, A.D., Gardner, K.H., Eds.; American Chemical Society: Washington, DC, USA, 1980; pp. 459–482.

5. Putseys, J.A.; Lamberts, L.; Delcour, J.A. Amylose-inclusion complexes: Formation, identity and physico-chemical properties. *J. Cereal Sci.* **2010**, *51*, 238–247. [CrossRef]

6. Shogren, R.L. Complexes of starch with telechelic poly(epsilon-caprolactone) phosphate. *Carbohydr. Polym.* **1993**, *22*, 93–98. [CrossRef]

7. Shogren, R.L.; Greene, R.V.; Wu, Y.V. Complexes of starch polysaccharides and poly(ethylene coacrylic acid)—structure and stability in solution. *J. Appl. Polym. Sci.* **1991**, *42*, 1701–1709. [CrossRef]

8. Star, A.; Steuerman, D.W.; Heath, J.R.; Stoddart, J.F. Starched carbon nanotubes. *Angew. Chem. Int. Ed.* **2002**, *41*, 2508–2512. [CrossRef]

9. Ikeda, M.; Furusho, Y.; Okoshi, K.; Tanahara, S.; Maeda, K.; Nishino, S.; Mori, T.; Yashima, E. A luminescent poly(phenylenevinylene)-amylose composite with supramolecular liquid crystallinity. *Angew. Chem. Int. Ed.* **2006**, *45*, 6491–6495. [CrossRef] [PubMed]

10. Kumar, K.; Woortman, A.J.J.; Loos, K. Synthesis of amylose-polystyrene inclusion complexes by a facile preparation route. *Biomacromolecules* **2013**, *14*, 1955–1960. [CrossRef] [PubMed]

11. Kobayashi, S.; Uyama, H.; Kimura, S. Enzymatic polymerization. *Chem. Rev.* **2001**, *101*, 3793–3818. [CrossRef] [PubMed]

12. Shoda, S.; Izumi, R.; Fujita, M. Green process in glycotechnology. *Bull. Chem. Soc. Jpn.* **2003**, *76*, 1–13. [CrossRef]

13. Kobayashi, S.; Makino, A. Enzymatic polymer synthesis: An opportunity for green polymer chemistry. *Chem. Rev.* **2009**, *109*, 5288–5353. [CrossRef] [PubMed]

14. Kadokawa, J. Precision polysaccharide synthesis catalyzed by enzymes. *Chem. Rev.* **2011**, *111*, 4308–4345. [CrossRef] [PubMed]

15. Shoda, S.; Uyama, H.; Kadokawa, J.; Kimura, S.; Kobayashi, S. Enzymes as green catalysts for precision macromolecular synthesis. *Chem. Rev.* **2016**, *116*, 2307–2413. [CrossRef] [PubMed]

16. Kadokawa, J. α-Glucan phosphorylase: A useful catalyst for precision enzymatic synthesis of oligo- and polysaccharides. *Curr. Org. Chem.* **2017**, *21*, 1192–1204. [CrossRef]

17. Seibel, J.; Jordening, H.J.; Buchholz, K. Glycosylation with activated sugars using glycosyltransferases and transglycosidases. *Biocatal. Biotransform.* **2006**, *24*, 311–342. [CrossRef]

18. Ziegast, G.; Pfannemüller, B. Phosphorolytic syntheses with di-functional, oligo-functional and multifunctional primers. *Carbohydr. Res.* **1987**, *160*, 185–204. [CrossRef]

19. Fujii, K.; Takata, H.; Yanase, M.; Terada, Y.; Ohdan, K.; Takaha, T.; Okada, S.; Kuriki, T. Bioengineering and application of novel glucose polymers. *Biocatal. Biotransform.* **2003**, *21*, 167–172. [CrossRef]

20. Yanase, M.; Takaha, T.; Kuriki, T. α-Glucan phosphorylase and its use in carbohydrate engineering. *J. Sci. Food Agric.* **2006**, *86*, 1631–1635. [CrossRef]

21. Kitamura, S. Starch polymers, natural and synthetic. In *The Polymeric Materials Encyclopedia, Synthesis, Properties and Applications*; Salamone, C., Ed.; CRC Press: New York, NY, USA, 1996; Volume 10, pp. 7915–7922.

22. Kaneko, Y.; Kadokawa, J. Vine-twining polymerization: A new preparation method for well-defined supramolecules composed of amylose and synthetic polymers. *Chem. Rec.* **2005**, *5*, 36–46. [CrossRef] [PubMed]

23. Kaneko, Y.; Kadokawa, J. Synthesis of nanostructured bio-related materials by hybridization of synthetic polymers with polysaccharides or saccharide residues. *J. Biomater. Sci. Polym. Ed.* **2006**, *17*, 1269–1284. [CrossRef] [PubMed]

24. Kaneko, Y.; Beppu, K.; Kadokawa, J. Amylose selectively includes a specific range of molecular weights in poly(tetrahydrofuran)s in vine-twining polymerization. *Polym. J.* **2009**, *41*, 792–796. [CrossRef]

25. Kadokawa, J. Preparation and applications of amylose supramolecules by means of phosphorylase-catalyzed enzymatic polymerization. *Polymers* **2012**, *4*, 116–133. [CrossRef]

26. Kadokawa, J. Architecture of amylose supramolecules in form of inclusion complexes by phosphorylase-catalyzed enzymatic polymerization. *Biomolecules* **2013**, *3*, 369–385. [CrossRef] [PubMed]

27. Kadokawa, J. Chemoenzymatic synthesis of functional amylosic materials. *Pure Appl. Chem.* **2014**, *86*, 701–709. [CrossRef]

28. Kadokawa, J. Hierarchically fabrication of amylosic supramolecular nanocomposites by means of inclusion complexation in phosphorylase-catalyzed enzymatic polymerization field. In *Eco-Friendly Polymer Nanocomposites: Processing and Properties*; Thakur, K.V., Thakur, K.M., Eds.; Springer India: New Delhi, India, 2015; pp. 513–525.

29. Kadokawa, J.; Kaneko, Y.; Tagaya, H.; Chiba, K. Synthesis of an amylose-polymer inclusion complex by enzymatic polymerization of glucose 1-phosphate catalyzed by phosphorylase enzyme in the presence of polythf: A new method for synthesis of polymer-polymer inclusion complexes. *Chem. Commun.* **2001**, 449–450. [CrossRef]

30. Kadokawa, J.; Kaneko, Y.; Nagase, S.; Takahashi, T.; Tagaya, H. Vine-twining polymerization: Amylose twines around polyethers to form amylose—polyether inclusion complexes. *Chem. Eur. J.* **2002**, *8*, 3321–3326. [CrossRef]

31. Kadokawa, J.; Kaneko, Y.; Nakaya, A.; Tagaya, H. Formation of an amylose-polyester inclusion complex by means of phosphorylase-catalyzed enzymatic polymerization of a-D-glucose 1-phosphate monomer in the presence of poly(e-caprolactone). *Macromolecules* **2001**, *34*, 6536–6538. [CrossRef]

32. Kadokawa, J.; Nakaya, A.; Kaneko, Y.; Tagaya, H. Preparation of inclusion complexes between amylose and ester-containing polymers by means of vine-twining polymerization. *Macromol. Chem. Phys.* **2003**, *204*, 1451–1457. [CrossRef]

33. Nomura, S.; Kyutoku, T.; Shimomura, N.; Kaneko, Y.; Kadokawa, J. Preparation of inclusion complexes composed of amylose and biodegradable poly(glycolic acid-*co*-e-caprolactone) by vine-twining polymerization and their lipase-catalyzed hydrolysis behavior. *Polym. J.* **2011**, *43*, 971–977. [CrossRef]

34. Kaneko, Y.; Beppu, K.; Kadokawa, J. Preparation of amylose/polycarbonate inclusion complexes by means of vine-twining polymerization. *Macromol. Chem. Phys.* **2008**, *209*, 1037–1042. [CrossRef]

35. Kaneko, Y.; Beppu, K.; Kadokawa, J. Amylose selectively includes one from a mixture of two resemblant polyethers in vine-twining polymerization. *Biomacromolecules* **2007**, *8*, 2983–2985. [CrossRef] [PubMed]

36. Kaneko, Y.; Beppu, K.; Kyutoku, T.; Kadokawa, J. Selectivity and priority on inclusion of amylose toward guest polyethers and polyesters in vine-twining polymerization. *Polym. J.* **2009**, *41*, 279–286. [CrossRef]

37. Kaneko, Y.; Ueno, K.; Yui, T.; Nakahara, K.; Kadokawa, J. Amylose's recognition of chirality in polylactides on formation of inclusion complexes in vine-twining polymerization. *Macromol. Biosci.* **2011**, *11*, 1407–1415. [CrossRef] [PubMed]

38. Gotanda, R.; Yamamoto, K.; Kadokawa, J.-I. Amylose stereoselectively includes poly(D-alanine) to form inclusion complex in vine-twining polymerization: A novel saccharide-peptide supramolecular conjugate. *Macromol. Chem. Phys.* **2016**. [CrossRef]

39. Kaneko, Y.; Fujisaki, K.; Kyutoku, T.; Furukawa, H.; Kadokawa, J. Preparation of enzymatically recyclable hydrogels through the formation of inclusion complexes of amylose in a vine-twining polymerization. *Chem. Asian J.* **2010**, *5*, 1627–1633. [CrossRef] [PubMed]

40. Kadokawa, J.; Nomura, S.; Hatanaka, D.; Yamamoto, K. Preparation of polysaccharide supramolecular films by vine-twining polymerization approach. *Carbohydr. Polym.* **2013**, *98*, 611–617. [CrossRef] [PubMed]

41. Kadokawa, J.; Tanaka, K.; Hatanaka, D.; Yamamoto, K. Preparation of multiformable supramolecular gels through helical complexation by amylose in vine-twining polymerization. *Polym. Chem.* **2015**, *6*, 6402–6408. [CrossRef]

42. Tanaka, T.; Sasayama, S.; Nomura, S.; Yamamoto, K.; Kimura, Y.; Kadokawa, J. An amylose-poly(L-lactide) inclusion supramolecular polymer: Enzymatic synthesis by means of vine-twining polymerization using a primer-guest conjugate. *Macromol. Chem. Phys.* **2013**, *214*, 2829–2834. [CrossRef]

43. Tanaka, T.; Tsutsui, A.; Gotanda, R.; Sasayama, S.; Yamamoto, K.; Kadokawa, J. Synthesis of amylose-polyether inclusion supramolecular polymers by vine-twining polymerization using maltoheptaose-functionalized poly(tetrahydrofuran) as a primer-guest conjugate. *J. Appl. Glycosci.* **2015**, *62*, 135–141. [CrossRef]

Polymers **2017**, *9*, 729

44. Tanaka, T.; Gotanda, R.; Tsutsui, A.; Sasayama, S.; Yamamoto, K.; Kimura, Y.; Kadokawa, J. Synthesis and gel formation of hyperbranched supramolecular polymer by vine-twining polymerization using branched primer-guest conjugate. *Polymer* **2015**, *73*, 9–16. [CrossRef]

45. Tanaka, T.; Sasayama, S.; Yamamoto, K.; Kimura, Y.; Kadokawa, J. Evaluating relative chain orientation of amylose and poly(L-lactide) in inclusion complexes formed by vine-twining polymerization using primer–guest conjugates. *Macromol. Chem. Phys.* **2015**, *216*, 794–800. [CrossRef]

polymers

【MDPI】

Review

Aliphatic Polyester Nanofibers Functionalized with Cyclodextrins and Cyclodextrin-Guest Inclusion Complexes

Ganesh Narayanan [1,*], Jialong Shen [1], Ramiz Boy [2], Bhupender S. Gupta [1,3] and Alan E. Tonelli [1,3,*]

[1] Fiber and Polymer Science Program, North Carolina State University, Raleigh, NC 27695, USA; jshen3@ncsu.edu (J.S.); bgupta@ncsu.edu (B.S.G.)
[2] Department of Textile Engineering, Namık Kemal University, Corlu/Tekirdag 59860, Turkey; rboy@nku.edu.tr
[3] Department of Textile Engineering Chemistry and Science, North Carolina State University, Raleigh, NC 27695, USA
* Correspondence: gnaraya@ncsu.edu (G.N.); atonelli@ncsu.edu (A.E.T.); Tel.: +1-973-204-3039 (G.N.); +1-919-515-6588 (A.E.T.)

Received: 27 February 2018; Accepted: 4 April 2018; Published: 11 April 2018

Abstract: The fabrication of nanofibers by electrospinning has gained popularity in the past two decades; however, only in this decade, have polymeric nanofibers been functionalized using cyclodextrins (CDs) or their inclusion complexes (ICs). By combining electrospinning of polymers with free CDs, nanofibers can be fabricated that are capable of capturing small molecules, such as wound odors or environmental toxins in water and air. Likewise, combining polymers with cyclodextrin-inclusion complexes (CD-ICs), has shown promise in enhancing or controlling the delivery of small molecule guests, by minor tweaking in the technique utilized in fabricating these nanofibers, for example, by forming core–shell or multilayered structures and conventional electrospinning, for controlled and rapid delivery, respectively. In addition to small molecule delivery, the thermomechanical properties of the polymers can be significantly improved, as our group has shown recently, by adding non-stoichiometric inclusion complexes to the polymeric nanofibers. We recently reported and thoroughly characterized the fabrication of polypseudorotaxane (PpR) nanofibers without a polymeric carrier. These PpR nanofibers show unusual rheological and thermomechanical properties, even when the coverage of those polymer chains is relatively sparse (~3%). A key advantage of these PpR nanofibers is the presence of relatively stable hydroxyl groups on the outer surface of the nanofibers, which can subsequently be taken advantage of for bioconjugation, making them suitable for biomedical applications. Although the number of studies in this area is limited, initial results suggest significant potential for bone tissue engineering, and with additional bioconjugation in other areas of tissue engineering. In addition, the behaviors and uses of aliphatic polyester nanofibers functionalized with CDs and CD-ICs are briefly described and summarized. Based on these observations, we attempt to draw conclusions for each of these combinations, and the relationships that exist between their presence and the functional behaviors of their nanofibers.

Keywords: poly(lactic acid); poly(ε-caprolactone); cyclodextrins; cyclodextrin-inclusion complexes; controlled drug delivery; rapid dissolution; pseudorotaxanes; mechanical properties

---

1. Introduction

Electrospinning is a widely utilized technique to fabricate nanofibers with intricate architectures, making this process highly suitable for tissue engineering and environmental, energy, sensor, textile, food packaging, and agricultural applications [1–3]. Reasons for the widespread use of nanofibers include their large surface to volume ratio with high porosities, and very low weight, compared to bulk fibrous materials [4,5]. In addition to their nano-scale dimensions, electrospun nanofibers are of special interest in biomedical applications, as they imitate the native extracellular matrix of physiologic tissues [4,6]. Another advantage of nanofibers is their capability of delivering drugs that are otherwise impermeable to physiologic tissues [7]. Aliphatic polyesters belong to the biodegradable polymers that are currently widely used in various biomedical applications, including for treating orthopaedic, oral and maxillofacial, and neurological ailments [4,8]. Despite their potential in various medical applications, aliphatic polyesters lack surface epitopes, have poor biomechanical properties (low modulus or brittleness, for example), and longer degradation times (polycaprolactone, for example), thereby necessitating the incorporation of additional molecules to modulate these characteristics [4].

Cyclodextrins (CDs) are cyclic oligosaccharides, essentially composed of α-D glucopyranose units joined by α-1,4 glycosidic bonds [9]. The most commonly used CDs are α-, β-, and γ-CDs. These have 6–8 glucopyranose units and all CD types have truncated cone-like structures with a hollow cavity [10]. Because of the presence of hydrophobic hollow cavities (Figure 1), CDs can encapsulate both small and large molecules, forming inclusion or, in the case of polymer guests, non-stoichiometric inclusion complexes, depending on the stoichiometry used between the host CD and guest molecules, resulting in the improved stability of the included guest molecules against heat, light, and other environmental conditions [11,12]. In addition, native CDs has been modified, resulting in modified CDs, such as random methyl, hydroxypropyl, peracetyl, sulfobutylether, and monochlorotriazinyl CDs, with improved water solubility compared to native CDs [11,13]. Both native as well as modified CDs have hydrophilic outer surfaces, in addition to hydrophobic cavities, facilitating secondary interactions with the included guest molecules [11].

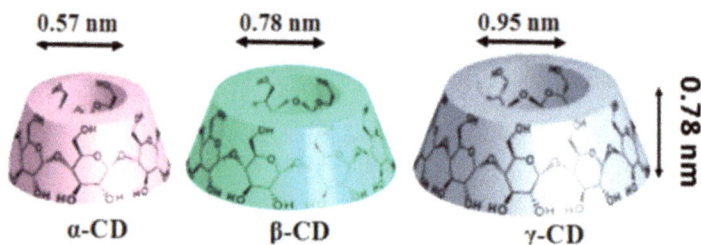

Figure 1. Schematic representations of the native cyclodextrins (CDs) (α-, β-, and γ-CDs). All three native CDs have a similar depth or height (7.9 Å); however, a key difference lies in their inner diameters, with α-, β- and γ-CDs having inner diameters of 5.7, 7.8, and 9.5 Å, respectively. Images adapted and reproduced from [11]. Copyright 2017 Elsevier Ltd.

Their unique structures make CDs capable of encapsulating smaller, as well as larger guest molecules (for example, polymers), resulting in the successful encapsulation of drugs, food ingredients (vitamins and fish oil, for example), light and air sensitive ingredients, as well as antibacterial/microbial agents [14,15]. Currently, a vast number of CDs are used in pharmaceutical formulations to improve the water solubility of hydrophobic drugs or to improve delivery rates to physiologic tissues, and/or in regenerative medicine [16–23]. Combining aliphatic polyesters with CDs, in particular, in nanofibrous form, can lead to interesting properties that can improve the versatility of these polymers for a variety of applications [24]. In this review, we summarize the various aliphatic polyester-based nanofibers, in particular those of biomedically relevant polymers, such as poly(ε-caprolactone) (PCL) and poly (lactic acid) (PLA), whose properties have been modified by either CDs or CD-inclusion complexes (CD-ICs), and that have been reported in the past decade. Although other biomedically relevant polymers, such as chitosan [25,26] and poly(butylene succinate-co-terephthalate) (PBST) [27], have also been modified with CDs or ICs. As those studies are far fewer, they are not addressed in this manuscript.

2. Cyclodextrin Functionalized Aliphatic Polyester Nanofibers

2.1. Poly(ε-caprolactone) (PCL) Nanofibers Functionalized with Cyclodextrins or Small Molecule Inclusion Complexes

The CDs were simply combined with aliphatic polyester solutions (for example, poly(ε-caprolactone) (PCL) and poly(lactic acid) (PLA)), and subsequently electrospun, resulting in nanofibers with improved hydrophilicity and/or crystallinity profiles. Although this procedure has been employed for electrospinning various other polymer/CD combinations, such as polymethyl methacrylate [28–30], polystyrene [31–33], polyurethane [34,35], chitosan [36], poly(vinyl alcohol) [37–39], poly(ethylene terephthalate) [40], and polyethylene oxide [41], the key difference in aliphatic polyester/CD combinations is the lack of a suitable common solvent (for example, dimethyl formamide (DMF)), necessitating the use of binary solvent mixtures, such as chloroform or dichloromethane with DMF. However, the use of binary solvent mixtures is not without limitations, as such mixtures could lead to the inadvertent formation of ICs between polymers and CDs. It should also be noted that a common route for preparing ICs is by first dissolving CDs in water and guest molecules in organic solvents, and subsequently combining the two solutions. As binary solvent mixtures necessitate the use of organic solvents and water for dissolving guest and host molecules, respectively, mixing the solutions would inadvertently lead to the formation of ICs.

Despite these challenges, our research group first reported the successful fabrication of PCL nanofibers containing α- and γ-CDs [42], and PCL with β-CDs [43]. In our first study, electrospinning of PCL with α- and γ-CDs was conducted, and subsequently the thermal, morphological, and water contact angle properties of the CD functionalized polymeric nanofibers were evaluated. With the incorporation of CDs, bead-free nano-sized fibers were obtained up to a loading of 15% (both α- and γ-CDs), with respect to PCL mass (Figure 2) [42]. Beyond the 15% addition, some thicker micro-sized fibers were obtained, although average fiber sizes remained at ~800 nm. With the addition of CD, major differences in the thermal properties of the PCL/CD composites, especially higher crystallinity and crystallization temperatures, were observed [42]. For instance, the crystallinity of PCL (from films), which is usually ~45%, increased dramatically to 53% and 65%, respectively, with the addition of 10% and 30% γ-CD. Furthermore, addition of CDs also resulted in a drastic reduction in the water contact angle (WCA) values of the composites. This effect of CD addition influencing WCA values was more prominent in PCL/α-CD composites at lower concentrations (5%, 10%, and 15%), and in PCL/γ-CD composites at higher concentrations (30% and 40%) [42] (Figure 3). Notwithstanding the thermal and morphological improvements, in addition to WCA values, thermogravimetric (TGA) and wide-angle diffraction (WAXD) analyses also showed the presence of some inclusion complexation between PCL and CDs. By virtue of lower weight loss at 370 °C, α-CD containing PCL composites illustrated the formation of IC between PCL and α-CD, compared to PCL/γ-CD at similar CD loadings [42]. In accordance with the TGA results, WAXD analyses of PCL/α-CD composites (10% CD loading) showed the presence of a diffraction peak at $2\theta = 20°$, in addition to PCL peaks at $2\theta = 22°$ (110 reflection) and 24° (220 reflection), indicating the presence of PCL/α-CD in the complexed form, rather than in the neat state (Figure 4) [42]. In summary, this promising study reported the fabrication of PCL/α-CD composite nanofibers, with minimal formation of ICs between PCL and α-CD.

The formation of ICs between PCL and CDs (α- and γ-CD) can, to some degree, be expected, as both α- and γ-CD have shown an affinity towards IC formation [44–49]. However, due to the mismatch in the size between PCL and β-CD, such IC formation is not expected [50,51], leading to the formation of IC-free CD functionalized nanofibers. By employing the identical solutions and processing parameters used for fabricating PCL/α-CD and PCL/γ-CD nanofibers [42], PCL/β-CD nanofibers were fabricated via electrospinning [43]. Unlike PCL/α- or γ-CD nanofibers, CD loading of up to 50% was feasible without detrimental effects on fiber size (570 ± 310 nm) and morphology (bead-free nanofibers) (Figure 5). In addition, the absence of IC peaks from WAXD and higher mass loss in TGA analyses, showed a lack of IC formation between PCL and β-CD. More importantly, when incubated with wound odor emulating solutions containing butyric and propionic acids, X-ray photoelectron spectroscopy (XPS) and TGA analyses demonstrated higher volume capture of these molecules in PCL nanofibers containing β-CDs. This effect was observed, in particular, in those nanofibers containing higher loadings of β-CDs (>20%), compared to the control nanofibers (PCL), demonstrating the potential for these CD-containing nanofibers in wound dressings [43].

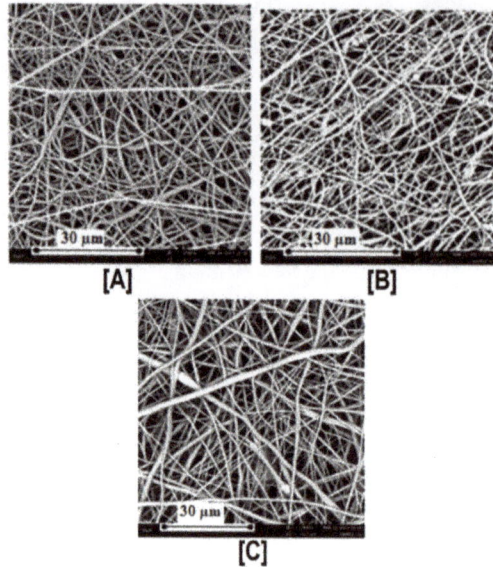

Figure 2. Scanning Electron Micrographs (SEM) of PCL and PCL/γ-CD functionalized nanofibers. By simply adjusting the PCL concentration from 14% to 12%, an increase in γ-CD loading (from 15% to upwards of 30%) was made possible without significant increases in the beaded structures in the nanofibers (**A–C**). Images adapted and reproduced with permission from [42]. Copyright American Chemical Society 2014.

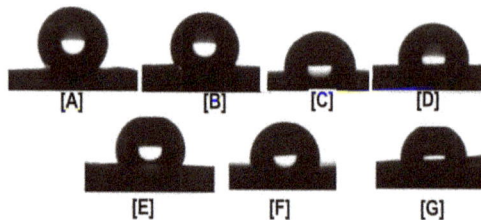

Figure 3. Water Contact Angle (WCA) values of PCL, PCL/α-CD, and PCL/γ-CD functionalized nanofibers. The neat PCL showed hydrophobic characteristics (WCA of 140°), while the WCA of functionalized nanofibers showed a marked decrease in hydrophobic characteristics, even with the addition of as little as 5% CD (WCA of PCL/5% α-CD and PCL/5% γ-CD are ~120°). With further addition, WCA plateaued at ~100°, depending on the CD used. Images adapted and reproduced with permission from [43]. Copyright American Chemical Society 2014.

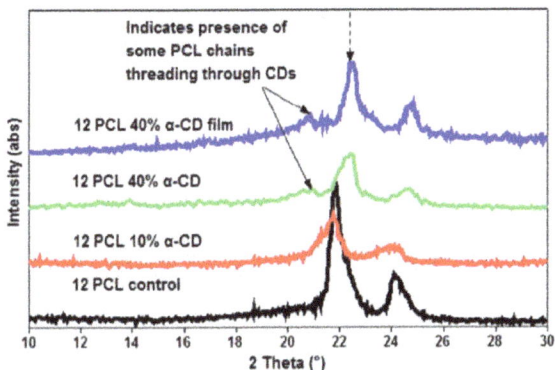

Figure 4. Wide-angle diffraction (WAXD) patterns of the electrospun PCL, α-CD functionalized electrospun PCL nanofibers, and PCL-α-CD film cast from solution. Neat PCL showed two peaks at 22° and 24° 2θ corresponding to (110) and (200) reflections. While the lack of any additional peaks in 10% α-CD containing PCL nanofibers, the presence of an additional peak at 20° 2θ and shifts in the PCL peaks, indicate the presence of some inclusion complex between PCL and α-CD. Images adapted and reproduced from [42]. Copyright American Chemical Society 2014.

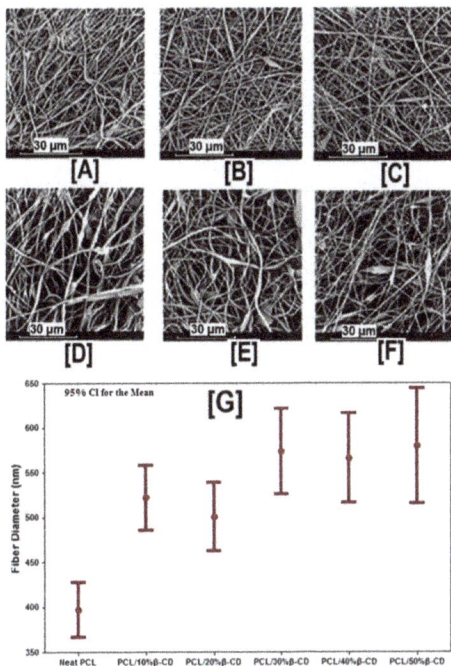

Figure 5. SEM images and average fiber diameters of β-CD functionalized PCL nanofibers at various β-CD loadings (0% to 50%). SEM images indicate the absence of beaded structures in PCL and PCL/β-CD nanofibers. While a marginal increase in fiber size was observed with the addition of β-CD (520 ± 200 nm for 10% β-CD vs. 400 ± 160 nm for neat PCL), further increases in β-CD concentration did not lead to further increases in fiber diameter. Images adapted and reproduced with permission from [43]. Copyright John Wiley and Sons 2015.

The encapsulation of small guest molecules in CD cavities enhances their stability, and in addition, facilitates melt or solution processing, irrespective of the nature of the guest molecules. Our research group has shown successful encapsulation of various guest molecules (triclosan, neomycin sulfate, antiblaze RD-1, nonoxynol-9, and azo-dyes) and subsequent melt processing with polymers [52–56]. By employing such a methodology, not only can the stability and controlled release of the guest molecules be ensured, but it also facilitates the fabrication of facile commercial products containing small guest molecules [56]. This approach is particularly advantageous when combined with nanofibers, as such products would, in addition, offer large surface areas with low weight, unlike those observed with bulk composite materials.

The unique advantages of solid CD-ICs and nanofibers via electrospinning have been exploited in the past decade, resulting in materials suitable for a wide range of applications. For instance, α-tocopherol (α-TC), a potent form of vitamin-E is frequently used in wound dressings to promote wound healing. However, α-TC is not only poorly water soluble, but also susceptible to degrade under light and oxygen, for example [57]. When α-TC was encapsulated in β-CD, electrospun and carried by PCL nanofibers, it showed improvements in photostability (~6% higher than un-encapsulated form) and antioxidant activity, despite only a marginal increase in fiber diameter (345 ± 140 vs. 205 ± 115 nm for PCL/α-TC nanofibers without β-CD) [57].

In addition to enhancing stability against various environmental factors, enhancing the delivery of drugs, in particular hydrophobic ones, is also another critical factor that has been explored using cyclodextrins as a carrier [58]. Naproxen and sulfisoxazole are hydrophobic drugs that are widely used for relieving pain. Due to their poor solubility, both drugs suffer from slow dissolution and, therefore, are not significantly available at the target site. One common way to improve their solubility and availability is by safely encapsulating them inside CD cavities [59–62]. When naproxen (NAP) was encapsulated in β-CD-IC cavities via traditional IC formation means, and was subsequently embedded in a PCL nanofibrous matrix, high performance liquid chromatography (HPLC) revealed two-fold improvements in the release rates of NAP, compared to PCL nanofibers with neat NAP (Figure 6A) [63]. This is not particularly surprising considering that UV-vis spectroscopy also showed significantly higher intensities in NAP-β-CD-ICs, compared to NAP alone in water solution (Figure 6B) [63]. Similar to NAP and sulfisoxazole, tetracycline, a biocidal drug, which has poor solubility, has been embedded in PCL nanofibers and encapsulated in β-CD cavities to regenerate periodontal ligaments [64]. Microbiological tests against the commonly found oral pathogens *Aggregatibacter actinomycetemcomitans* (*A.a*) and *Porphyromonas gingivalis* (*P.g*) revealed significantly higher halos of bacterial inhibition against both oral bacteria in PCL nanofibers containing tetracycline/β-CD, compared to PCL nanofibers with (28 ± 4 and 26 ± 3 mm) and without tetracycline (0 and 0 mm) [64]. In addition, PCL nanofibers containing tetracycline/β-CD also showed higher inhibition than chlorhexidine (26 ± 4 vs. 27 ± 3 mm for chlorhexidine), a clinically recommended oral disinfectant [64], illustrating the potential for novel, highly impactful solutions for periodontitis. Ciprofloxacin is yet another hydrophobic biocidal drug whose application has been impaired due to poor solubility and stability [65,66]. By combining ciprofloxacin/β-CD with PCL nanofiber technology, large amounts of ciprofloxacin can be encapsulated, especially when prepared under sonic energy, and subsequently carried in PCL nanofibers [67]. An increase in the released amount of ciprofloxacin *via* this technique has been found in stimulated physiologic environments (pH 7.2) when delivered from β-CD/PCL nanofibers, compared to α-CD/PCL nanofibers [67].

Another way to control the solubility of a drug is by utilizing a sandwich of nanofibrous layers containing water soluble and insoluble matrices, as shown recently by Aytac et al. [68]. The authors prepared ICs of sulfisoxazole (SFS) and HP-β-CD, which were subsequently electrospun and carried by hydrophilic, hydroxypropyl cellulose (HPC) nanofibers. As both the carrier polymer and the CDs were highly soluble in water, by sandwiching the HPC/SFS-HP-β-CD nanofibers, release of SFS could be significantly modulated [68]. By employing such a strategy, a slow release of SFS, with the highest amount released over time (~720 min), was realized, demonstrating the existence of several facile strategies to improve/modulate the release of hydrophobic drugs.

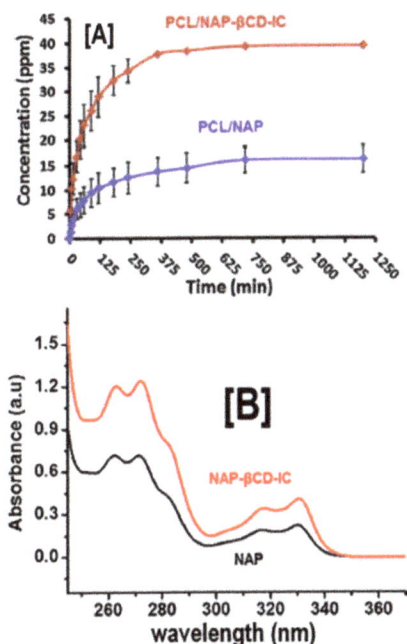

Figure 6. The release and solubility profiles of naproxen (NAP) from PCL and PCL/β-CD-IC nanofibers. The release profile study (**A**) by HPLC revealed significant release of NAP from PCL/NAP-β-CD-IC, compared to the PCL/NAP nanofibers, owing to improved solubility of encapsulated NAP, compared to pristine NAP. These results were further corroborated by solubility analyses (UV-vis) (**B**), which showed significantly enhanced solubility, as observed by increases in the intensity. Images adapted and reproduced from [63]. Copyright Elsevier Ltd., 2014.

In addition to enhanced retention, solubility, and release of hydrophobic drugs, newer applications of CD functionalized PCL nanofibers are as biocatalysts. It has been recently reported that immobilized biocatalysts on suitable substrates, compared to simple coatings, significantly increase catalytic efficiency and significantly decrease cost, compared to conventional catalysts [69–72]. More recently, α-amylase, a digestive enzyme immobilized on HP-β-CD, has shown a significant increase in stability, solubility, and enzymatic activity, compared to other routes of administration, such as therapeutic proteins [73]. The large surface to area ratios, high porosity and small pore sizes of nanofibers typically obtained via the electrospinning process give another suitable substrate for biocatalyst immobilization [74]. Recently, electrospun nanofibers possessing pendant CD molecules have been prepared for immobilizing biocatalysts and have been subsequently evaluated for catalytic activities [74]. For instance, catalase, an anti-free radical enzyme, has been successfully immobilized onto PEO nanofibers containing γ-CDs (in complexed and uncomplexed forms), sandwiched between PCL nanofibers [75]. Catalase enzymatic activity results showed the positive influence of CDs on enzyme activity and the stability of the biocatalysts (catalase) over prolonged periods of time [75]. Similar to the catalase enzyme, laccase enzymes immobilized on γ-CD/PCL nanofibers showed higher catalytic activity (96.48 U/mg), compared to enzymes immobilized on PCL nanofibers (without CDs) (23.2 U/mg) or γ-CD/laccase physical mixtures in PCL nanofibers (71.6 U/mg) [76]. In addition to higher enzymatic activity, similar to previous studies on catalase enzyme immobilization, no discernible loss of enzyme activity was observed, indicating a synergistic effect in enhancing catalytic activity by combining polymeric nanofibers and CD-IC formation [76].

2.2. PCL Nanofibers Functionalized with Cyclodextrin-Large Molecule Inclusion Complexes

The key advantage of CDs is their facile capability to form ICs with both small as well as large molecules. Even though polymer-CD-ICs may not be used directly, as they neither melt nor are soluble in solvents, after coalescing the polymers from the ICs, the coalesced polymers can be processed using conventional means. These coalesced polymers elicit properties that are unique from their neat native samples, making them attractive for a plethora of applications. For example, in the past, we have shown the possibility of synthesizing polymers, such as polystyrene, with well-defined architectures, through a greener route in aqueous solution [77,78]. Coalescing the polymers from the CD cavity results in an intimate blend of immiscible homopolymers or copolymers, as evidenced by single glass transition temperatures, suppressed microphase separation, elevated melting temperatures, and specific interactions observed by DSC and FTIR, respectively [79–86]. In addition, degradation behaviors of these coalesced biodegradable copolymers (PCL-*b*-PLLA) have been observed to be unique, with accelerated enzymatic degradation, in particular during the early phase of degradation, compared to copolymer or physical homopolymer blends [86]. This effect has been observed particularly in the PCL block phase, which could partially be attributed to the degradation environment (*R. arrhizus* in phosphate buffer solution at 37 °C and pH 7) [86]. In the case of homopolymers, even with slow crystallizing polymers, such as polyethylene terephthalate (PET), upon coalescence the ethylene glycol units have been known to adopt a highly extended g $\pm$ tg $\mp$ kink conformation (where g $\pm$ are gauche and t is all trans conformations), and PET chains isolated from neighboring chains quickly convert to the fully extended, all-trans crystalline PET conformer [87–89]. Even faster crystallizing semi-crystalline polymers, such as PCL, show substantial increases in crystallization temperatures (ΔT_cs) (~25 °C), Young's modulus (~100%), and hardness [90].

While the traditional processing of completely covered polymer chains in CD cavities is not feasible, the processing of polymers can be facilitated if coverage is less than complete. Chains dangling out of the CD cavities are also able to undergo crystallization and melting [91]. In addition, depending on the stoichiometry between the cyclodextrin host and guest polymer, the lengths of dangling polymer chains and their molecular weights, crystallizability (T_cs and T_ms) increases significantly, irrespective of the crystalline nature of the dangling polymer [50,92,93]. Besides improvements in thermal properties, the additions of such non-stoichiometric ICs to bulk polymers have also resulted in significant increases in elastic modulus values and tensile strength, with drastic reductions in the loss tangent and elongation, indicating significant improvements in the stiffness and elasticity of the composites [94]. This is particularly interesting, as the evidenced improvements occurred over a wide range of stoichiometries (the more dangling chains the better) and temperatures (−50 to 50 °C) [94].

Recently, our group took advantage of non-stoichiometric inclusion complexes (PCL-α-CD-ICs) by using them as nucleating agents for neat bulk PCL nanofibers [95], as well as synthesizing pseudorotaxane nanofibers (PCL-α-CD-ICs) without a polymeric carrier [96,97]. In the first instance, three ICs were synthesized with stoichiometries of 1:1 (complete coverage), 4:1, and 6:1 (PCL:CD molar ratio) [95]. The as-synthesized ICs were characterized by ^{1}H-NMR (for stoichiometry) [98], wide angle X-ray diffraction, FTIR, and DSC, respectively [95]. The ICs (5, 10, and 15% wt loading) were then added to PCL solutions (in 60:40 chloroform/*N*,*N* dimethyl formamide (CFM/DMF)), and were electrospun into nanofibers (PCL/PCL-α-CD-ICs). In the first heating cycle, DSC analyses showed elevated melting temperatures (T_m), and in TGA analyses, higher degradation temperatures of the composites were observed, indicating the presence of intact ICs in the composite nanofibers (Figure 7A,B) [95]. More interestingly, quasi-static tensile testing of PCL/PCL-α-CD-ICs indicated significant increases in modulus values, with corresponding decreases in elongation values at all wt % IC loadings, compared to control nanofibers (PCL and PCL/α-CD) [95]. Within the wt % loading and stoichiometry evaluated, 10% IC-6 (6:1 stoichiometric ratio) loading resulted in a high tensile modulus value (15.2 ± 5.0 MPa) and low elongation (170 ± 40%), compared to 2.6 ± 0.9 MPa and 390 ± 70%, respectively, observed for neat PCL nanofibers [95].

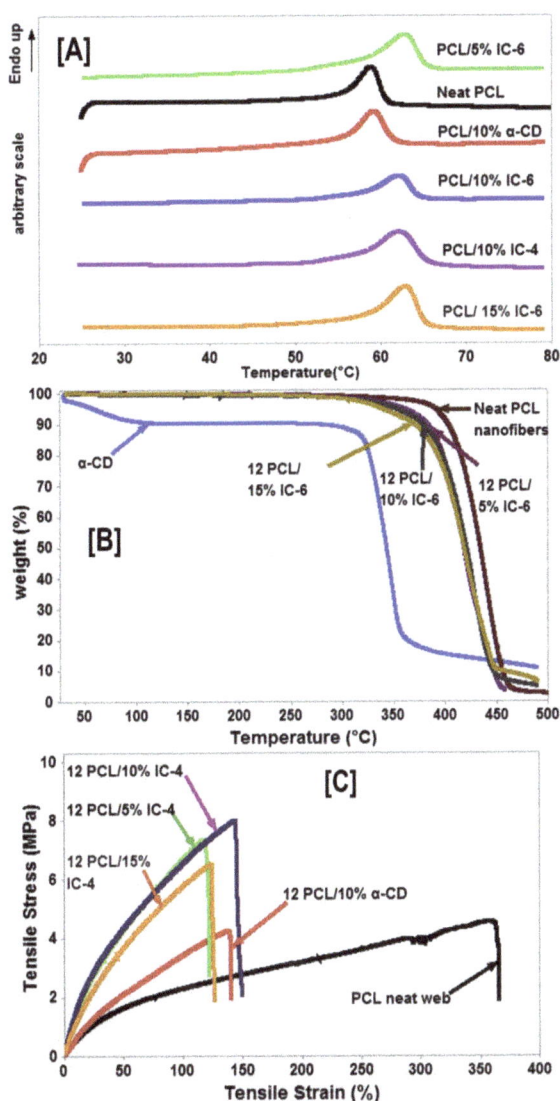

Figure 7. Structure–property relationships (thermal and mechanical), due to the presence of intact non-stoichiometric inclusion complexes (n-s PCL-α-CD-IC) in PCL nanofibers. The presence of intact (n-s PCL-α-CD-IC) caused increases (~5 °C) in PCL melting temperature (**A**), irrespective of the stoichiometric ratio of the IC or the wt % loading of those ICs in the nanofibers. Likewise, the degradation temperature of the α-CD phase significantly increased in those composites, indicating the presence of α-CD in complexed state (**B**). Finally, intact ICs caused changes in the mechanical behavior (increased modulus value and decreased elongation) of the composites. Images adapted with permission from [95]. Copyright Elsevier Ltd., 2015.

An identical trend was also observed in IC-4 (4:1 stoichiometric ratio) nucleated PCL composites, with a high tensile modulus value (10.2 ± 1.2 MPa) and low elongation (170 ± 20%) observed for composites containing 10% IC-4 [95] (Figure 7C), indicating the necessity of optimal stoichiometry, as well as the wt % loading for improving the mechanical properties of PCL nanofibers for potential applications in tissue engineering.

By employing a similar approach, our group also fabricated intact PCL-α-CD-IC nanofibers with coverages ranging from 3% to 12% from CFM solution, facilitated by the addition of benzyl triethylammonium chloride salt (BTEAC) to increase solution conductivity [96]. Preliminary FTIR analyses provided qualitative assessment of the varying stoichiometries in the nanofibers by decreasing carbonyl peaks (bending vibration at 1027 cm^{-1}) and increasing hydroxyl peaks (3351 cm^{-1}), with increases in the coverage of PCL chains by α-CD [96]. Similar to the previous study that reported PCL/PCL-α-CD-IC composites, increases in thermal stability and the channel-like crystal structure of IC nanofibers were observed by TGA and WAXD analyses, respectively, and provided conclusive evidence as to the presence of intact rotaxanated structures [96]. Finally, just like the PCL/PCL-α-CD-IC study, mechanical testing showed a substantial increase in modulus values (16.0 ± 4.9 vs. 8.0 ± 1.7 MPa for neat PCL nanofibers) [96].

To further understand the structure–property relationships that exist in liquid (polymer solutions) and solid phases (nanofibers), a subsequent study was carried out to evaluate the rheology of solutions, and 2D-WAXD, selected area electron diffraction (SAED), and dynamic mechanical analyses were performed on the nanofibers [97]. The frequency sweep of PCL and pseudorotaxane solutions showed significant differences in terms of elastic and loss moduli and cross-over patterns within the studied frequency ranges. While the neat PCL solution had both lower elastic, as well as loss moduli, the PCL solutions also displayed several cross over points, indicating the separation of PCL chains from one another, as would be expected for viscoelastic materials [97].

In P-3 solutions (3% coverage by CDs), a solitary cross-over point was observed at higher frequencies, indicating longer relaxation times for the excluded portions of the PCL chains. In P-12 solutions (12% coverage by CDs), this cross-over was observed at even higher frequencies, leading to Rouse–Zimm-like behavior, which would be typically expected of unfettered polymer chains in dilute solutions, leading to significant increases in the elastic moduli (Figure 8B,C) [97]. Further 2D-WAXD and SAED experiments (Figure 8D–K) demonstrated the presence of crystalline domains in control PCL nanofibers (both aligned and random) corresponding to the (110) and (200) reflections, while all the pseudorotaxane nanofibers showed an absence of reflections, indicating the randomness of the crystal lattice arrangement along the fiber axis [97]. Contrastingly, both quasi-static tensile testing, as well as dynamic mechanical analyses (DMA), provided further insights into the deformation behavior of the composite and control nanofibers. With an increase in the coverage of PCL chains by CDs, linear trends of increasing modulus values (14 ± 5 vs. 16 ± 6 vs. 29 ± 5 MPa for P-3, P-6, and P-12, respectively) and decreasing elongation (103 ± 34 vs. 62 ± 12% for P-3, and P-12, respectively) were observed [97]. Just like quasi-static tensile testing, DMA analyses also showed a similar trend, with high elastic and loss moduli values observed for P-12 > P-6 > P-3 > neat PCL nanofibers, at most studied temperatures, with the key difference being a gradual decrease in the moduli values, compared to a sharp decrease observed for neat PCL nanofibers, indicating a strong intermolecular interaction between pseudorotaxane nanofibers [97].

One of the foremost advantages of polymer coverage by CDs is their potential ability to immobilize proteins/growth factors/biomolecules, owing to the presence of abundant hydroxyl groups present on the external rims of the CDs [11,99]. This is typically accomplished by a variety of post-modification reactions on hydroxyl groups, facilitating intermolecular interactions, such as ionic interactions [100,101] and click chemistries [102], potentially resulting in the immobilization of bioactive molecules. Although our studies did not focus on further development of the functionalized PCL, Elisseeff [103] and Schlatter [104,105] have reported various bioactive molecule immobilization on PCL-α-CD-IC nanofibers.

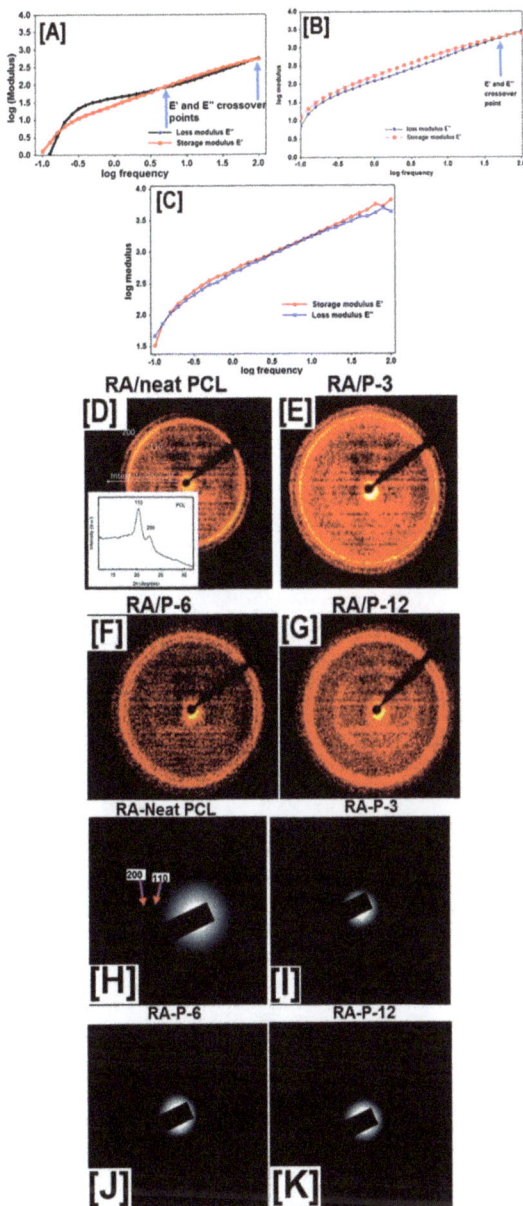

Figure 8. Establishing structure property relationships between pseudorotaxane (PpR) solution rheology and the resultant molecular arrangement of PpR in the nanofibers. Solution rheology of neat PCL solutions at higher concentrations (12%) showed low modulus values (elastic and loss) with multiple cross-over points, typical of viscoelastic material. On the other hand, PpR solutions showed frequent cross-over points, indicating a Rouse–Zimm effect (**A–C**). However, both 2D-wide angle X-ray diffraction (**D–G**) and selected area electron diffraction analyses (**H–K**) showed the presence of crystalline domains in both randomly aligned, as well as aligned PCL nanofibers, while no such peaks were evidently observed in PpR nanofibers. Images adapted and reproduced with permission from [97]. Copyright The Royal Society of America 2016.

Elisseeff's research group, in proof-of-concept experiments, showed the conjugation of fluorescamine onto PCL-α-CD-IC nanofibers upon activation of the hydroxyl groups by *N,N'*-carbonyldiimidazole (CDI) (Figure 9A–E) [103].

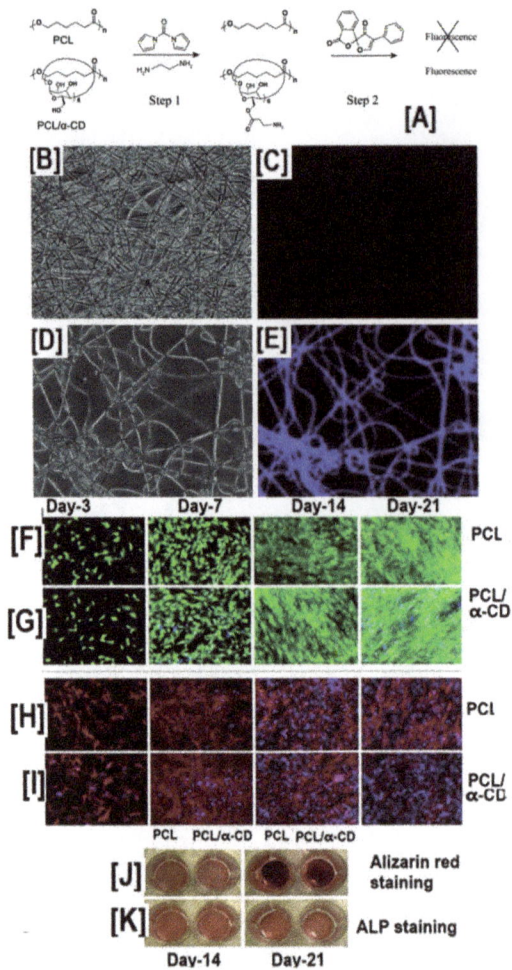

Figure 9. Schematic illustrating the immobilization of fluorescent molecules and the effect of hydroxyl groups of cyclodextrins on the osteogenic differentiation of human adipose-derived stem cells. The Neat PCL does not have surface epitopes facilitating immobilization of bioactive molecules. Conversely, the pseudorotaxanes contain abundant hydroxyl groups, facilitating the immobilization of bioactive molecules, and a representative fluorescent molecule (fluorescamine) (**A**), whose presence was monitored by fluorescent spectroscopy (**B–E**). Although immobilization of active biomolecule was possible only in scaffolds containing CDs, live-dead cell assay and F-actin assay showed no statistical significance in the cell viability at any of the time point studied (**F–I**) In addition to facilitating the immobilization of bioactive molecules, PCL/α-CD nanofibers without immobilization of biomolecules also facilitated the osteogenic differentiation of human adipose derived stem cells (h-ADSCs), although the osteogenic marker levels showing similarities in the expression levels between the study groups were the same (**J,K**). Images adapted and reproduced with permission from [103]. Copyright Springer 2012.

Even though further studies on immobilizing bioactive molecules were not conducted, unconjugated PCL-α-CD-IC nanofibers were evaluated for their potential role in bone tissue engineering. Cell viability/proliferation, cytoskeleton formation, calcium mineralization, and gene expression assays, were carried out on PCL and PCL-α-CD-IC nanofiber scaffolds seeded with human adipose-derived stem cells (h-ADSCs) [103]. Despite cell viability or mineralization assays not showing significant differences over a 21-day period (cell viability: 96.5 ± 2.5% and 97.1 ± 1.5% live cells for PCL and PCL-α-CD-IC nanofibers) (Figure 9F,G), gene expression levels of osteogenic markers such as RunX2, osteopontin (OPN), COL1A1, COL3A1, COLXA1, and hydroxyproline levels were upregulated in PCL-α-CD-IC scaffolds compared to PCL nanofibers, indicating the potential application of PCL-α-CD-IC scaffolds in bone tissue engineering [103].

As we have shown in our PCL-α-CD-IC electrospinning experiments [96,97], the addition of DMF or dimethyl sulfoxide (DMSO) leads to the destabilization of the rotaxanated structure, leading to the de-threading of PCL from CD cavities. We overcame this difficulty by electrospinning pseudorotaxanes (PpR) from CFM and improved solution conductivity with the addition of BTEAC salt. Another easier method was reported by Schlatter's research group, who reported on the fabrication of core–shell nanofibers comprised of four branched star-PpR or mikto-arm PpR for shell and neat PCL for core regions, respectively [104,105]. Small angle neutron scattering measurements (SANS) showed temperature-dependent shell stability/core instability of PpR in DMSO, based on which stable PpR solution was prepared for electrospinning (35 °C) (Figure 10A). WAXD analyses of PCL/PpR nanofibers (10:2 and 10:3 ratios) showed a $2\theta = 20°$ peak, indicating the presence of intact rotaxane crystals in the coaxial spun nanofibers [104]. The intact PpR was further taken advantage of by immobilizing fluorescein isothiocyanate (FITC) onto star PpR *via* immersing the scaffolds in acetonitrile, a water (1:1) solution containing FITC [104]. Interestingly, confocal microscopy experiments showed fluorescence in core–shell fibers containing star-PpR in the shell region (Figure 10B,C), whereas in the core–shell nanofibers containing linear PpR, such a fluorescent effect was not observed due to the dethreading and subsequent leaching of CDs [104].

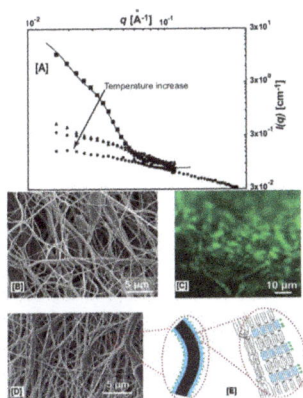

Figure 10. Temperature-dependent stability of pseudorotaxanes (PpR) using DMSO and an electron micrograph of electrospun core–shell nanofibers with PpR in the shell region, which was subsequently conjugated with FITC. The SANS analyses of the PpR (shell region) in DMSO showed decreases in scattering intensity, indicating dethreading of PpRs into α-CD and PCL phases. Whereas at 35 °C, higher scattering intensity indicated the presence of a PpR structure, and in addition, a parallelepiped model was developed which showed CD assemblies approximating 11 α-CDs in length and 20 α-CDs in width (**A**). Additionally, conjugation of fluorescamine was only possible onto hydroxyl groups present in PpR (**B–E**), whereas in neat PCL nanofibers, conjugation was not possible. Images adapted and reproduced with permission from [104]. Copyright John Wiley and Sons 2015.

Using a similar approach, FITC was immobilized onto both star-PpR, as well as mikto-arm PpR (two of the four ends were covered by CDs), with both nanofibers showing fluorescent activity. However, some damage to the mikto-arm PpR containing nanofibers was observed, indicating some degree of PpR structure destabilization in nanofibers by DMSO [105]. In addition to FITC, a red fluorescent molecule, Megastokes 673 was immobilized onto PCL, star-, and mikto-arm PpRs, with no fluorescent activity observed in PCL nanofibers, again demonstrating the necessity of functional groups on the nanofibers surface for bioconjugation [105].

2.3. Cyclodextrins or Inclusion Complex Functionalized Poly(lactic acid) Nanofibers

Just like PCL, poly(lactic acid) (PLA) belongs to the class of aliphatic polyesters, also known as α-hydroxy esters, that are renowned for their excellent bioresorbable and biocompatible characteristics, making them suitable for pharmaceutical and biomedical implants. Unlike PCL, which is a fast crystallizing polymer, PLA is a slow crystallizing, semi-crystalline polymer with crystallinity, melting, and glass transition temperature values ranging from 40–50 °C, 55–80 °C, and 170–180 °C, respectively [4]. Another difference lies in the chirality of PLA, enabling PLA to exist in three enantiomeric states, namely L-Lactide (PLLA), D-Lactide (PDLA), and *meso*-lactide, with PLLA and PDLA commonly used as biomedical implants, in particular in orthopaedic and dental implants [4]. To further improve degradation characteristics, material properties (for example, thermal and mechanical), and processing behavior, lactic acid monomers are copolymerized with various monomers, including other aliphatic polyester monomers such as ε-caprolactone and glycolic acid, or with monomers such as ethylene glycol and glutamic acid, making PLA a versatile biopolymer [4].

Even though some of the surface properties of these bulk copolymers were enhanced, uncontrolled detrimental effects were observed on other properties, necessitating the development of a newer "softer" approach in functionalizing PLA-based materials. Just like PCL, PLA-based nanofibers have been functionalized with cyclodextrins or their inclusion complexes, leading to the development of materials suitable for anti-cancer therapies, modulating drug release, anti-microbial packaging, and wound dressing [106].

In one study, slow degradation characteristics (compared to hydrophilic polymers) of poly(lactic-*co*-glycolic acid) (PLGA) were taken advantage of by forming sequential layers of hydrophilic PVA nanofibers containing haloperidol, encapsulated in randomly methylated β-CD (RM-β-CD) cavities, resulting in the slow, controlled released of haloperidol [107]. While the release of haloperidol was also low from PVA nanofibers without β-CD encapsulation, the drug release was instantaneous and the fibers dissolved rapidly when encapsulated in RM-β-CD cavities. On the other hand, in sequential layers containing PLGA nanofibrous layers, a more controlled release of haloperidol over several days was observed, with some initial burst of the drug, along with improved sensitivity towards heat and light, and low water solubility [107].

While this study evaluated the release of haloperidol from single- or multi-layered nanofibrous films, structural changes, which determine the efficacy of the drug, were not evaluated. Xie et al., for instance, evaluated both the release rates, as well as the form of released hydroxycamptothecin (HCPT), an active drug prescribed for treating malignant tumors, from poly(ethylene glycol)–poly(dl-lactic acid) (PELA) copolymer nanofibers [108]. With the addition of HCPT in encapsulated form in 2-hydroxypropyl-β-CD (HP-β-CD), greater release of HCPT from HCPT/PELA fibers was observed, depending on the HP-β-CD concentration (5% > 2.5% > 1.5%). For example, at HP-β-CD concentrations of 1.5% and 2.5%, respectively, $50.5 \pm 3.8\%$ and $58.2 \pm 2.0\%$ of HCPT was released from the PELA fibers after 20-days of inoculation, attributed to the enhancement of drug dissolution by hydrophilic HP-β-CD [108]. Also, with the addition of HCPT in complexed form, the half-life of the active drug form (lactone) improved from 24 to 40 mins in physiologic conditions, and in addition, the half maximal inhibitory concentration (IC_{50}) of the drug improved dramatically, by 7 times compared to free HCPT [108], indicating the therapeutic potential of drug-CD embedded PLA nanofibers.

Similar to the study of HCPT for treating cancer, curcumin (CUR) encapsulated in β-CD carried by poly(lactide-*co*-caprolactone) PLCL nanofibers, in conjunction with other biologics, such as aloe vera (AV) and magnesium oxide (MgO) nanoparticles, have been evaluated for treating breast cancer [109]. In vitro proliferation evaluation of the composite fibers of breast cancer cell line, MCF-7 (Michigan Cancer Foundation-7), by MTS assay (3-(4,5-dimethylthiazol-2-yl)-5-(3-carboxymethoxyphenyl)-2-(4-sulfophenyl)-2H-tetrazolium, inner salt) showed a significant reduction in cell numbers ($p < 0.01$) after 6- and 9-days, with either free or encapsulated CUR (PLCL/AV/MgO/CUR and PLCL/AV/MgO/CUR/β-CD scaffolds) (Figure 11B,C) [109]. These results were corroborated by laser confocal scanning microscopy images of fluorescent dye 5-chloromethylfluoresceindiacetate (CMFDA) (chloromethyl derivatives)-labelled MCF-7 cells, which showed decreased cell densities in scaffolds containing free or encapsulated curcumin, compared to the scaffolds that did not contain curcumin (PLCL/AV, PLCL/MgO, and controls) (Figure 11D–G) [109]. In addition, CUR-containing scaffolds (PLCL/AV/MgO/CUR and PLCL/AV/MgO/CUR/β-CD) showed a drastic reduction in collagen secretion, indicating poor proliferation of MCF-7 cells, due to the presence of CUR/β-CD, and showing a synergistic effect of CD/drugs delivered via PLA-based nanofibers (Figure 11H,I) [109].

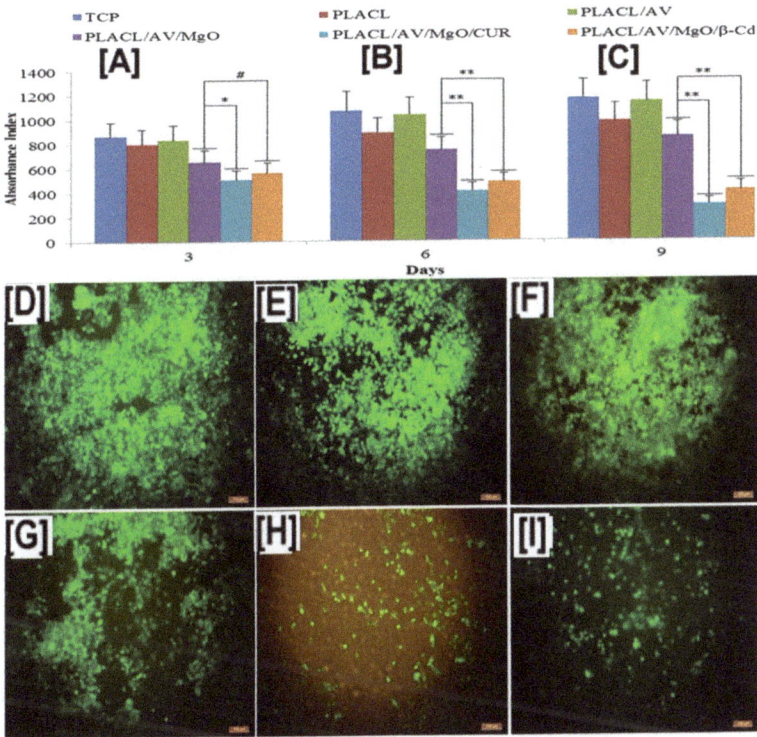

Figure 11. Proliferation and cell densities of Michigan Cancer Foundation-7 (MCF-7) cells on poly(lactide-*co*-caprolactone) (PLCL or PLACL) scaffolds and PLCL scaffolds containing aloe vera (AV), magnesium oxide (MgO), and free curcumin (CUR), or encapsulated in β-CD. The addition of AV alone did not cause MCF-7 cell death at all studied time points; however, the addition of CUR, in either free or encapsulated form, resulted in significant cell death of the proliferating MCF-7 cells (**A**). These results were further corroborated by laser confocal scanning microscopy measurements, which showed low cell densities in scaffolds containing CUR and MgO/CUR (11 B: (**E,F**)). Images adapted and reproduced with permission from [109]. Copyright Elsevier Ltd., 2017.

Given the fact that release of drugs or other small molecule guests are significantly enhanced with encapsulation in CD cavities, drug release can also be controlled by simply encapsulating the complex in the core region of PLA nanofibers. Using such a facile strategy, CUR/HP-β-CD ICs were safely embedded in the core region, with the PLA forming the shell region of the coaxial nanofibers [110]. Although the rate of CUR dissolution from core–shell nanofibers was slow, the total release of CUR from core–shell nanofibers was significantly higher at simulated physiologic conditions (pH 1 and 7.4). In addition, release of CUR from PLA nanofibers and core–shell nanofibers were evaluated in methanol and methanol–water (1:1), which showed higher release of CUR from PLA nanofibers into the methanol solution [110]. On the other hand, CUR released from core–shell nanofibers was higher in methanol–water, owing to improved dissolution in aqueous solution, due to the presence of HP-β-CD [110]. In accordance with the release study, antioxidant activity of CUR in methanol and methanol–water, also showed a similar trend, with PLA/CUR nanofibers and core–shell nanofibers, showing improved activity in methanol and methanol–water solutions, respectively [110].

Just like the delivery (rapid or controlled) of hydrophobic anti-cancer drugs is critical for treating cancer, improving the efficiency of antimicrobial agents, such as cinnamaldehyde (CA), is critical for treating wounds. Recently, CA encapsulated in β-CDs delivered by PLA nanofibers was evaluated for bactericidal effects against *Escherichia coli* (*E. coli*) and *Staphylococcus aureus* (*S. aureus*), and for cytocompability against CCC-HSF-1 human skin fibroblasts (HSFs) [111]. With the addition of CA-β-CDs in PLA, concentration dependent effects of CA-β-CDs on average fiber diameter (the higher the loading, the higher the fiber diameter), were observed. While this trend was reversed for tensile modulus values and surface hydrophilicity of the nanofibers (the higher the loading, the lower the modulus value and surface hydrophilicity) [111]. In addition, similar to the release rates observed in anti-cancer agents, the cumulative release rate of CA also depended on the concentration of CA-β-CDs (12% > 6% > 3% > 1.5%) used in the nanofiber formulation (Figure 12A). Notwithstanding the fiber diameters or the enhanced release rate of CA, a concentration (of CA-β-CD) independent effect on cell death of *S. aureus* and *E. coli* with the addition of CA-β-CD was observed. In other words, the addition of CA-β-CD resulted in poor viability (>95% cell death) of *S. aureus* and *E. coli* at all studied times (20, 40, and 60 h) [111]. Another advantage of utilizing CD-ICs over free CAs was observed in cell viability (Figure 12B), where the toxicity of CA was significantly decreased by its encapsulation, indicating the possibility of modulating drug response against bacteria and physiologic cells by varying the concentration of ICs [111].

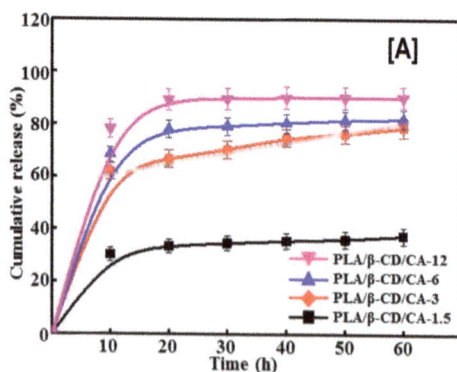

Figure 12. *Cont.*

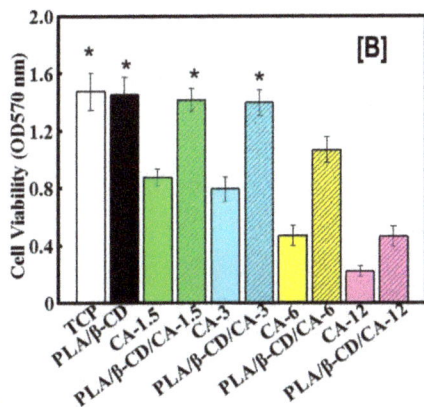

Figure 12. The effects of poly(lactic acid) nanofibers containing β-CD/cinnamaldehyde (CA) inclusion complex on the cumulative release of CA and cell viability of human skin fibroblasts. The cumulative release of CA depended on the β-CD/CA concentration, i.e., the higher the concentration, the higher the release rate of CA, possibly owing to improved dissolution of CA from β-CD complex (**A**). In comparison, low cell densities of *S. aureus* and *E. coli* were observed with increasing β-CD/CA complex concentrations. Similar trends were also seen with the proliferation of human skin fibroblasts cells, when exposed to CA containing PLA nanofibers or free CA (**B**). Images adapted and reproduced with permission from [111] Copyright Liu et al., 2017.

Exploding population growth requires increased food production, as well as the prevention of food spoilage due to microbial action. For these reasons, there is an immediate need to curtail the growth of food borne microbes and, thereby, extend the shelf life of food products, without significant loss of nutritional and sensory characteristics. In addition, prevention of active ingredients against volatility, light, heat, and air, are other factors that need to be actively addressed, such as through encapsulation inside CD cavities. Several active ingredients, such as gallic acid, triclosan, and essential oils (cinnamon), have been investigated, in conjunction with PLA nanofibers, for their capability preventing microbial growth. For instance, triclosan (TR) has been encapsulated in β- and γ-CDs (not possible in α-CD cavity), which were subsequently electrospun with a polymeric carrier (PLA), resulting in nanofibers with fiber sizes ranging from 640 ± 480 and 940 ± 500 nm for PLA/TR/β-CD-IC and PLA/TR/γ-CD-IC nanofibers, respectively [112]. Successful complexation was confirmed by DSC and TGA analyses, which showed an absence of TR melting (typically seen at ~60 °C) and the enhanced thermal stability of TR, indicating successful IC formation. An antibacterial test of the electrospun nanowebs (PLA/TR, PLA/TR/β-CD-IC, and PLA/TR/γ-CD-IC) showed significant increases in the inhibition zones against *S. aureus* and *E. coli* for the PLA/TR/CD system over the PLA/TR system [112]. In particular, the TR/β-CD-IC system showed significant inhibition zones (3.50 ± 0.00 and 3.7 ± 0.15 cm) against *S. aureus* and *E. coli*, compared to the PLA/TR/γ-CD-IC system, which showed inhibition zones of 2.90 ± 0.17 and 3.3 ± 0.00 cm, respectively, that in turn, were significantly higher than PLA/TR (2.83 ± 0.28 and 3.0 ± 0.20 cm) [112].

Unlike TR, which is a synthetic compound, due to health and environmental considerations, antimicrobial agents from natural sources, such as gallic acid and cinnamon oil (CEO), are also being actively investigated for their antimicrobial and antioxidant characteristics in food packaging. Similar to PLA/TR/CD systems, PLA/CEO showed enhanced inhibition against *S. aureus* and *E. coli*, with quantitative evaluation demonstrating a drastic reduction in the viability of microbes for up to 9 days [113]. In addition to antimicrobial activity, some of these agents are also antioxidants that can scavenge free radicals from the system. Gallic acid (GA), an antimicrobial/antioxidant agent derived from plants and herbs, has been encapsulated in HP-β-CD, and subsequently fabricated into bead-free

PLA/GA-HP-β-CD nanofibers [114]. While the loading was obviously low in PLA/GA-HP-β-CD nanofibers (75 ± 0.7%) compared to PLA/GA nanofibers (80 ± 0.7%), the cumulative release of GA from these systems was significantly higher in ethanol solutions (10% and 95%) (Figure 13A). In addition, GA from PLA/GA-HP-β-CD nanofibers retained excellent antioxidant characteristics, as was evidenced by DPPH (2,2-diphenyl-1-picrylhydrazyl) assay (Figure 13C) [114].

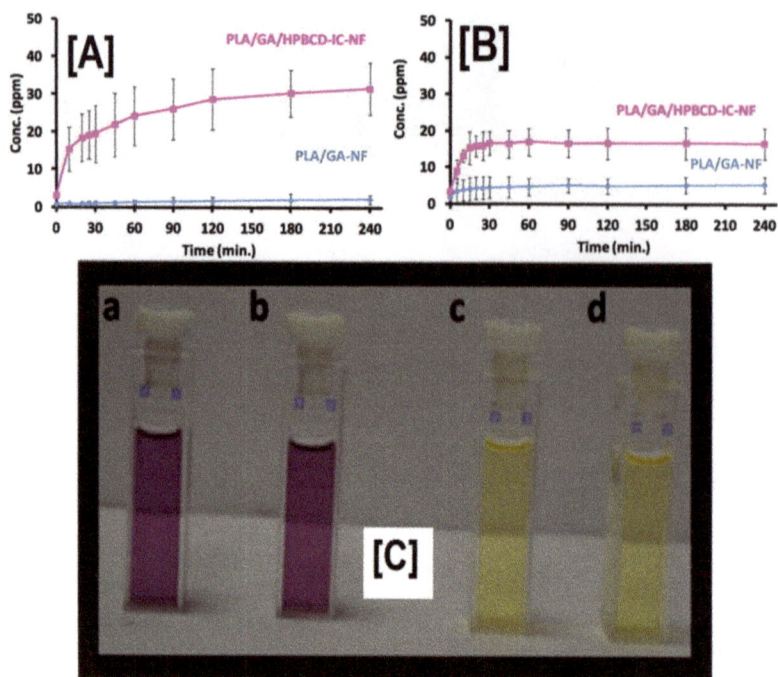

Figure 13. Total cumulative release and antioxidant characteristics of gallic acid (GA) released from PLA/GA-HP-β-CD nanofibers. The total cumulative release of GA from PLA/GA-HP-β-CD nanofibers was always higher in water (**A**) and 10% ethanol solutions (**B**), than those released from PLA/GA nanofibers, owing to the enhanced dissolution characteristics of GA/HP-β-CD compared to free GA. In addition, the DPPH assay showed that the presence of GA in both PLA/GA, as well as PLA/GA-HP-β-CD nanofibers, retained its antioxidant characteristics (**C**). Images adapted and reproduced with permission from [114]. Copyright Elsevier Ltd., 2016.

3. Future Outlook

The aliphatic polyesters discussed in this review are well-studied materials for biomedical applications and are known for their excellent biocompatible and bioresorbable characteristics. Despite these features, both PCL and PLA are known for their hydrophobicity, rendering their utility in pharmaceutical applications, where the controlled or rapid delivery of drugs is desirable. In addition, despite being a fast crystallizing, semi-crystalline material, PCL suffers from poor mechanical properties (low modulus value, strength, and high elongation). Modulating small molecule delivery/capture using a facile approach significantly increases its appeal for these applications.

Although CDs were originally developed predominantly for the rapid delivery of drugs, which are mostly hydrophobic, CDs have also been utilized to improve the properties of polymers in several ways, for example, by reorganizing polymer chains [115]. As seen from this review, aliphatic polyesters can be combined with CDs or CD-ICs in numerous ways and can be fabricated into nanofibers; however, significant research needs to be carried out. For example, to date, no in vivo studies have been carried out to explore the safety and efficacy of aliphatic polyester/CD or IC nanofibers. In addition, almost all studies deal with nanofibers in non-woven mats; however, additional processing, such as textile braiding, would further improve the prospects of these fibers for biomedical applications.

There is a lack of literature detailing the use of other aliphatic polyesters, for example, poly(glycolic acid) and poly(lactic-*co*-glycolic) acid, which are well-known as suture and drug delivery materials, owing to their fast degradation nature. Not only could combining these polymers with CD-based strategies lead to novel materials with improved properties, but based on further in vivo experiments, it is quite likely that application-specific or tailorable materials could be developed for a wide range of biomedical applications. In addition, these polymers are approved as "generally regarded as safe" (GRAS) by the Food and Drug Administration for a variety of applications. In combination with the GRAS status of CDs, this is expected to accelerate the utility of these materials in biomedical applications.

4. Conclusions

We have presented some of the combinations of aliphatic polyester nanofibers functionalized with cyclodextrins or their inclusion complexes, that have been reported predominantly in the past five years. A striking observation is that various strategies are now available to fabricate polymeric nanofibers containing free cyclodextrins or their small molecule inclusion complexes or even large molecule (polymer) inclusion complexes, with or without even a polymer carrier. In addition, if added cyclodextrins encapsulate guest molecules, it is now possible to tailor/modulate release rates by using, for example, multilayered sheets or core–shell structures. Just like in solid powder form, encapsulating results in rapid dissolution of hydrophobic molecules in aqueous solution (and possibly into physiologic tissues in vivo). Additionally, the structures of the drug molecules remain intact, meaning they retain their pharmacologic, antimicrobial, and antioxidant effects, which may not be possible without a cyclodextrin carrier.

Author Contributions: G.N. and A.E.T. conceived the idea and formulated the plan to write the manuscript. G.N., J.S., and R.B. drafted the manuscript under the direct supervision of A.E.T. and B.S.G. G.N. and A.E.T. revised the manuscript and communicated with the editors.

Conflicts of Interest: The authors declare no conflict of interest.

Abbreviations

A.a	*Aggregatibacter actinomycetemcomitans*
AV	Aloe vera
BTEAC	Benzyl triethylammonium chloride salt
CA	Cinnamaldehyde
CDs	Cyclodextrins
CDI	*N,N'*-carbonyldiimidazole
CD-ICs	CD-inclusion complexes
CFM	Chloroform
CMFDA	5-chloromethylfluoresceindiacetate
CEO	Cinnamon essential oil
CUR	Curcumin
DMA	Dynamic mechanical analyses
DMSO	Dimethyl sulfoxide
DMF	Dimethyl formamide
DPPH	2,2-diphenyl-1-picrylhydrazyl
DSC	Differential scanning calorimetry
E. coli	*Escherichia coli*
FITC	Fluorescein isothiocyanate

GA	Gallic acid
hADSCs	Human adipose-derived stem cells
HP-β-CD	2-hydroxypropyl-β-CD
HPC	Hydroxypropyl cellulose
HPLC	High performance liquid chromatography
HCPT	Hydroxycamptothecin
HSF	Human skin fibroblasts
IC$_{50}$	Half maximal inhibitory concentration
MCF-7	Michigan Cancer Foundation-7
MgO	Magnesium oxide
NAP	Naproxen
OPN	Osteopontin
PCL	Poly(ε-caprolactone)
PDLA	Poly(D-lactic acid)
PELA	Poly(ethylene glycol)–poly(dl-lactic acid)
P.g	*Porphyromonas gingivalis*
PLCL	Poly(lactide-*co*-caprolactone)
PLLA	Poly(L-lactic acid)
PpR	Pseudorotaxanes
PLA	Poly lactic acid
RM-β-CD	Randomly methylated β-CD
S. aureus	*Staphylococcus aureus*
SAED	Selected area electron diffraction
SFS	Sulfisoxazole
α-TC	α-tocopherol
TGA	Thermogravimetric analyses
TR	Triclosan
WAXD	Wide angle-X-ray diffraction
WCA	Water contact angle
XPS	X-ray photoelectron spectroscopy

References

1. Kny, E.; Uyar, T.; Kny, E. *Electrospun Materials for Tissue Engineering and Biomedical Applications*; Kny, E., Uyar, T., Eds.; Woodhead Publishing: Sawston, UK, 2017; pp. i–iii.
2. Anu Bhushani, J.; Anandharamakrishnan, C. Electrospinning and electrospraying techniques: Potential food based applications. *Trends Food Sci. Technol.* **2014**, *38*, 21–33. [CrossRef]
3. Wang, W.; Huang, H.; Li, Z.; Zhang, H.; Wang, Y.; Zheng, W.; Wang, C. Zinc Oxide Nanofiber Gas Sensors via Electrospinning. *J. Am. Ceram. Soc.* **2008**, *91*, 3817–3819. [CrossRef]
4. Narayanan, G.; Vernekar, V.N.; Kuyinu, E.L.; Laurencin, C.T. Poly (lactic acid)-based biomaterials for orthopaedic regenerative engineering. *Adv. Drug Deliv. Rev.* **2016**, *107*, 247–276. [CrossRef] [PubMed]
5. Ramakrishna, S.; Fujihara, K.; Teo, W. E.; Yong, T.; Ma, Z.; Ramaseshan, R. Electrospun nanofibers: Solving global issues. *Mater. Today* **2006**, *9*, 40–50. [CrossRef]
6. Muerza-Cascante, M.L.; Shokoohmand, A.; Khosrotehrani, K.; Haylock, D.; Dalton, P.D.; Hutmacher, D.W.; Loessner, D. Endosteal-like extracellular matrix expression on melt electrospun written scaffolds. *Acta Biomater.* **2017**, *52*, 145–158. [CrossRef] [PubMed]
7. Mendes, A.C.; Gorzelanny, C.; Halter, N.; Schneider, S.W.; Chronakis, I.S. Hybrid electrospun chitosan-phospholipids nanofibers for transdermal drug delivery. *Int. J. Pharm.* **2016**, *510*, 48–56. [CrossRef] [PubMed]
8. Woodruff, M.A.; Hutmacher, D.W. The return of a forgotten polymer—Polycaprolactone in the 21st century. *Prog. Polym. Sci.* **2010**, *35*, 1217–1256. [CrossRef]
9. Uyar, T.; El-Shafei, A.; Wang, X.; Hacaloglu, J.; Tonelli, A.E. The Solid Channel Structure Inclusion Complex Formed between Guest Styrene and Host γ-Cyclodextrin. *J. Incl. Phenom. Macrocycl. Chem.* **2006**, *55*, 109–121. [CrossRef]

10. Uyar, T.; Hunt, M.A.; Gracz, H.S.; Tonelli, A.E. Crystalline Cyclodextrin Inclusion Compounds Formed with Aromatic Guests: Guest-Dependent Stoichiometries and Hydration-Sensitive Crystal Structures. *Cryst. Growth Des.* **2006**, *6*, 1113–1119. [CrossRef]

11. Narayanan, G.; Boy, R.; Gupta, B.S.; Tonelli, A.E. Analytical techniques for characterizing cyclodextrins and their inclusion complexes with large and small molecular weight guest molecules. *Polym. Test.* **2017**, *62*, 402–439. [CrossRef]

12. Tonelli, A.E. Molecular Processing of Polymers with Cyclodextrins. In *Inclusion Polymers*; Wenz, G., Ed.; Springer: Berlin/Heidelberg, Germany, 2009; pp. 55–77.

13. Croft, A.P.; Bartsch, R.A. Synthesis of chemically modified cyclodextrins. *Tetrahedron* **1983**, *39*, 1417–1474. [CrossRef]

14. Tonelli, A.E. Restructuring polymers via nanoconfinement and subsequent release. *Beil. J. Organ. Chem.* **2012**, *8*, 1318. [CrossRef] [PubMed]

15. Saha, S.; Roy, M.N. *Encapsulation of Vitamin C into β-Cyclodextrin for Advanced and Regulatory Release*; Hamza, A.H., Vitamin, C., Eds.; InTech: Rijeka, Croatian, 2017; p. Ch. 07.

16. Alvarez-Lorenzo, C.; García-González, C.A.; Concheiro, A. Cyclodextrins as versatile building blocks for regenerative medicine. *J. Control. Release* **2017**, *268*, 269–281. [CrossRef] [PubMed]

17. Alvarez-Lorenzo, C.; Moya-Ortega, M.D.; Loftsson, T.; Concheiro, A.; Torres-Labandeira, J.J. Cyclodextrin-Based Hydrogels. In *Cyclodextrins in Pharmaceutics, Cosmetics, and Biomedicine*; John Wiley & Sons, Inc.: Hoboken, NJ, USA, 2011; pp. 297–321.

18. Costoya, A.; Concheiro, A.; Alvarez-Lorenzo, C. Electrospun Fibers of Cyclodextrins and Poly(cyclodextrins). *Molecules* **2017**, *22*, 230. [CrossRef] [PubMed]

19. Loftsson, T.; Moya-Ortega, M.D.; Alvarez-Lorenzo, C.; Concheiro, A. Pharmacokinetics of cyclodextrins and drugs after oral and parenteral administration of drug/cyclodextrin complexes. *J. Pharm. Pharmacol.* **2016**, *68*, 544–555. [CrossRef] [PubMed]

20. Brackman, G.; Garcia-Fernandez, M.J.; Lenoir, J.; De Meyer, L.; Remon, J.-P.; De Beer, T.; Concheiro, A.; Alvarez-Lorenzo, C.; Coenye, T. Dressings Loaded with Cyclodextrin–Hamamelitannin Complexes Increase Staphylococcus aureus Susceptibility Toward Antibiotics Both in Single as well as in Mixed Biofilm Communities. *Macromol. Biosci.* **2016**, *16*, 859–869. [CrossRef] [PubMed]

21. Simoes, S.M.N.; Rey-Rico, A.; Concheiro, A.; Alvarez-Lorenzo, C. Supramolecular cyclodextrin-based drug nanocarriers. *Chem. Commun.* **2015**, *51*, 6275–6289. [CrossRef] [PubMed]

22. Simões, S.M.N.; Veiga, F.; Torres-Labandeira, J.J.; Ribeiro, A.C.F.; Concheiro, A.; Alvarez-Lorenzo, C. Poloxamine-Cyclodextrin-Simvastatin Supramolecular Systems Promote Osteoblast Differentiation of Mesenchymal Stem Cells. *Macromol. Biosci.* **2013**, *13*, 723–734. [CrossRef] [PubMed]

23. Concheiro, A.; Alvarez-Lorenzo, C. Chemically cross-linked and grafted cyclodextrin hydrogels: From nanostructures to drug-eluting medical devices. *Adv. Drug Deliv. Rev.* **2013**, *65*, 1188–1203. [CrossRef] [PubMed]

24. Malikmammadov, E.; Tanir, T.E.; Kiziltay, A.; Hasirci, V.; Hasirci, N. PCL and PCL-based materials in biomedical applications. *J. Biomater. Sci. Polym. Ed.* **2017**, *29*, 863–893. [CrossRef] [PubMed]

25. Flores, C.; Lopez, M.; Tabary, N.; Neut, C.; Chai, F.; Betbeder, D.; Herkt, C.; Cazaux, F.; Gaucher, V.; Martel, B.; et al. Preparation and characterization of novel chitosan and β-cyclodextrin polymer sponges for wound dressing applications. *Carbohydr. Polym.* **2017**, *173*, 535–546. [CrossRef] [PubMed]

26. Tabuchi, R.; Anraku, M.; Iohara, D.; Ishiguro, T.; Ifuku, S.; Nagae, T.; Uekama, K.; Okazaki, S.; Takeshita, K.; Otagiri, M.; et al. Surface-deacetylated chitin nanofibers reinforced with a sulfobutyl ether β-cyclodextrin gel loaded with prednisolone as potential therapy for inflammatory bowel disease. *Carbohydr. Polym.* **2017**, *174*, 1087–1094. [CrossRef] [PubMed]

27. Wei, Z.; Pan, Z.; Li, F.; Yu, J. Poly(butylene succinate-co-terephthalate) nanofibrous membrane composited with cyclodextrin polymer for superhydrophilic property. *RSC Adv.* **2018**, *8*, 1378–1384. [CrossRef]

28. Uyar, T.; Balan, A.; Toppare, L.; Besenbacher, F. Electrospinning of cyclodextrin functionalized poly(methyl methacrylate) (PMMA) nanofibers. *Polymer* **2009**, *50*, 475–480. [CrossRef]

29. Tamer, U.; Yusuf, N.; Jale, H.; Flemming, B. Electrospinning of functional poly(methyl methacrylate) nanofibers containing cyclodextrin-menthol inclusion complexes. *Nanotechnology* **2009**, *20*, 125703.

30. Uyar, T.; Havelund, R.; Nur, Y.; Balan, A.; Hacaloglu, J.; Toppare, L.; Besenbacherac, F.; Kingshotta, P. Cyclodextrin functionalized poly(methyl methacrylate) (PMMA) electrospun nanofibers for organic vapors waste treatment. *J. Membr. Sci.* **2010**, *365*, 409–417. [CrossRef]

31. Tamer, U.; Rasmus, H.; Jale, H.; Xingfei, Z.; Flemming, B.; Peter, K. The formation and characterization of cyclodextrin functionalized polystyrene nanofibers produced by electrospinning. *Nanotechnology* **2009**, *20*, 125605.

32. Uyar, T.; Hacaloglu, J.; Besenbacher, F. Electrospun polystyrene fibers containing high temperature stable volatile fragrance/flavor facilitated by cyclodextrin inclusion complexes. *React. Funct. Polym.* **2009**, *69*, 145–150. [CrossRef]

33. Uyar, T.; Havelund, R.; Hacaloglu, J.; Besenbacher, F.; Kingshott, P. Functional Electrospun Polystyrene Nanofibers Incorporating α-, β-, and γ-Cyclodextrins: Comparison of Molecular Filter Performance. *ACS Nano* **2010**, *4*, 5121–5230. [CrossRef] [PubMed]

34. Akçakoca Kumbasar, E.P.; Akduman, Ç.; Çay, A. Effects of β-cyclodextrin on selected properties of electrospun thermoplastic polyurethane nanofibres. *Carbohydr. Polym.* **2014**, *104*, 42–49. [CrossRef] [PubMed]

35. Akduman, C.; Kumbasar, E.P.A.; Morsunbul, S. Electrospun nanofiber membranes for adsorption of dye molecules from textile wastewater. *Mater. Sci. Eng.* **2017**, *254*, 102001. [CrossRef]

36. Burns, N.A.; Burroughs, M.C.; Gracz, H.; Pritchard, C.Q.; Brozena, A.H.; Willoughby, J.; Khan, S.A. Cyclodextrin facilitated electrospun chitosan nanofibers. *RSC Adv.* **2015**, *5*, 7131–7137. [CrossRef]

37. Sun, X.; Yu, Z.; Cai, Z.; Yu, L.; Lv, Y. Voriconazole Composited Polyvinyl Alcohol/Hydroxypropyl-β-Cyclodextrin Nanofibers for Ophthalmic Delivery. *PLoS ONE* **2016**, *11*, e0167961. [CrossRef] [PubMed]

38. Zhou, J.; Wang, Q.; Lu, H.; Zhang, Q.; Lv, P.; Wei, Q. Preparation and characterization of electrospun polyvinyl alcoholstyrylpyridinium/β-cyclodextrin composite nanofibers: Release behavior and potential use for wound dressing. *Fibers Polym.* **2016**, *17*, 1835–1841. [CrossRef]

39. Lemma, S.M.; Scampicchio, M.; Mahon, P.J.; Sbarski, I.; Wang, J.; Kingshott, P. Controlled Release of Retinyl Acetate from β-Cyclodextrin Functionalized Poly(vinyl alcohol) Electrospun Nanofibers. *J. Agric. Food Chem.* **2015**, *63*, 3481–3488. [CrossRef] [PubMed]

40. Kayaci, F.; Uyar, T. Electrospun polyester/cyclodextrin nanofibers for entrapment of volatile organic compounds. *Polym. Eng. Sci.* **2014**, *54*, 2970–2978. [CrossRef]

41. Uyar, T.; Besenbacher, F. Electrospinning of cyclodextrin functionalized polyethylene oxide (PEO) nanofibers. *Eur. Polym. J.* **2009**, *45*, 1032–1037. [CrossRef]

42. Narayanan, G.; Gupta, B.S.; Tonelli, A.E. Poly(ε-caprolactone) Nanowebs Functionalized with α- and γ-Cyclodextrins. *Biomacromolecules* **2014**, *15*, 4122–4133. [CrossRef] [PubMed]

43. Narayanan, G.; Ormond, B.R.; Gupta, B.S.; Tonelli, A.E. Efficient wound odor removal by β-cyclodextrin functionalized poly (ε-caprolactone) nanofibers. *J. Appl. Polym. Sci.* **2015**, *132*, 42782. [CrossRef]

44. Huang, L.; Allen, E.; Tonelli, A.E. Study of the inclusion compounds formed between α-cyclodextrin and high molecular weight poly(ethylene oxide) and poly(ε-caprolactone). *Polymer* **1998**, *39*, 4857–4865. [CrossRef]

45. Wei, M.; Shuai, X.; Tonelli, A.E. Melting and Crystallization Behaviors of Biodegradable Polymers Enzymatically Coalesced from Their Cyclodextrin Inclusion Complexes. *Biomacromolecules* **2003**, *4*, 783–792. [CrossRef] [PubMed]

46. Lu, J.; Mirau, P.A.; Tonelli, A.E. Dynamics of Isolated Polycaprolactone Chains in Their Inclusion Complexes with Cyclodextrins. *Macromolecules* **2001**, *34*, 3276–3284. [CrossRef]

47. Rusa, C.C.; Luca, C.; Tonelli, A.E. Polymer–Cyclodextrin Inclusion Compounds: Toward New Aspects of Their Inclusion Mechanism. *Macromolecules* **2001**, *34*, 1318–1322. [CrossRef]

48. Shuai, X.; Porbeni, F.E.; Wei, M.; Shin, I.D.; Tonelli, A.E. Formation of and Coalescence from the Inclusion Complex of a Biodegradable Block Copolymer and α-Cyclodextrin: A Novel Means To Modify the Phase Structure of Biodegradable Block Copolymers. *Macromolecules* **2001**, *34*, 7355–7361. [CrossRef]

49. Shuai, X.; Porbeni, F.E.; Wei, M.; Bullions, T.; Tonelli, A.E. Inclusion Complex Formation between α,γ-Cyclodextrins and a Triblock Copolymer and the Cyclodextrin-Type-Dependent Microphase Structures of Their Coalesced Samples. *Macromolecules* **2002**, *35*, 2401–2405. [CrossRef]

50. Williamson, B.R.; Tonelli, A.E. Constrained polymer chain behavior observed in their non-stoichiometric cyclodextrin inclusion complexes. *J. Incl. Phenom. Macrocycl. Chem.* **2012**, *72*, 71–78. [CrossRef]

51. Chan, S.-C.; Kuo, S.-W.; Chang, F.-C. Synthesis of the Organic/Inorganic Hybrid Star Polymers and Their Inclusion Complexes with Cyclodextrins. *Macromolecules* **2005**, *38*, 3099–3107. [CrossRef]

52. Lu, J.; Hill, M.A.; Hood, M.; Greeson, D.F., Jr.; Horton, J.R.; Orndorff, P.E.; Herndon, A.S.; Tonelli, A.E. Formation of antibiotic, biodegradable polymers by processing with Irgasan DP300R (triclosan) and its inclusion compound with β-cyclodextrin. *J. Appl. Polym. Sci.* **2001**, *82*, 300–309. [CrossRef]

53. Huang, L.; Gerber, M.; Lu, J.; Tonelli, A.E. Formation of a flame retardant-cyclodextrin inclusion compound and its application as a flame retardant for poly(ethylene terephthalate). *Polym. Degrad. Stab.* **2001**, *71*, 279–284. [CrossRef]

54. Huang, L.; Taylor, H.; Gerber, M.; Orndorff, P.E.; Horton, J.R.; Tonelli, A. Formation of antibiotic, biodegradable/bioabsorbable polymers by processing with neomycin sulfate and its inclusion compound with β-cyclodextrin. *J. Appl. Polym. Sci.* **1999**, *74*, 937–947. [CrossRef]

55. Suk Whang, H.; Hunt, M.A.; Wrench, N.; Hockney, J.E.; Farin, C.E.; Tonelli, A.E. Nonoxynol-9-α-cyclodextrin inclusion compound and its application for the controlled release of nonoxynol-9 spermicide. *J. Appl. Polym. Sci.* **2007**, *106*, 4104–4109. [CrossRef]

56. Tonelli, A.E. Nanostructuring and functionalizing polymers with cyclodextrins. *Polymer* **2008**, *49*, 1725–1736. [CrossRef]

57. Aytac, Z.; Uyar, T. Antioxidant activity and photostability of α-tocopherol/β-cyclodextrin inclusion complex encapsulated electrospun polycaprolactone nanofibers. *Eur. Polym. J.* **2016**, *79*, 140–149. [CrossRef]

58. Siafaka, P.; Üstündağ Okur, N.; Mone, M.; Giannakopoulou, S.; Er, S.; Pavlidou, E.; Karavas, E.; Bikiaris, D.N. Two Different Approaches for Oral Administration of Voriconazole Loaded Formulations: Electrospun Fibers versus β-Cyclodextrin Complexes. *Int. J. Mol. Sci.* **2016**, *17*, 282. [CrossRef] [PubMed]

59. Banik, A.; Gogoi, P.; Saikia, M.D. Interaction of naproxen with β-cyclodextrin and its derivatives/polymer: Experimental and molecular modeling studies. *J. Incl. Phenom. Macrocycl. Chem.* **2012**, *72*, 449–458. [CrossRef]

60. Blanco, J.; Vila-jato, J.L.; Otero, F.; Anguiano, S. Influence of Method of Preparation on Inclusion Complexes of Naproxen with Different Cyclodextrins. *Drug Dev. Ind. Pharm.* **1991**, *17*, 943–957. [CrossRef]

61. Szafran, B.; Pawlaczyk, J. Preparation and characterization of the β-cyclodextrin inclusion complex with sulfafurazole. *J. Incl. Phenom. Mol. Recognit. Chem.* **1995**, *23*, 277–288. [CrossRef]

62. Gladys, G.; Claudia, G.; Marcela, L. The effect of pH and triethanolamine on sulfisoxazole complexation with hydroxypropyl-β-cyclodextrin. *Eur. J. Pharm. Sci.* **2003**, *20*, 285–293. [CrossRef]

63. Canbolat, M.F.; Celebioglu, A.; Uyar, T. Drug delivery system based on cyclodextrin-naproxen inclusion complex incorporated in electrospun polycaprolactone nanofibers. *Colloids Surf. B Biointerfaces* **2014**, *115*, 15–21. [CrossRef] [PubMed]

64. Monteiro, A.P.; Rocha, C.M.; Oliveira, M.F.; Gontijo, S.M.; Agudelo, R.R.; Sinisterra, R.D.; Cortés, M.E. Nanofibers containing tetracycline/β-cyclodextrin: Physico-chemical characterization and antimicrobial evaluation. *Carbohydr. Polym.* **2017**, *156*, 417–426. [CrossRef] [PubMed]

65. Sousa, J.; Alves, G.; Oliveira, P.; Fortuna, A.; Falcão, A. Intranasal delivery of ciprofloxacin to rats: A topical approach using a thermoreversible in situ gel. *Eur. J. Pharm. Sci.* **2017**, *97*, 30–37. [CrossRef] [PubMed]

66. Sahoo, S.; Chakraborti, C.K.; Mishra, S.C.; Naik, S. Ftir and Xrd Investigations of Some Fluoroquinolones. *Int. J. Pharm. Pharm. Sci.* **2011**, *3*, 165–170.

67. Masoumi, S.; Amiri, S.; Bahrami, S.H. PCL-based nanofibers loaded with ciprofloxacin/cyclodextrin containers. *J. Text. Inst.* **2017**, *54*. [CrossRef]

68. Aytac, Z.; Sen, H.S.; Durgun, E.; Uyar, T. Sulfisoxazole/cyclodextrin inclusion complex incorporated in electrospun hydroxypropyl cellulose nanofibers as drug delivery system. *Colloids Surf. B Biointerfaces* **2015**, *128*, 331–338. [CrossRef] [PubMed]

69. Schrader, J.; Etschmann, M.M.W.; Sell, D.; Hilmer, J.-M.; Rabenhorst, J. Applied biocatalysis for the synthesis of natural flavour compounds—Current industrial processes and future prospects. *Biotechnol. Lett.* **2004**, *26*, 463–472. [CrossRef] [PubMed]

70. Kourist, R.; Domínguez de María, P.; Bornscheuer, U.T. Enzymatic Synthesis of Optically Active Tertiary Alcohols: Expanding the Biocatalysis Toolbox. *ChemBioChem* **2008**, *9*, 491–498. [CrossRef] [PubMed]

71. Kelley, B. Very Large Scale Monoclonal Antibody Purification: The Case for Conventional Unit Operations. *Biotechnol. Prog.* **2007**, *23*, 995–1008. [CrossRef] [PubMed]

72. Betancor, L.; Luckarift, H.R. Bioinspired enzyme encapsulation for biocatalysis. *Trends Biotechnol.* **2008**, *26*, 566–572. [CrossRef] [PubMed]

73. Aleti, S.; Karaturi, H.; Subrahmanyam, C.V.S.; Narasu, M.L. Complexation and Characterization of -Amylase with Hydroxypropyl-Cyclodextrin. *Int. J. Pharm. Phytopharmacol. Res.* **2012**, *1*, 375–378.

74. Lee, J.H.; Hwang, E.T.; Kim, B.C.; Lee, S.-M.; Sang, B.-I.; Choi, Y.S.; Kim, J.; Gu, M.B. Stable and continuous long-term enzymatic reaction using an enzyme–nanofiber composite. *Appl. Microbiol. Biotechnol.* **2007**, *75*, 1301–1307. [CrossRef] [PubMed]

75. Canbolat, M.F.; Savas, H.B.; Gultekin, F. Improved catalytic activity by catalase immobilization using γ-cyclodextrin and electrospun PCL nanofibers. *J. Appl. Polym. Sci.* **2017**. [CrossRef]

76. Canbolat, M.F.; Savas, H.B.; Gultekin, F. Enzymatic behavior of laccase following interaction with γ-CD and immobilization into PCL nanofibers. *Anal. Biochem.* **2017**, *528*, 13–18. [CrossRef] [PubMed]

77. Hunt, M.A.; Jung, D.W.; Shamsheer, M.; Uyar, T.; Tonelli, A.E. Polystyrenes in channels. *Polymer* **2004**, *45*, 1345–1347. [CrossRef]

78. Uyar, T.; Gracz, H.S.; Rusa, M.; Shin, I.D.; El-Shafei, A.; Tonelli, A.E. Polymerization of styrene in γ-cyclodextrin channels: Lightly rotaxanated polystyrenes with altered stereosequences. *Polymer* **2006**, *47*, 6948–6955. [CrossRef]

79. Wei, M.; Tonelli, A.E. Compatiblization of Polymers via Coalescence from Their Common Cyclodextrin Inclusion Compounds. *Macromolecules* **2001**, *34*, 4061–4065. [CrossRef]

80. Wei, M.; Shin, I.D.; Urban, B.; Tonelli, A.E. Partial miscibility in a nylon-6/nylon-66 blend coalesced from their common α-cyclodextrin inclusion complex. *J. Polym. Sci. Part B Polym. Phys.* **2004**, *42*, 1369–1378. [CrossRef]

81. Rusa, C.C.; Wei, M.; Shuai, X.; Bullions, T.A.; Wang, X.; Rusa, M.; Uyar, T.; Tonelli, A.E. Molecular mixing of incompatible polymers through formation of and coalescence from their common crystalline cyclodextrin inclusion compounds. *J. Polym. Sci. Part B Polym. Phys.* **2004**, *42*, 4207–4224. [CrossRef]

82. Rusa, C.C.; Wei, M.; Bullions, T.A.; Rusa, M.; Gomez, M.A.; Porbeni, F.E.; Gomez, M.A.; Porbeni, F.E.; Wang, X.; Shin, I.D.; et al. Controlling the Polymorphic Behaviors of Semicrystalline Polymers with Cyclodextrins. *Cryst. Growth Des.* **2004**, *4*, 1431–1441. [CrossRef]

83. Porbeni, F.E.; Shin, I.D.; Shuai, X.; Wang, X.; White, J.L.; Jia, X.; Tonelli, A.E. Morphology and dynamics of the poly(ε-caprolactone)-*b*-poly(L-lactide) diblock copolymer and its inclusion compound with α-cyclodextrin: A solid-state ^{13}C NMR study. *J. Polym. Sci. Part B Polym. Phys.* **2005**, *43*, 2086–2096. [CrossRef]

84. Rusa, C.C.; Rusa, M.; Peet, J.; Uyar, T.; Fox, J.; Hunt, M.A.; Wang, X.; Balik, C.M.; Tonell, A.E. The Nano-threading of Polymers. *J. Incl. Phenom. Macrocycl. Chem.* **2006**, *55*, 185–192. [CrossRef]

85. Rusa, C.C.; Shuai, X.; Shin, I.D.; Bullions, T.A.; Wei, M.; Porbeni, F.E.; Lu, L.; Huang, L.; Fox, J.; Tonelli, A.E. Controlling the Behaviors of Biodegradable/Bioabsorbable Polymers with Cyclodextrins. *J. Polym. Environ.* **2004**, *12*, 157–163. [CrossRef]

86. Shuai, X.; Wei, M.; Porbeni, F.E.; Bullions, T.A.; Tonelli, A.E. Formation of and Coalescence from the Inclusion Complex of a Biodegradable block copolymer and α-cyclodextrin. 2: A novel way to regulate the biodegradation behavior of biodegradable block copolymers. *Biomacromolecules* **2002**, *3*, 201–207. [CrossRef] [PubMed]

87. Wei, M.; Bullions, T.A.; Rusa, C.C.; Wang, X.; Tonelli, A.E. Unique morphological and thermal behaviors of reorganized poly(ethylene terephthalates). *J. Polym. Sci. Part B Polym. Phys.* **2004**, *42*, 386–394. [CrossRef]

88. Pang, K.; Schmidt, B.; Kotek, R.; Tonelli, A. Reorganization of the chain packing between poly(ethylene isophthalate) chains via coalescence from their inclusion compound formed with γ-cyclodextrin. *J. Appl. Polym. Sci.* **2006**, *102*, 6049–6053. [CrossRef]

89. Vedula, J.; Tonelli, A.E. Reorganization of poly(ethylene terephthalate) structures and conformations to alter properties. *J. Polym. Sci. Part B Polym. Phys.* **2007**, *45*, 735–746. [CrossRef]

90. Williamson, B.R.; Krishnaswamy, R.; Tonelli, A.E. Physical properties of poly(ε-caprolactone) coalesced from its α-cyclodextrin inclusion compound. *Polymer* **2011**, *52*, 4517–4527. [CrossRef]

91. Dong, T.; Shin, K.-M.; Zhu, B.; Inoue, Y. Nucleation and Crystallization Behavior of Poly(butylene succinate) Induced by Its α-Cyclodextrin Inclusion Complex: Effect of Stoichiometry. *Macromolecules* **2006**, *39*, 2427–2428. [CrossRef]

92. Mohan, A.; Joyner, X.; Kotek, R.; Tonelli, A.E. Constrained/Directed Crystallization of Nylon-6. I. Nonstoichiometric Inclusion Compounds Formed with Cyclodextrins. *Macromolecules* **2009**, *42*, 8983–8991. [CrossRef]

93. Tonelli, A. Non-Stoichiometric Polymer-Cyclodextrin Inclusion Compounds: Constraints Placed on Un-Included Chain Portions Tethered at Both Ends and Their Relation to Polymer Brushes. *Polymers* **2014**, *6*, 2166–2185. [CrossRef]

94. Dong, T.; Mori, T.; Pan, P.; Kai, W.; Zhu, B.; Inoue, Y. Crystallization behavior and mechanical properties of poly(ε-caprolactone)/cyclodextrin biodegradable composites. *J. Appl. Polym. Sci.* **2009**, *112*, 2351–2357. [CrossRef]

95. Narayanan, G.; Gupta, B.S.; Tonelli, A.E. Enhanced mechanical properties of poly (ε-caprolactone) nanofibers produced by the addition of non-stoichiometric inclusion complexes of poly (ε-caprolactone) and α-cyclodextrin. *Polymer* **2015**, *76*, 321–330. [CrossRef]

96. Narayanan, G.; Aguda, R.; Hartman, M.; Chung, C.-C.; Boy, R.; Gupta, B.S.; Tonelli, A.E. Fabrication and Characterization of Poly(ε-caprolactone)/α-Cyclodextrin Pseudorotaxane Nanofibers. *Biomacromolecules* **2016**, *17*, 271–279. [CrossRef] [PubMed]

97. Narayanan, G.; Chung, C.-C.; Aguda, R.; Boy, R.; Hartman, M.; Mehraban, N.; Guptaa, B.S.; Tonelli, A.E. Correlation of the stoichiometries of poly(ε-caprolactone) and α-cyclodextrin pseudorotaxanes with their solution rheology and the molecular orientation, crystallite size, and thermomechanical properties of their nanofibers. *RSC Adv.* **2016**, *6*, 111326–111336. [CrossRef]

98. Narayanan, G.; Gupta, B.S.; Tonelli, A.E. Estimation of the poly (ε-caprolactone) [PCL] and α-cyclodextrin [α-CD] stoichiometric ratios in their inclusion complexes [ICs], and evaluation of porosity and fiber alignment in PCL nanofibers containing these ICs. *Data Brief* **2015**, *5*, 1048–1055. [CrossRef] [PubMed]

99. Narayanan, G.; Boy, R.; Gupta, B.S.; Tonelli, A.E. Functional Nanofibers Containing Cyclodextrins. In *Polysaccharide-Based Fibers and Composites: Chemical and Engineering Fundamentals and Industrial Applications*; Lucia, L., Ayoub, A., Eds.; Springer: Berlin, Germany, 2018; pp. 29–62.

100. Narayanan, G.; Bhattacharjee, M.; Nair, L.S.; Laurencin, C.T. Musculoskeletal Tissue Regeneration: The Role of the Stem Cells. *Regen. Eng. Transl. Med.* **2017**, *3*, 133–165. [CrossRef]

101. Narayanan, G.; Nair, L.S.; Laurencin, C.T. Regenerative Engineering of the Rotator Cuff of the Shoulder. *ACS Biomater. Sci. Eng.* **2017**. [CrossRef]

102. Hein, C.D.; Liu, X.-M.; Wang, D. Click Chemistry, A Powerful Tool for Pharmaceutical Sciences. *Pharm. Res.* **2008**, *25*, 2216–2230. [CrossRef] [PubMed]

103. Zhan, J.; Singh, A.; Zhang, Z.; Huang, L.; Elisseeff, J.H. Multifunctional aliphatic polyester nanofibers for tissue engineering. *Biomatter* **2012**, *2*, 202–212. [CrossRef] [PubMed]

104. Oster, M.; Hébraud, A.; Gallet, S.; Lapp, A.; Pollet, E.; Avérous, L.; Schlatter, G. Star-Pseudopolyrotaxane Organized in Nanoplatelets for Poly(ε-caprolactone)-Based Nanofibrous Scaffolds with Enhanced Surface Reactivity. *Macromol. Rapid Commun.* **2015**, *36*, 292–297. [CrossRef] [PubMed]

105. Oster, M.; Schlatter, G.; Gallet, S.; Baati, R.; Pollet, E.; Gaillard, C.; Avérous, L.; Fajolles, C.; Hébraud, A. The study of the pseudo-polyrotaxane architecture as a route for mild surface functionalization by click chemistry of poly(ε-caprolactone)-based electrospun fibers. *J. Mater. Chem. B* **2017**, *5*, 2181–2189. [CrossRef]

106. Wang, J.; Qiu, C.; Narsimhan, G.; Jin, Z. Preparation and Characterization of Ternary Antimicrobial Films of β-Cyclodextrin/Allyl Isothiocyanate/Polylactic Acid for the Enhancement of Long-Term Controlled Release. *Materials* **2017**, *10*, 1210. [CrossRef] [PubMed]

107. Fathi-Azarbayjani, A.; Chan, S.Y. Single and Multi-Layered Nanofibers for Rapid and Controlled Drug Delivery. *Chem. Pharm. Bull.* **2010**, *58*, 143–146. [CrossRef] [PubMed]

108. Xie, C.; Li, X.; Luo, X.; Yang, Y.; Cui, W.; Zou, J.; Zhou, S. Release modulation and cytotoxicity of hydroxycamptothecin-loaded electrospun fibers with 2-hydroxypropyl-β-cyclodextrin inoculations. *Int. J. Pharm.* **2010**, *391*, 55–64. [CrossRef] [PubMed]

109. Sudakaran, S.V.; Venugopal, J.R.; Vijayakumar, G.P.; Abisegapriyan, S.; Grace, A.N.; Ramakrishna, S. Sequel of MgO nanoparticles in PLACL nanofibers for anti-cancer therapy in synergy with curcumin/β-cyclodextrin. *Mater. Sci. Eng. C* **2017**, *71*, 620–628. [CrossRef] [PubMed]

110. Aytac, Z.; Uyar, T. Core-shell nanofibers of curcumin/cyclodextrin inclusion complex and polylactic acid: Enhanced water solubility and slow release of curcumin. *Int. J. Pharm.* **2017**, *518*, 177–184. [CrossRef] [PubMed]

111. Liu, Y.; Liang, X.; Zhang, R.; Lan, W.; Qin, W. Fabrication of Electrospun Polylactic Acid/Cinnamaldehyde/β-Cyclodextrin Fibers as an Antimicrobial Wound Dressing. *Polymers* **2017**, *9*, 464. [CrossRef]

112. Kayaci, F.; Umu, O.C.O.; Tekinay, T.; Uyar, T. Antibacterial Electrospun Poly(lactic acid) (PLA) Nanofibrous Webs Incorporating Triclosan/Cyclodextrin Inclusion Complexes. *J. Agric. Food Chem.* **2013**, *61*, 3901–3908. [CrossRef] [PubMed]

113. Wen, P.; Zhu, D.-H.; Feng, K.; Liu, F.-J.; Lou, W.-Y.; Li, N.; Zong, M.-H.; Wu, H. Fabrication of electrospun polylactic acid nanofilm incorporating cinnamon essential oil/β-cyclodextrin inclusion complex for antimicrobial packaging. *Food Chem.* **2016**, *196*, 996–1004. [CrossRef] [PubMed]
114. Aytac, Z.; Kusku, S.I.; Durgun, E.; Uyar, T. Encapsulation of gallic acid/cyclodextrin inclusion complex in electrospun polylactic acid nanofibers: Release behavior and antioxidant activity of gallic acid. *Mater. Sci. Eng. C* **2016**, *63*, 231–239. [CrossRef] [PubMed]
115. Gurarslan, A.; Joijode, A.; Shen, J.; Narayanan, G.; Antony, G.; Li, S.; Caydamli, Y.; Tonelli, A.E. Reorganizing Polymer Chains with Cyclodextrins. *Polymers* **2017**, *9*, 673. [CrossRef]

polymers

Article

Enhancement of Wound Healing in Normal and Diabetic Mice by Topical Application of Amorphous Polyphosphate. Superior Effect of a Host–Guest Composite Material Composed of Collagen (Host) and Polyphosphate (Guest)

Werner E. G. Müller [1,*], **Dinko Relkovic** [2], **Maximilian Ackermann** [3], **Shunfeng Wang** [1], **Meik Neufurth** [1], **Andrea Paravic Radicevic** [2], **Hiroshi Ushijima** [4], **Heinz C. Schröder** [1] and **Xiaohong Wang** [1,*]

[1] ERC Advanced Investigator Grant Research Group at the Institute for Physiological Chemistry, University Medical Center of the Johannes Gutenberg University, Mainz, Duesbergweg 6, 55128 Mainz, Germany; Shunwang@uni-mainz.de (S.W.); mneufurt@uni-mainz.de (M.N.); hschroed@uni-mainz.de (H.C.S.)

[2] Fidelta Ltd., Prilaz baruna Filipovića 29, 10000 Zagreb, Croatia; Dinko.Relkovic@glpg.com (D.R.); andrea.paravicradicevic@glpg.com (A.P.R.)

[3] Institute of Functional and Clinical Anatomy, University Medical Center of the Johannes Gutenberg University, Johann Joachim Becher Weg 13, D-55099 Mainz, Germany; maximilian.ackermann@uni-mainz.de

[4] Division of Microbiology, Department of Pathology and Microbiology, Nihon University School of Medicine, 30-1 Oyaguchi Kamicho, Itabashi-ku, Tokyo 173-8610, Japan; ushijima-hiroshi@jcom.home.ne.jp

*   Correspondence: wmueller@uni-mainz.de (W.E.G.M.); wang013@uni-mainz.de (X.W.); Tel.: +49-6131-392-5910 (W.E.G.M. & X.W.)

Received: 2 July 2017; Accepted: 20 July 2017; Published: 22 July 2017

Abstract: The effect of polyphosphate (polyP) microparticles on wound healing was tested both in vitro and in a mice model in vivo. Two approaches were used: pure salts of polyphosphate, fabricated as amorphous microparticles (MPs, consisting of calcium and magnesium salts of polyP, "Ca–polyp-MPs" and "Mg–polyp-MPs"), and host–guest composite particles, prepared from amorphous collagen (host) and polyphosphate (guest), termed "col/polyp-MPs". Animal experiments with polyP on healing of excisional wounds were performed using both normal mice and diabetic mice. After a healing period of 7 days "Ca–polyp-MP" significantly improved re-epithelialization in normal mice from 31% (control) to 72% (polyP microparticle-treated). Importantly, in diabetic mice, particularly the host–guest particles "col/polyp-MP", increased the rate of re-epithelialization to ≈40% (control, 23%). In addition, those particles increased the expression of *COL-I* and *COL-III* as well as the expression the α-smooth muscle actin and the plasminogen activator inhibitor-1. We propose that "Ca–polyp-MPs", and particularly the host–guest "col/polyp-MPs" are useful for topical treatment of wounds.

Keywords: polyphosphate; microparticles; delayed wound healing; collagen; PAI-1; re-epithelialization; diabetic mice

1. Introduction

Acute and, in particular, chronic wounds are a global health problem, with over 10 million people affected and ≈300,000 people hospitalized every year alone in the United States [1]. Non-healing wounds affect 3–6 million people in the United States; persons aged over 65 years represent 85% of these patients, with costs representing more than $3 billion per year [2]. Wound healing has been staged into the following phases; coagulation/inflammation, formation of granulation tissue, production of new structures and tissue, and finally, remodeling [3]. These complex processes are regulated by cytokines and growth factors and are decisively modulated by systemic conditions, e.g., diabetes. During the coagulation/inflammatory phase, blood platelets adhere to damaged blood vessels and initiate a release reaction resulting in the initiation of the blood-clotting cascade [1]. Blood platelets release an array of growth factors and cytokines as well as survival or apoptosis-inducing agents. Major factors involved are the platelet-derived growth factor and the transforming growth factors A1 and A2. In turn, inflammatory cells, including macrophages and leukocytes, are attracted, and release antimicrobial reactive oxygen species and proteases which remove non-self-bacteria and cell debris. Finally, the inflammatory phase is terminated by apoptosis during which inflammatory cells are eliminated. In the following proliferative/granulation phase, tissue repair starts a process that is controlled by growth factors produced by invading epidermal and dermal cells that via autocrine, paracrine, and juxtacrine pathways induce and maintain cellular proliferation and cellular migration [4]. The formation of granulation tissue allows epithelialization within the wound. During the final matrix remodeling and scar formation phase normal blood supply to the regenerated tissue is completed providing a suitable microenvironment for epidermal and dermal cell proliferation and migration, contributing to wound re-epithelialization and restoration.

The supply of metabolic energy to the regeneration zone of the wound is a critical parameter that influences wound-healing kinetics. This becomes overt on studying chronic wound metabolism. As an example, venous ulcers are attributed as a consequence of venous valvular insufficiency both in the deep and the superficial veins. This impairment results in blood backflow followed by an increased venous pressure [5] and increased blood vessel wall permeabilities. Similarly, complications during aging and diabetes show serious consequences, with both arterial and venous insufficiencies. In diabetic patients (diabetic ulcers) vascular impairment and cellular deficiencies are frequently parallel to microvascular pathologies [6]. The biochemical consequences of impaired vascularization are reduced cell metabolism and supply of oxygen required for energy production by means of ATP [7]. It is well thematized and proven that intracellular ATP levels are directly correlated with the efficiency of wound healing and tissue regeneration. However, until recently it remained obscure as to which metabolic energy storages exist in the extracellular space of the tissue. Now it has become more and more obvious that polyphosphate (polyP), comprising up to several hundreds of phosphate units linked with "high-energy" phosphoanhydride bonds, contributes as a "metabolic fuel" [8,9] to the establishment and maintenance of the extracellular structural and functional organization. ATP is present (if at all) solely in minute concentrations in the extracellular space [10] where it acts as a signaling molecule only [11]. In this place ATP is prone to hydrolytic cleavage through the alkaline phosphatase (ALP) [12]. It is exactly this enzyme that also hydrolyzes polyP [13]. Consequently—as in any biochemical cleavage reaction of energy-rich phosphoanhydride bonds—enzymatic hydrolysis of polyP by the ALP will release metabolically useful chemical energy [14] as well as some heat (see scheme in Figure 1). It has been shown that metabolic chemical energy, in the form of ATP, is required for endothelial cells to grow and differentiate—especially during the maintenance and regulation of angiogenesis [15]. For the preparation of Ca^{2+} polyP particles we introduced a new technology; the amorphous microparticles were obtained by co-precipitation of the highly soluble Na^+–polyP in the presence of a large surplus of $CaCl_2$ [8]. Very recently we also succeeded in preparing a collagen (host) and Ca–polyP (guest) hybrid material which displayed pronounced potencies for the induction of cell proliferation of primary human osteoblasts under in vitro conditions [16].

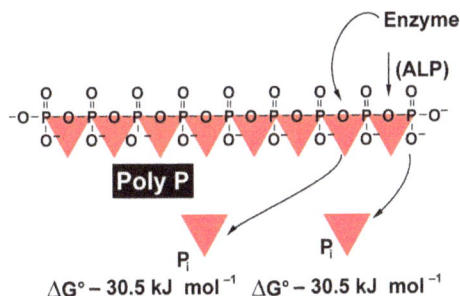

Figure 1. The linear inorganic polyphosphate (poly P) is composed of many phosphate (Pi) residues which are linked by high-energy phosphoanhydride bonds. If those "high-energy" phosphate bonds are cleaved, e.g., by enzymes (like the alkaline phosphatase), the "stored" energy is released (ΔG—30.5 kJ·mol^{-1}).

polyP is a remarkable inorganic macromolecule. It exists as a polymer of around 100 phosphate units in many cells and also in the blood, in especially high concentrations in blood platelets [17]. This physiological polymer has been implicated in the synthesis of amorphous calcium phosphate from amorphous calcium carbonate after its enzymatic hydrolysis to ortho-phosphate during bone formation [18]. In addition, polyP as a (potential) metabolic fuel acts as a suitable bio-ink in three-dimensional tissue printing [9]. The major polyP stores in mammals are the blood platelets which play a crucial role during all phases of wound healing, including coagulation, immune cell recruitment/inflammation, and angiogenesis as well as in remodeling. The ALP activity is highest during the phase of granulation/tissue formation [19]. Hence, there was an obvious need to clarify the effect of polyP, incorporated as a topical wound dressing, on wound-healing kinetics. However, a formulation for polyP needed to be developed in which polyP is released in a slower rate, which prevents rapid enzymatic hydrolysis. Furthermore, those salts must be amorphous, in order to be biologically active. This task was achieved by the fabrication of this polymer as an amorphous Ca^{2+} complex in 100- to 300-nm large microparticles [20]. In this form polyP retains its propensity to undergo enzymatic hydrolysis and shows morphogenetic activity.

Furthermore, fibrillar collagen remodeling during wound healing, especially during the transition from granulation tissue to scar tissue, is dependent on a continuous synthesis and catabolism of the respective collagen types [4]. In a previous contribution we could demonstrate that the expression of the fibrillar collagen genes, of types I, II, and III, are strongly upregulated in vitro by polyP microparticles, especially if they are enriched with retinol [21]. In addition, for the genes encoding for collagens, the steady-state-expression levels of aggrecan [22] (a major regulatory proteoglycan also during wound healing) and of ALP (an enzyme found in the first stage of acute wound healing) [23], increase in response to polyP [24]. These results prove the morphogenetic activity of polyP, as summarized [9] (Figure 2). Hence, it is the aim of the present study to test both in vitro and in vivo to evaluate if polyP in a suitable formulation(s) elicits a beneficial effect on experimental wound healing in mice. Three different forms of polyP microparticles (MPs) have been fabricated, polyP in the form of a Ca^{2+} salt (Ca–polyP-MPs) and a Mg^{2+} salt (Mg–polyp-MPs) as well as a hybrid material together with collagen (col/polyP-MPs). Collagen has been selected as additional component (as host) since this structural macromolecule has been proven to act as a scaffold for wound dressings and facilitates cell attachment, growth and differentiation [25]. The animal experiments were performed both with normal (C57BL/6) and diabetic mice. As a morphogenetic parameter for the effect of the polyP particles in vivo, the potency to induce gene expression of collagen has been chosen. More specifically, the expression of *collagen type I* (dominant in skin and vascular ligature) and the reticulate *type III* (reticular fibers) was determined; these types of collagen are critical for wound healing [26]. In addition, the steady-state-expression levels for *α-smooth muscle actin* (α-SMA) [27] and for *plasminogen activator inhibitor-1* (PAI-1) [28], both genes that are strongly upregulated during wound healing, were determined. Based on the data obtained it is concluded that both the "Ca–polyP-MPs" and the "col/polyP-MPs" positively accelerate the kinetics of wound healing and are (potentially) beneficial for the topical treatment of wounds in humans as well.

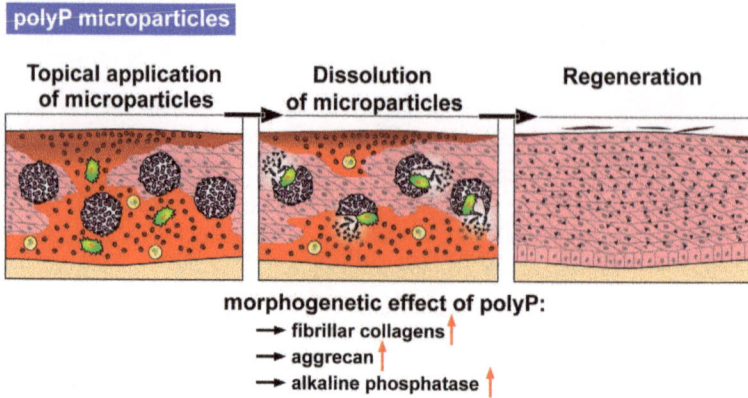

Figure 2. Proposed action mode of polyP, in the form of polyP microparticles, as a morphogenetically active and wound-healing-promoting physiological inorganic polymer. After topical application of the polyP microparticles (with collagen as host and polyP as guest) they undergo dissolution in the impaired tissue, as a result of ALP enzymatic hydrolysis. Then, polyP is known to activate the genes e.g., those encoding fibrillar collagens, aggrecan and the ALP. The experimental data given verify that these polyP microparticles accelerate wound healing and tissue regeneration.

2. Materials and Methods

2.1. Materials

Na–polyphosphate (Na–polyP) with an average chain length of 40 phosphate units was from Chemische Fabrik Budenheim (Budenheim, Germany). The general formula of this Na–polyP is $[NaPO_3]_n[NaPO_3(OH)]_2$, where n is $\approx$40. Rat tail collagen type I was obtained from Shenzhen Lando Biomaterials Co., Ltd. (Shenzhen, China).

2.2. Amorphous PolyP Microparticles

Amorphous polyP particles were prepared from Na–polyP in the presence of a surplus of divalent ions (Ca^{2+} or Mg^{2+}). The final stoichiometric ratio between these ions and the polyP polymer (with reference to one charged phosphate unit) is close to 2. The procedure is outlined in brief.

Mg–polyP particles were basically prepared as described [29]. In short, 3.86 g of $MgCl_2 \cdot 6H_2O$ (#A537.1, Roth, Karlsruhe, Germany) was dissolved in 25 mL of distilled water and added dropwise to 1 g of Na–polyP, likewise in 25 mL of distilled water. During particle formation the suspension was kept at pH 10 and then stirred for 12 h. The microparticles formed were collected by filtration, washed with ethanol and dried at 50 °C. The sample was termed "Mg–polyP-MP".

Ca–polyP particles were fabricated from 2.8 g of $CaCl_2 \cdot 2H_2O$ (#223506; Sigma-Aldrich, Taufkirchen, Germany) (25 mL of distilled water) and 1 g of Na–polyP (25 mL water) at room temperature [20]. The microparticles are formed in the pH 10 suspension during the 12-h stirring period. Then the microparticles were collected by filtration, washed with ethanol and dried at 50 °C. They are termed "Ca–polyP-MP".

Collagen-polyP host–guest hybrid particles were prepared as follows: A suspension of 20 mL of collagen (containing 0.2 g of fibrous collagen material) was added to 50 mL of aqueous Na–polyP solution (containing 0.2 g of solid salt). The developing suspension was kept at pH 9 (with NaOH) and stirred for 4 h at room temperature. The resulting suspension was filled into a syringe (aperture of 2 mm), clamped into an automatic pump, and injected with a speed of 1 mL/min into a $CaCl_2$ bath (3 g of $CaCl_2 \cdot 2H_2O$ per 100 mL) composed of ethanol:acetone (1:2 v/v). Ethanol and acetone prevented shrinkage of the particles, reduced the surface tension and extracted water. Under those conditions the organic solvents are considered not to alter the biological properties of collagen [30]. The suspension was stirred overnight. The particles were collected by filtration and washed twice in acetone. Then the particles, termed "col/polyP-MP" (host–guest particles), were dried at room temperature. A schematic outline of this procedure, and preparation of the host (collagen)–guest (Ca–polyP microparticles) hybrid material, has been given previously [16].

Prior to use all particles were sieved through a 500-μm mesh net.

2.3. Microstructure Analyses

Scanning electron microscopic (SEM) imaging was performed using a Hitachi SU-8000 electron microscope (Hitachi High-Technologies Europe GmbH, Krefeld, Germany). Reflection electron microscope (REM) was performed in a Philips XL30 microscope (Philips, Eindhoven, The Netherlands) at 15 KeV and 21 μA. The samples were coated with 20–25 Å gold in an argon atmosphere.

For energy-dispersive X-ray (EDX) spectroscopy an EDAX Genesis EDX System attached to the scanning electron microscope (Nova 600 Nanolab, FEI, Eindhoven, The Netherlands) was used, using the operation mode 10 kV and a collection time of 30–45 s.

2.4. Cultivation of MC3T3-E1 Cells

Mouse calvaria cells MC3T3-E1 (ATCC-CRL-2593, #99072810) were cultivated in α-MEM medium (Gibco/Invitrogen, Darmstadt, Germany) enriched with 20% fetal calf serum (FCS, Gibco, Schwerte, Germany). In addition, the medium contained 2 mM L-glutamine, 1 mM Na-pyruvate and 50 μg/mL of gentamicin. For the gene expression studies the cells were cultivated in 24-well plates (Greiner Bio-One, Frickenhausen, Germany). The cells were seeded at a density of 5×10^3 cells/well. After an incubation period of 4 days the cells were harvested and subjected to PCR analysis.

Na–polyP was added to the culture/serum, after stoichiometric complexation with Ca^{2+} (molar ratio of 2:1/phosphate monomer: Ca^{2+}). Prior to addition of the microparticles to the culture they were washed twice in medium for 3 min.

2.5. Animals

Genetically diabetic male mice, BKS.Cg-m + Lepr[db]/+Lepr[db] (db/db), aged 6 and 7 weeks at arrival (Charles River, Calco, Italy), and a common inbred strain of laboratory mouse, male C57BL/6, aged 7 weeks at arrival were used. The diabetic mice were markedly hyperglycemic (mean blood glucose: 527 ± 9 mg/dL), compared to non-diabetic animals (205 ± 9 mg/dL) with all details given previously [31]. For acclimatization, a minimum of 5 consecutive days prior to the experiments in the laboratory animal house was chosen. All animals received a detailed physical examination from the resident veterinarian to confirm that the animals were in a good, adequate state of health. Daily observations were performed at the time of delivery of the animals, during the total period of acclimatization and also throughout the duration of the study.

Housing of the animals: Mice were kept in solid bottomed cages (polysulfone type III H; Tecniplast, Buguggiate, Italy) with dimensions of 425 mm × 266 mm × 185 mm. The animals were kept on 3–4 cm thick ALPHA-dri dust free bedding (pure cellulose fiber, uniform particle size 5 mm sq, highly absorbent; LBS Serving (RH6 0UW, Horley, UK)). Each cage was provided with a nestlet and Des Res Standard (for mice) 16 cm long × 12 cm wide × 8 cm high; LBS Serving. The animals were maintained under standard laboratory conditions (temperature 22 ± 2 °C, relative humidity (55 ± 10)%, 15–20 air changes per h, 12 h artificial lighting/12 h darkness per day (7.00 a.m. lights on–7.00 p.m. lights off)).

The animals had free access to food VRF1 (P) (Akronom KfT, Budapest, Hungary) and water distributed in bottles (Tecniplast) and filled with drinking water from the municipal water supply.

2.6. Permissions

All animal-related research was conducted in accordance with 2010/63/EU and national legislation regulating the use of laboratory animals in scientific research and for other purposes (Official Gazette 55/13). An Institutional Committee on Animal Research Ethics (CARE-Zg) had overseen that animal-related procedures did not compromise animal welfare. All experiments conducted in studies described herein were performed under the institutional ethics committee approval number CAREZG_13-06-14_49 EP/2016 (SP-167-15 and SP-167-16). The approval number from the Ministry of Agriculture, Republic of Croatia was KLASA: UP/I-322-01/15-01/108, URBROJ: 525-10/0255-16-8.

2.7. Experimental Procedures in Wound-Healing Studies

In each study, mice were divided into groups (six animals per group). The interscapular region was shaved, and depilatory cream Veet (Slough, UK) was applied onto the shaved region and removed 2 min after application on day 2. The interscapular region was disinfected and, utilizing strictly aseptic procedures, a single full-thickness excisional wound 8-mm in diameter was inserted midline with a sterile, disposable biopsy punch, thus exposing the underlying fascia muscularis as described.

The test samples were applied in the powder form (100%) and directly spread into the wound beds immediately post-wounding at day 0. The wound was covered by Tegaderm Wound dressing (3M, St. Paul, MN, USA) which remained there until the end of the study.

Postoperative pain control included daily s.c. injections of 4 mg/kg carprofen (5% Norocarp (Pfizer, New York, NY, USA)–50 mg/mL; a 1:20 dilution was made in sterile, deionised water; 30 µL were injected into an animal) for two days post-wounding at day 0 until day 2. Skin samples were collected at time points specified in the study. Prior to skin sampling all animals were humanely killed with an overdose of ketamine (Taj Pharmaceuticals, Newcastle, UK)/xylasine (KHBoddin, Hamburg, Germany) administered via the intraperitoneal route.

2.8. Sample Analysis: Wound Excision and Morphometry

After terminating the experiments, the wound areas and the surrounding healthy tissues were excised post mortem (1 cm × 2.5 cm, rectangular shape), clamped onto a strip of paper and stored in 10% formalin for histological assessment.

Morphometry: Slides cut from paraffin blocks were stained with hematoxylin–eosin [32]. Morphometric evaluation of the degree of re-epithelialization was performed using a Zeiss Axioskop 2 Plus microscope and an Axiovision program (Zeiss, Oberkochen; Germany). The magnification used was 100×. Re-epithelialization is expressed as length of newly formed epithelium (given in mm) and percentage of the wound diameter covered with a new epithelial layer.

2.9. Gene Expression Studies

Animals: The technique of quantitative real-time reverse transcription polymerase chain reaction (qRT-PCR) was applied to determine semi-quantitatively the effect of the polyP particles on wound healing. Formalin-fixed paraffin-embedded (FFPE) tissue samples were used and RNA was extracted from those [33]. The reactions were performed in 1.5 mL Eppendorf tubes and all incubation steps were performed in a thermal cycler (Thermomixer comfort, Eppendorf, Hamburg, Germany). Random primers were used in concentrations of 250 ng/reaction. RNA was mixed with primers together with 1 µL of 10 mM dNTPS, incubated for 5 min at 65 °C and then cooled on ice for 1 min. To each sample the following components were added to the reaction: 4 µL 5 × buffer (Thermo Fisher Scientific, Langenselbold, Germany), 1 µL 0.1 M dithiothreitol (DTT), 1 µL RNaseOUT (RNAse inhibitor, Thermo Fisher Scientific, Schwerte, Germany) and 1 µL Superscript III reverse transcriptase (Thermo Fisher Scientific). The mixture was incubated as follows: 5 min at 25 °C, 60 min at 50 °C and 15 min at

70 °C. When cooled down the mixture was diluted 5 times with RNase free water and used for qPCR. The following primer pairs for mouse genes were used: for *procollagen type Iα* (COL-I; AK075707) Fwd: 5′-AGGCTGACACGAACTGAGGT-3′ and Rev: 5′-ATGCACATCAATGTGGAGGA-3′; *collagen type III αI* (COL-III; P08121) Fwd: 5′-GCTGTTTCAACCACCCAATACAGG-3′ and Rev: 5′-CTGGTGAATGAGTATGACCGTTGC-3′; *α-smooth muscle actin* (α-SMA; NC_000074.6) Fwd: 5′-CAGGGAGTAATGGTTGGAAT-3′ and Rev: 5′-TCTCAAACATAATCTGGGTCA-3′; *plasminogen activator inhibitor-1* (PAI-1; NC_000071.6) Fwd: 5′-CTGCAGATGACCACAGCGGG-3′ and Rev: 5′-AGCTGGCGGAGGGCATGA-3′. The *hypoxanthine phosphoribosyltransferase 1* gene was used as house-keeping gene (HPRT1; NM_013556.2) Fwd: 5′-TGGATACAGGCCAGACTTTG-3′ and Rev: 5′-GTACTCATTATAGTCAAGGGCATAT-3′. All reactions were carried out at least in duplicate and the results were analyzed by a $2^{-\Delta\Delta CT}$ method [33].

MC3T3-E1 cells: The cells were harvested and RNA was extracted. Then the RNA was subjected to qRT-PCR analysis for *collagen type I, α 1* (COL-I, NM_007742) using the primer pair Fwd: 5′-TACATCAGCCCGAACCCCAAG-3′ and Rev: 5′-GGTGGACATTAGGCGCAGGAAG-3′ and for *collagen type III, α 1* (COL-III; NM_009930) the pair Fwd: 5′-GCTGTTTCAACCACCCAATACAGG-3′ and Rev: 5′-CTGGTGAATGAGTATGACCGTTGC-3′. In this series of experiments the expression of *glyceraldehyde 3-phosphate dehydrogenase* (GAPDH; NM_008084) was chosen as a reference gene; the primer pair Fwd: 5′-TCACGGCAAATTCAACGGCAC-3′ and Rev: 5′-AGACTCCACGACATACTCAGCAC-3′ was chosen. The determinations were performed in an iCycler (Bio-Rad, Hercules, CA, USA) with the respective iCycler software.

2.10. Statistical Analyses

The gene expression studies for the incubation experiments in vitro as well as for the re-epithelialization determinations in vivo are expressed as mean (± standard error of mean; independent two-sample Student's *t*-test; Mann–Whitney U-test) [34]. The PCR studies forming the in vivo experiments are given as Box plot analyses [35].

3. Results

3.1. Amorphous PolyP Microparticles

For both the in vivo and the in vitro tests Na–polyP powder and the microparticles "Mg–polyP-MP", "Ca–polyP-MP" and "col/polyP-MP" were prepared. The size of the globular "Ca–polyP-MP" particles varied between 100 nm and 800 nm with an average of 450 ± 170 nm (n = 60) (Figure 3A–C). The surface of the particles showed pores of sizes around 10 nm only at a higher magnification.

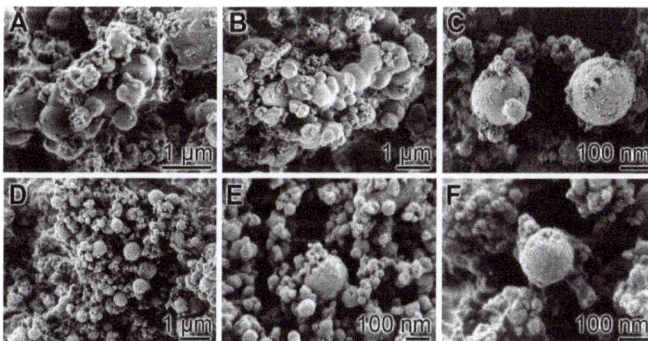

Figure 3. Morphology of "Ca–polyP-MPs" and "Mg–polyP-MPs"; SEM analysis. (**A–C**) "Ca–polyP-MP" and (**D–F**) "Mg–polyP-MP".

A similar morphology is characteristic for the likewise globular "Mg–polyP-MP" microparticles (Figure 3D–F). These particles are more homogeneous than the "Ca–polyP-MP" with an average of 170 ± 65 nm. Broken particles revealed that the particles are close to compact.

The "col/polyP-MP" are less spherical than the other two particles. The particle size distribution is fairly homogeneous with an average diameter of 450 ± 0.190 μm (Figure 4A,B). The morphology varies from almost globular to close to disc-like. At REM magnification it is already seen that the intact particles show ball-like protrusions (Figure 4C) that proved to be gas bubbles (Figure 4D). At the higher SEM magnification it is revealed that the solid material that surrounds the gas vesicles is built of a collagen scaffold around which the polyP particles are arranged (Figure 4E). At a higher magnification the size of the polyP nanoparticles can be determined with $\approx$30–50 nm (Figure 4F). In those images it can be seen that basically the distribution of the collagen scaffold within the particles is homogeneous (Figure 4). The Ca–polyP particles become associated with the collagen fibers during the precipitation with CaCl$_2$. The distribution of those particles onto the surface of the fibers is dense (Figure 4E,F).

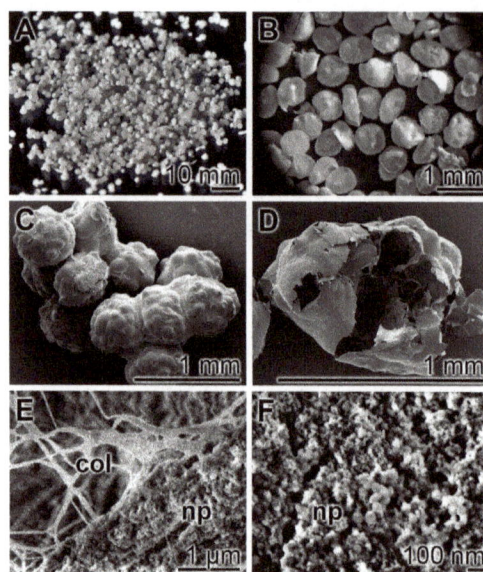

Figure 4. Electron microscopic images of the host–guest "col/polyP-MPs"; (**A**) to (**D**): Reflection electron microscope (REM) and (**E**) and (**F**): SEM. (**A**) and (**B**) At a lower magnification (REM) the particles appear as globular to disc-like particles. (**C**) The intact particles have on their surfaces ball like protrusions which (**D**) proved to be gas bubbles in broken particles. (**E**) and (**F**) At SEM magnification it can be identified that the scaffold material is built of a collagen (col) fiber framework around which nanoparticles (NPs), within the microparticles, are arranged. (**F**) At the higher magnification it can be resolved that the polyP nanoparticles, formed at the collagen scaffold, have a homogeneous morphology with a diameter of $\approx$ 30 nm.

For the analysis of the elements, present in the "col/polyP-MP" particles, EDX spectroscopy was applied. As documented in Figure 5, the spectrum shows the characteristic signals for C, N, O, P and Ca. The C, N O signal peaks can be attributed to the collagen framework and the P and Ca peaks to Ca–polyP. In previous studies it is reported that both the "Ca–polyP-MP" [20] and the "Mg–polyP-MP" particles are amorphous [28]. It has also been verified that the "col/polyP-MP" have this state (data not shown).

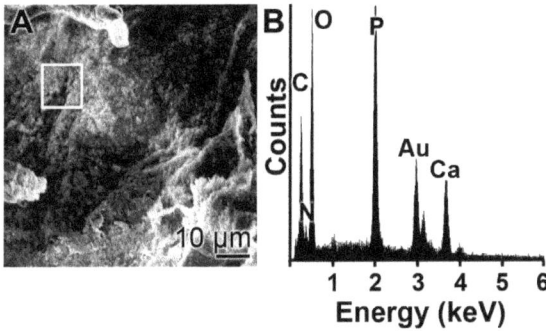

Figure 5. Energy-dispersive X-ray (EDX) analysis of "col/polyP-MP". (**A**) SEM analysis and (**B**) EDX spectrum. The prominent element peaks (C, N, O, P and Ca) are marked. The Au peak originates from the gold surface after sputtering. The white square in (**A**) delimits the area selected for the EDX analysis.

3.2. Potency of polyP to Change the Steady-State Expression of Collagen Genes in MC3T3-E1 Cells In Vitro

Mouse calvaria MC3T3-E1 cells were cultivated in medium/serum as described under "Materials and Methods"; the seeding cell concentration was 5×10^3 cells/well [24-well plates]. The cultures received either no additional component (controls) or were exposed to 50 µg/mL of "Na–polyP", "Mg–polyP-MPs", "Ca–polyP-MPs" or "col/polyP-MPs" and incubated for 4 days. Then, the cells were harvested, and RNA was extracted which was subjected to PCR analysis to assess the expression levels for *collagen type I* and *collagen type III*. The data are summarized in Figure 6. It is striking that, under the conditions used, "Na–polyP" did not change the expression levels for either collagen type. However, a significant upregulation of the steady-state-expressions to about 2-fold was measured for *collagen type I* and also for *collagen type III* in the experiments with "Mg–polyP-MPs", "Ca–polyP-MPs" and "col/polyP-MPs".

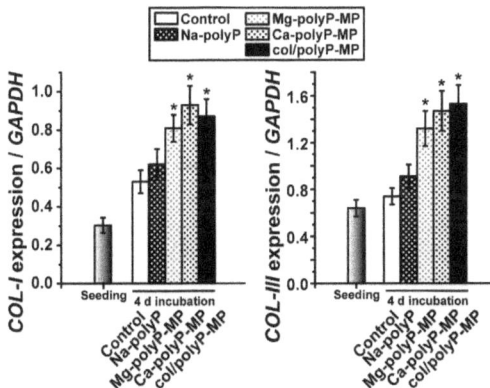

Figure 6. Gene expression levels of the two collagen types, *collagen type I* (COL-I), and *collagen type III* (COL-III) in MC3T3-E1 cells exposed to 50 µg/mL of "Na–polyP", "Mg–polyP-MPs", "Ca–polyP-MPs" or "col/polyP-MPs"; the controls received no additional component. After the 4-day incubation period the cells were harvested, their RNA was extracted and the steady-state levels of collagen expression were determined by qRT-PCR. The expression levels at time 0 (seeding) and time 4 days are correlated with the one of the housekeeping genes *GAPDH*. * $p < 0.01$ ($n = 6$ parallel experiments; Student's *t*-test) with respect to the values of the controls.

3.3. Effect of polyP Application on Re-epithelialization in C57BL/6 and db/db Mice

Groups of six specimens each of C57BL/6 and db/db mice were used for the study. Each mouse received one defined wound with a diameter of 8 mm. The wounds were supplied once with 3 mg of the respective samples immediately after setting the wound. After a healing period of 7 days (Figure 7) or 13 days (Figure 8) the experimental animals were analyzed for the degree of re-epithelialization.

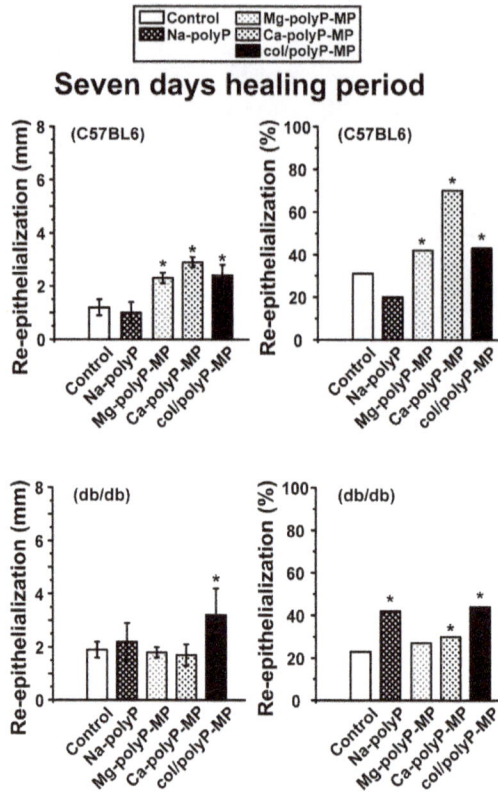

Figure 7. Effects of different polyP formulations on wound-healing kinetics (7 days after setting the wound), measured on the basis of re-epithelialization (in percent) which is calculated by dividing the degree of re-epithelialization (in mm) by the wound diameter (mm) × 100. For the study both wild-type (C57BL/6) and db/db mice were included (experiment #SP-013-16). The values for both the controls (without polyP) and the different polyP formulations, "Na–polyP", "Mg–polyP-MP", "Ca–polyP-MP" and "col/polyP-MP", are given. The data are presented as mean ± SEM. The significance is * $p < 0.05$; $n = 6$ animals (Student's *t*-test) [* $p < 0.05$ versus negative control according to the Mann–Whitney U-test].

In the present study a delayed wound healing in diabetic (db/db) mice is described in comparison to normal mice. The results revealed that after a 7-day healing period (Figure 7) the C57BL/6 mice recovered with a 31% score of re-epithelialization (controls), while the db/db mice showed only 23% re-epithelialization. At day 7 some polyP formulations already showed a distinct beneficial effect on the wound healing kinetics. Statistical significant is the effect of "Ca–polyP-MP" in C57BL/6 mice with a score of re-epithelialization of 72%; somewhat lower is the positive wound healing effect of "Mg–polyP-MP" (40%) and "col/polyP-MP" (44%). In contrast, "Na–polyP" proved to be ineffective.

Very distinctive is also the beneficial effect of polyP in db/db mice. In this model "Na–polyP" increased the rate of wound healing to 42%, if checked by the rate of re-epithelialization. Likewise significant is the effect of "Ca–polyP-MP" (30% re-epithelialization) and "col/polyP-MP" (44%) in diabetic mice.

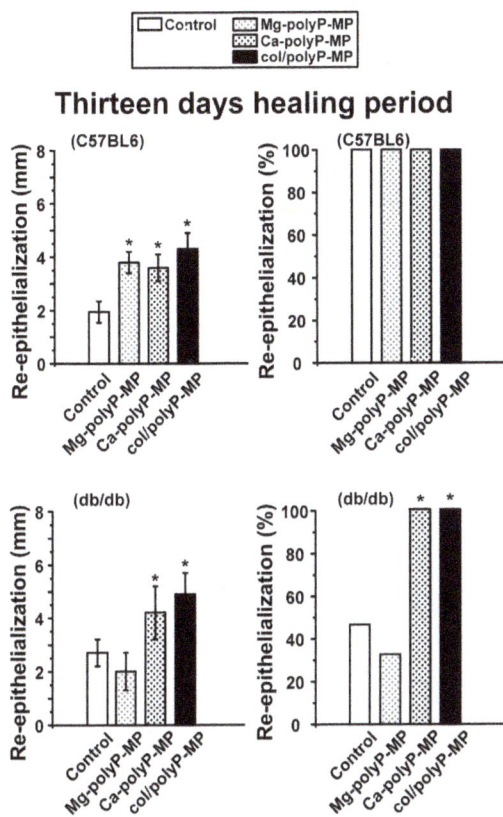

Figure 8. The effects of the tested materials on wound re-epithelialization in C57BL/6 and db/db mice on day 13. The data are presented as mean ± SEM.

The wound healing velocity after the 13-day healing period is also very pronounced. While in the experiments with normal mice both the controls and the polyP-treated wounds had already re-epithelialized 100%, the diabetic mice (controls) showed only 48% re-epithelialization (Figure 8). In this series of experiments with both "Ca–polyP-MP" (score of 100%) and "col/polyP-MP" (100%) the epithelium totally covered the wound.

It is worth mentioning that the Ca–polyP microparticles are more efficient under the in vivo conditions than Na–polyP or Mg–polyP, as demonstrated previously by showing that the Ca-polyP particles are less soluble compared to Na–polyP/Mg–polyP and hence are less susceptible to the ALP [20], an enzyme which is highly expressed during the granulation phase [19]. This implies also that the release kinetics of Ca–polyP from the particles is much slower compared to the polymer from Na–polyP or Mg–polyP and hence is longer effective.

3.4. Expression of Selected Genes in Tissue from the Regenerating Wound by qRT-PCR

RNA was extracted from FFPE-fixed tissue samples and then subjected to qRT-PCR. As a marker for cell migration and tissue re-organization the expression of the genes encoding for *procollagen type Iα*

(COL-I) and *collagen type III αI* (COL-III) have been chosen. As a measure for the granulation efficiency the genes *α-smooth muscle actin* (α-SMA) and *plasminogen activator inhibitor-1* (PAI-1) have been selected. The studies were performed with wild type mice. As summarized in Figure 9 it became evident that after the 7-day incubation period the transcript levels for the *collagen* marker gene *COL-III* and for *PAI-1* significantly upregulate in tissue taken from the wound area that has been treated with powder composed of polyP microparticles ("Ca–polyP-MP" and "col/polyP-MP"). With respect to the α-*SMA* gene or *COL-I* gene no significant change is seen.

Seven days healing period

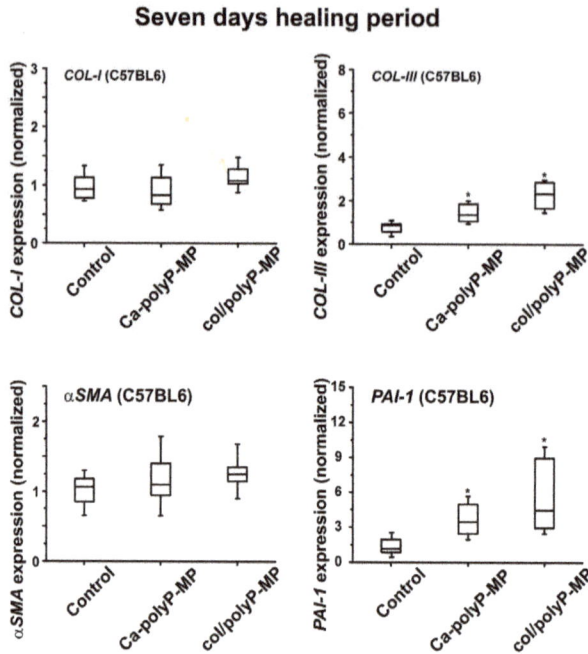

Figure 9. Upregulation of the collagen genes, *procollagen type Iα* (COL-I) and *collagen type III αI* (COL-III), in tissue from wild-type animals treated with the microparticles "Ca-polyP-MP" and "col/polyP-MP" (experiment #SP-013-15). The healing period was 7 days. The genes encoding *procollagen type Iα* (COL-I) and *α-smooth muscle actin* (α-SMA) remained unchanged. The significance (Box plot characteristics) are given by comparing the values of controls (no polyP added) with the values measured for the "Ca–polyP-MP" and "col/polyP-MP" treated wounds. The horizontal bars within the boxes indicate the median value; * $p < 0.005$ ($n = 6$ animals).

After the longer healing period of 13 days the expression of all four marker genes for wound healing and tissue regeneration becomes upregulated by up to 1.6-fold (compared to the controls) for COL-I, and up to 4.6-fold for COL-III, 1.5-fold for α-*SMA* and 5.1-fold for *PAI-1* (Figure 10).

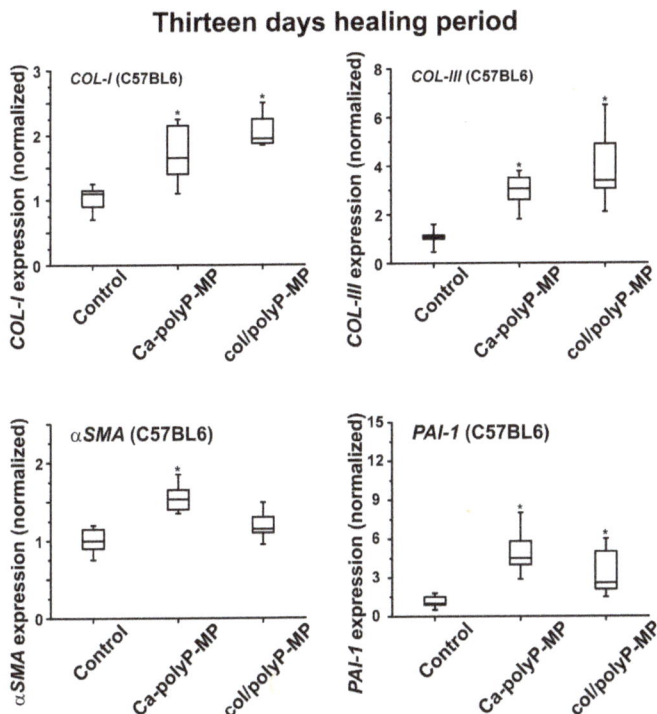

Figure 10. Changes of the steady-state expression of the genes *COL-I* and *COL-III* as well as of the genes α-*SMA* and *PAI-1* in wild-type animals after treatment with polyP microparticles. The healing period was 13 days. In all three series of experiments the expression levels of genes in tissue from regenerating wounds that were treated with polyP microparticles ("Ca–polyP-MP" and "col/polyP-MP") significantly increased compared to control; * $p < 0.005$ ($n = 6$ animals).

4. Discussion

It is very obvious that the therapy/care of both acute and chronic wounds has to begin with the understanding of the molecular and cellular etiology of the components present within each wound bed [1]. The failure in the venous and arterial blood flow and the balance of the levels of cytokines and growth facts in the wound area are a priority. In addition, systemic factors such as nutritional status, immunosuppression and infection contribute to the kinetics of wound healing. Also, the imbalance of the growth and differentiation state and rate of the cells involved, mainly fibroblasts, determine the functional quality of the hyperproliferative wound margin [2]. In addition, an imbalance of the pro-inflammatory cytokines and an uncontrolled enzymatic environment predisposes and causes delayed healing. Finally, local tissue hypoxia in concert with a repetitive ischemia–reperfusion injury accelerates pathogenesis in chronic wounds.

The first factors produced during the wound healing process originate from the blood platelets (platelet-derived growth factor) and the fibroblasts (fibroblast growth factor) (reviewed in [36]). One secret why blood platelets comprise a prime role during hemostasis is that they release coagulation factors that stop bleeding and cause clumping and clotting. However, as outlined in the "Introduction" they are filled with an inorganic polymer, polyP, that has recently been implicated in the extracellular energy supply [8]. In a very recent thorough review [37] the role of polyP as a signaling molecule in mammalian cells, causing an increased metabolism of mitochondria, has been highlighted. As examples, the signaling role of polyP between astroglial cells has been elucidated and the function as an amplifier proinflammatory response has been established [38]. Since the (potential) solution of the problem of the origin of polyP is the polymer delivering metabolic fuel to the cells we asked in the present study if the accelerating and anabolic effect of polyP on mineralization of bone cells can be applied also to cells, e.g., fibroblasts, controlling wound healing in vitro (fibroblasts) and in vivo (mouse wound healing model). Besides being a signaling molecule that migrates across the tiny spaces between adjacent cells, polyP has been proven to be a metabolic signal that regulates cell growth and differentiation, providing polyP with morphogenetically active properties. Examples are the induction of osteoinductive cytokine(s) and enzymes, like the ALP, whose dephosphorylation reactions result in an increased rate of diffusion of molecules into and inhibition of them to diffuse out the cells (scheme in Figure 2).

The major studies, summarized here have been performed both with wild-type (C57BL/6) mice and diabetic (db/db) mice in order to bring together in a comparative manner results from the topical effects of polyP during acute and chronic wound healing. It is interesting to note that db/db mice have a genetic mutation in their leptin production pathway and/or leptin receptor signaling in the hypothalamus and, by that, lose control of satiation [39]. Mice having a homozygous mutation of the leptin gene show a phenotype of hyperphagia, extreme obesity, diabetes, neuroendocrine abnormalities, and infertility [40]. This exact gene has been found to be induced by polyP in mouse calvaria cells MC3T3-E1 cells in vitro [41]. We introduced for the study, summarized here, in addition to the already proven amorphous Ca–polyP particles ("Ca–polyP-MP") [20] and the amorphous Mg–polyP microparticles ("Mg–polyP-MP"), a new hybrid polyP formulation with collagen as a scaffold, around which the Ca–polyP nanoparticles are arranged ("col/polyP-MP"). It is surprising that the size of the Ca–polyP nanoparticles that are formed onto the collagen scaffold is small ($\approx$30–50 nm), compared to that of the Ca–polyP particles (450 $\pm$ 170 nm), formed in the absence of the collagen scaffold. At present we attribute the smaller size of the collagen-associated particles to an alteration of the surface tensions existing in the multi-phase hybrid system (collagen–polyP) versus the one phase (polyP) [42].

EDX experiments revealed that this hybrid material is indeed composed of polyP and collagen. With this latter completely physiological two-component biomaterial we intended to combine both morphogenetic activity elicited from polyP and the structural guidance of the collagen in the particles as well as the biological property to attract cells to attach and to support migration and differentiation in the vicinity of this fibrous meshwork. The structure guidance of the cells via integrin receptors (present on the cell surface) and the RGD domains within the collagen fibers allow directed migration of the cells and additionally also modulate morphogenetic signaling within the cells. These properties provide the key advantage of the collagen/Ca–polyP hybrid material in comparison to the pure amorphous Ca–polyP particle. In the in vitro studies performed here we showed that the amorphous microparticles "Mg–polyP-MP" and "Ca–polyP-MP", but not the non-particulate Na–polyP ("Na–polyP"), significantly increase the expression of the collagen types *collagen-I* and *collagen-III*. A likewise significant upregulation of the steady-state-expressions of the two collagen genes is observed with the "col/polyP-MP" formulation.

For the in vivo wound healing experiments both with normal and diabetic mice the three polyP preparations and also the original Na–polyP sample were tested. At two time points, after 7 days (Figure 6) and 13 days (Figure 7), the topical effects of the polymers were analyzed. Previously it has been found that during the 7-day healing period of acute wounds in normal mice the process is completed to about 50%, while after 13 days the healing is complete [43]. In diabetic mice the healing kinetics is delayed by two-fold. In the present animal study, we found that the re-epithelialization process in normal mice was completed by 30% after a period of 7 days, while only 22% epithelialization was measured in diabetic mice (see Figures 6 and 7). In contrast to the wounds in normal mice, where an increase in the wound diameters is measured in response to a topical application of polyP, a significant reduction of the diameter is seen in diabetic mice using the same polymer samples. The increase in the wound size in normal mice could be attributed to a higher blood flow, perhaps caused by a local increase of NO, while the decrease in db/db mice could be indicative for an onset of fibroblast proliferation and differentiate into myofibroblasts [44]. In general, the topical treatment of wounds by polyP results in an increase of the wound healing. All polyP samples are significantly active. After 13 days the wound healing in normal mice is already complete, while in diabetic mice for "Ca–polyP-MP" and especially the host–guest "col/polyP-MP" groups, an increase of the re-epithelialization to 100% (complete cure) is measured. In contrast, the controls as well as the "Na–polyP" treated groups showed a similar wound healing state with around 30% re-epithelialization. This main part of the study here confers polyP, especially the "Ca–polyP-MP" and "col/polyP-MP" microparticles, a beneficial effect as an accelerator of wound healing especially in diabetic mice. This result provides a strong impact on future developments in topic wound healing therapy. In order to support these morphometric data by molecular biological experiments, qRT-PCR-based steady-state-expression determinations have been performed. As marker genes indicative for the wound healing progress the two collagen types (I and III) have been chosen; de-paraffinized tissue samples, taken from the wounds, have been applied. Those genes have been found to be upregulated in regenerating wound tissue [45]. The *collagen type I* gene and the additional marker gene *plasminogen activator inhibitor-1* (PAI-1; also termed endothelial plasminogen activator inhibitor, or serpin E1) were found to be significantly upregulated both in wild-type mice and db/db mice after a 7-day healing period. If the healing period was prolonged to 13 days, the transcript levels of all four genes were found to be upregulated if the wound area had been treated with "Ca–polyP-MPs" and "col/polyP-MPs" (host–guest particles).

In conclusion, the presented findings support the conclusion that polyP, in the form of microparticles, causes beneficial anabolic effects on cells involved in wound regeneration. The question might be discussed as to why polyP in the form of microparticles, especially as Ca^{2+}-salts, is superior to the highly soluble Na–polyP. As outlined above, that finding is on the basis of a delayed, beneficial release of the morphogenetically active Ca–polyP. Evidence has been presented that polyP is taken up by mammalian cells via endocytosis [8]. Since in the present study the size of the particles supplied to the cells is larger than the preferred particle size for the endocytic uptake machinery we postulate that only after ALP-mediated degradation of the microparticles are the nanoparticles generated taken up by the cells. In consideration of transfection studies it is established that the highly negatively charged DNA, like polyP, can be readily taken up by mammalian cells if entrapped into less charged particles, like liposomes [46]. Building on the presented in vitro and in vivo data we propose further safety studies in animals that might be followed by first human trials.

Acknowledgments: We thank Maren Müller (EDX analysis) and Gunnar Glaßer (SEM imaging) from the Max-Planck Institute for Polymer Research, Mainz (Germany) for their continuous support. Furthermore, we thank our colleagues Snjezana Čužić and Maja Antolić (histopathological analysis) as well as Paravić Radičević (some PCR experiments (animals)) from Fidelta Ltd. (10000 Zagreb, Croatia) for their expert contributions. Werner E.G. Müller is the holder of an ERC Advanced Investigator Grant (grant number 268476). This work was supported by grants from the European Commission (grant numbers: 604036 and 311848), the International Human Frontier Science Program and the BiomaTiCS research initiative of the University Medical Center, Mainz.

Author Contributions: Werner E.G. Müller, Dinko Relkovic, Maximilian Ackermann, Hiroshi Ushijima and Xiaohong Wang conceived and designed the experiments; Dinko Relkovic , Maximilian Ackermann, Shunfeng Wang, Meik Neufurth and Andrea Paravic Radicevic performed the experiments; Werner E.G. Müller, Dinko Relkovic, Maximilian Ackermann, Shunfeng Wang, Meik Neufurth, Andrea Paravic Radicevic, Hiroshi Ushijima., Heinz-C Schröder and Xiaohong Wang analyzed the data; Werner E.G. Müller, Dinko Relkovic, Maximilian Ackermann, Shunfeng Wang, Meik Neufurth, Andrea Paravic Radicevic, Hiroshi Ushijima, Heinz-C Schröder and Xiaohong Wang contributed reagents/materials/analysis tools; Werner E.G. Müller, Dinko Relkovic, Maximilian Ackermann, Shunfeng Wang, Meik Neufurth, Andrea Paravic Radicevic, Hiroshi Ushijima, Heinz-C Schröder and Xiaohong Wang wrote the paper.

References

1. Demidova-Rice, T.N.; Hamblin, M.R.; Herman, I.M. Acute and impaired wound healing: Pathophysiology and current methods for drug delivery, part 1: Normal and chronic wounds: biology, causes, and approaches to care. *Adv. Skin Wound Care* **2012**, *25*, 304–314. [CrossRef] [PubMed]
2. Menke, N.B.; Ward, K.R.; Witten, T.M.; Bonchev, D.G.; Diegelmann, R.F. Impaired wound healing. *Clin. Dermatol.* **2007**, *25*, 19–25. [CrossRef] [PubMed]
3. Clark, R.A. Fibrin and wound healing. *Ann. N. Y. Acad. Sci.* **2001**, *936*, 355–367. [CrossRef] [PubMed]
4. Singer, A.J.; Clark, R.A. Cutaneous wound healing. *N. Engl. J. Med.* **1999**, *341*, 738–746. [PubMed]
5. Kobrin, K.L.; Thompson, P.J.; van de Scheur, M.; Kwak, T.H.; Kim, S.; Falanga, V. Evaluation of dermal pericapillary fibrin cuffs in venous ulceration using confocal microscopy. *Wound Repair Regen.* **2008**, *16*, 503–506. [CrossRef] [PubMed]
6. Wang, J.; Wan, R.; Mo, Y.; Li, M.; Zhang, Q.; Chien, S. Intracellular delivery of adenosine triphosphate enhanced healing process in full-thickness skin wounds in diabetic rabbits. *Am. J. Surg.* **2010**, *199*, 823–832. [CrossRef] [PubMed]
7. Guo, S.; Dipietro, L.A. Factors affecting wound healing. *J. Dent. Res.* **2010**, *89*, 219–229. [CrossRef] [PubMed]
8. Müller, W.E.G.; Tolba, E.; Feng, Q.; Schröder, H.C.; Markl, J.S.; Kokkinopoulou, M.; Wang, X.H. Amorphous Ca^{2+} polyphosphate nanoparticles regulate ATP level in bone-like SaOS-2 cells. *J. Cell Sci.* **2015**, *128*, 2202–2207. [CrossRef] [PubMed]
9. Müller, W.E.G.; Tolba, E.; Schröder, H.C.; Wang, X.H. Polyphosphate: A morphogenetically active implant material serving as metabolic fuel for bone regeneration. *Macromol. Biosci.* **2015**, *15*, 1182–1197. [CrossRef] [PubMed]
10. Seminario-Vidal, L.; Lazarowski, E.R.; Okada, S.F. Assessment of extracellular ATP concentrations. *Methods Mol. Biol.* **2009**, *574*, 25–36. [PubMed]
11. Schwiebert, E.M.; Zsembery, A. Extracellular ATP as a signaling molecule for epithelial cells. *Biochim. Biophys. Acta* **2003**, *1615*, 7–32. [CrossRef]
12. Butterworth, P.J. Alkaline phosphatase—biochemistry of mammalian alkaline phosphatases. *Cell Biochem. Funct.* **1983**, *1*, 66–68. [CrossRef] [PubMed]
13. Lorenz, B.; Schröder, H.C. Mammalian intestinal alkaline phosphatase acts as highly active exopolyphosphatase. *Biochim. Biophys. Acta* **2001**, *1547*, 254–261. [CrossRef]
14. Wang, X.H.; Schröder, H.C.; Müller, W.E.G. Polyphosphate as a metabolic fuel in Metazoa: A foundational breakthrough invention for biomedical applications. *Biotechnol. J.* **2016**, *11*, 11–30. [CrossRef] [PubMed]
15. Stapor, P.; Wang, X.; Goveia, J.; Moens, S.; Carmeliet, P. Angiogenesis revisited - role and therapeutic potential of targeting endothelial metabolism. *J. Cell Sci.* **2014**, *127*, 4331–4341. [CrossRef] [PubMed]
16. Müller, W.E.G.; Neufurth, M.; Ackermann, M.; Tolba, E.; Wang, S.; Feng, Q.; Schröder, H.C.; Wang, X.H. Fabrication of a new physiological macroporous hybrid biomaterial/bioscaffold material based on polyphosphate and collagen by freeze-extraction. *J. Mater. Chem. B* **2017**, *5*, 3823–3835.
17. Morrissey, J.H.; Choi, S.H.; Smith, S.A. Polyphosphate: an ancient molecule that links platelets, coagulation, and inflammation. *Blood* **2012**, *119*, 5972–5979. [CrossRef] [PubMed]
18. Tolba, E.; Müller, W.E.G.; El-Hady, B.M.A.; Neufurth, M.; Wurm, F.; Wang, S.; Schröder, H.C.; Wang, X.H. High biocompatibility and improved osteogenic potential of amorphous calcium carbonate/vaterite. *J. Mater. Chem. B* **2016**, *4*, 376–386. [CrossRef]
19. Alpaslan, G.; Nakajima, T.; Takano, Y. Extracellular alkaline phosphatase activity as a possible marker for wound healing: A preliminary report. *J. Oral Maxillofac. Surg.* **1997**, *55*, 56–63. [CrossRef]
20. Müller, W.E.G.; Tolba, E.; Schröder, H.C.; Wang, S.; Glaßer, G.; Muñoz-Espí, R.; Link, T.; Wang, X.H. A new polyphosphate calcium material with morphogenetic activity. *Mater. Letters* **2015**, *148*, 163–166. [CrossRef]
21. Müller, W.E.G.; Tolba, E.; Schröder, H.C.; Diehl-Seifert, B.; Wang, X.H. Retinol encapsulated into amorphous Ca^{2+} polyphosphate nanospheres acts synergistically in MC3T3-E1 cells. *Eur. J. Pharm. Biopharm.* **2015**, *93*, 214–223. [CrossRef] [PubMed]
22. Ghatak, S.; Maytin, E.V.; Mack, J.A.; Hascall, V.C.; Atanelishvili, I.; Moreno Rodriguez, R.; Markwald, R.R.; Misra, S. Roles of proteoglycans and glycosaminoglycans in wound healing and fibrosis. *Int. J. Cell Biol.* **2015**, *2015*, 834893. [CrossRef] [PubMed]

23. Krötzsch, E.; Salgado, R.M.; Caba, D.; Lichtinger, A.; Padilla, L.; Di Silvio, M. Alkaline phosphatase activity is related to acute inflammation and collagen turnover during acute and chronic wound healing. *Wound Repair Regen.* **2008**, *13*, March 2008; abstract 162.

24. Müller, W.E.G.; Neufurth, M.; Wang, S.; Tolba, E.; Schröder, H.C.; Wang, X.H. Morphogenetically active scaffold for osteochondral repair (polyphosphate/alginate/N,O-carboxymethyl chitosan). *Eur. Cell Mater. J.* [*eCM*] **2016**, *31*, 174–190.

25. Brett, D. Review of collagen and collagen-based wound dressings. *Wounds* **2008**, *20*, 347–356. [PubMed]

26. Yates, C.C.; Hebda, P.; Wells, A. Skin wound healing and scarring: Fetal wounds and regenerative restitution. *Birth Defects Res. C* **2012**, *96*, 325–333. [CrossRef] [PubMed]

27. Hinz, B.; Celetta, G.; Tomasek, J.J.; Gabbiani, G.; Chaponnier, C. Alpha-smooth muscle actin expression upregulates fibroblast contractile activity. *Mol. Biol. Cell* **2001**, *12*, 2730–2741. [CrossRef] [PubMed]

28. Chan, J.C.; Duszczyszyn, D.A.; Castellino, F.J.; Ploplis, V.A. Accelerated skin wound healing in plasminogen activator inhibitor-1-deficient mice. *Am. J. Pathol.* **2001**, *159*, 1681–1688. [CrossRef]

29. Müller, W.E.G.; Ackermann, M.; Tolba, E.; Neufurth, M.; Wang, S.; Schröder, H.C.; Wang, X.H. A bio-imitating approach to fabricate an artificial matrix for cartilage tissue engineering using magnesium-polyphosphate and hyaluronic acid. *RSC Adv.* **2016**, *6*, 88559–88570. [CrossRef]

30. Rajan, N.; Habermehl, J.; Coté, M.F.; Doillon, C.J.; Mantovani, D. Preparation of ready-to-use, storable and reconstituted type I collagen from rat tail tendon for tissue engineering applications. *Nat. Protoc.* **2006**, *1*, 2753–2758. [CrossRef] [PubMed]

31. Tkalcević, V.; Cuzić, S.; Parnham, M.J.; Pasalić, I.; Brajsa, K. Differential evaluation of excisional non-occluded wound healing in db/db mice. *Toxicol. Pathol.* **2009**, *37*, 183–192. [CrossRef] [PubMed]

32. Lillie, R.D. *Histopathologic Technic and Practical Histochemistry*, 3rd ed.; McGraw-Hill Book Co.: New York, NY, USA, 1965.

33. Berg, D.; Malinowsky, K.; Reischauer, B.; Wolff, C.; Becker, K.F. Use of formalin-fixed and paraffin-embedded tissues for diagnosis and therapy in routine clinical settings. *Methods Mol. Biol.* **2011**, *785*, 109–122.

34. Petrie, A.; Watson, P. *Statistics for Veterinary and Animal Science*; Wiley-Blackwell: Oxford, UK, 2013; pp. 85–99.

35. Müller, W.E.G.; Schmidseder, R.; Rohde, H.J.; Zahn, R.K.; Scheunemann, H. Bleomycin-sensitivity test: Application for human squamous cell carcinoma. *Cancer* **1977**, *40*, 2787–2791. [CrossRef]

36. Demidova-Rice, T.N.; Hamblin, M.R.; Herman, I.M. Acute and impaired wound healing: Pathophysiology and current methods for drug delivery, part 2: Role of growth factors in normal and pathological wound healing: Therapeutic potential and methods of delivery. *Adv. Skin Wound Care* **2012**, *25*, 349–370. [CrossRef] [PubMed]

37. Angelova, P.R.; Baev, A.Y.; Berezhnov, A.V.; Abramov, A.Y. Role of inorganic polyphosphate in mammalian cells: From signal transduction and mitochondrial metabolism to cell death. *Biochem. Soc. Trans.* **2016**, *44*, 40–45. [CrossRef] [PubMed]

38. Dinarvand, P.; Hassanian, S.M.; Qureshi, S.H.; Manithody, C.; Eissenberg, J.C.; Yang, L.; Rezaie, A.R. Polyphosphate amplifies proinflammatory responses of nuclear proteins through interaction with receptor for advanced glycation end products and P2Y1 purinergic receptor. *Blood* **2014**, *123*, 935–945. [CrossRef] [PubMed]

39. Wang, B.; Chandrasekera, P.C.; Pippin, J.J. Leptin- and leptin receptor-deficient rodent models: Relevance for human type 2 diabetes. *Curr. Diabetes Rev.* **2014**, *10*, 131–145. [CrossRef]

40. Heymsfield, S.B.; Greenberg, A.S.; Fujioka, K.; Dixon, R.M.; Kushner, R.; Hunt, T.; Lubina, J.A.; Patane, J.; Self, B.; Hunt, P.; McCamish, M. Recombinant leptin for weight loss in obese and lean adults: A randomized, controlled, dose-escalation trial. *JAMA* **1999**, *282*, 1568–1575. [CrossRef] [PubMed]

41. Müller, W.E.G.; Tolba, E.; Dorweiler, B.; Schröder, H.C.; Diehl-Seifert, B.; Wang, X.H. Electrospun bioactive mats enriched with Ca-polyphosphate/retinol nanospheres as potential wound dressing. *Biochem. Biophys. Rep.* **2015**, *3*, 150–160. [CrossRef]

42. Landfester, K. Miniemulsions for nanoparticle synthesis. *Top. Curr. Chem.* **2003**, *227*, 75–123.

43. Frank, S.; Stallmeyer, B.; Kämpfer, H.; Kolb, N.; Pfeilschifter, J. Leptin enhances wound re-epithelialization and constitutes a direct function of leptin in skin repair. *J. Clin. Investig.* **2000**, *106*, 501–509. [CrossRef] [PubMed]

44. Reinke, J.M.; Sorg, H. Wound repair and regeneration. *Eur. Surg. Res.* **2012**, *49*, 35–43. [CrossRef] [PubMed]

45. Haukipuro, K. Synthesis of collagen types I and III in reincised wounds in humans. *Br. J. Surg.* **1991**, *78*, 708–712. [CrossRef] [PubMed]

46. Rizzo, W.B.; Schulman, J.D.; Mukherjee, A.B. Liposome-mediated transfer of simian virus 40 DNA and minichromosome into mammalian cells. *J. Gen. Virol.* **1983**, *64*, 911–919. [CrossRef] [PubMed]

MDPI

Article

Directional Alignment of Polyfluorene Copolymers at Patterned Solid-Liquid Interfaces

Xiaolu Pan [1], Hongwei Li [1,2] and Xinping Zhang [1,*]

[1] Institute of Information Photonics Technology and College of Applied Sciences,
Beijing University of Technology, Beijing 100124, China; xiaolupan@emails.bjut.edu.cn (X.P.);
hw0425@126.com (H.L.)

[2] Advanced Nano-Materials Division, Suzhou Institute of Nano-Tech and Nano-Bionics (SINANO),
Chinese Academy of Sciences (CAS), Suzhou 215123, China

* Correspondence: zhangxinping@bjut.edu.cn; Tel.: +86-10-67396371

Received: 26 June 2017; Accepted: 3 August 2017; Published: 11 August 2017

Abstract: Polyfluorene and its derivatives have been recognized as efficient light-emitting semiconductors. However, directional alignment of polyfluorene copolymers at a large scale has rarely been observed, in particular for the two relatively more amorphous members of poly-9,9-dioctylfluorene-co-bethiadisazole (F8BT) and poly-(9,9-dioctylfluorenyl-2,7-diyl)-*co*-(N,N0-diphenyl)-N,N′di(p-butyl-oxy-pheyl)-1,4-diamino-benzene) (PFB) molecules. Furthermore, the directional alignment of PFB has not been observed so far due to the triphenylamine units in its molecular structures. We present, in this work, a solution-processible method to achieve large-scale alignment of F8BT and PFB molecules into fibers as long as millimeters in a defined direction. Spin-coating the polymer film on to a glass substrate patterned by one-dimensional dielectric nano-grating structures through interference lithography and subsequent modification using 1,5-pentanediol have been used in all of the preparation procedures. Polymer fibers have been obtained in an arrangement parallel to the grating lines. The microscopic, spectroscopic, and photoconductive performances verified the formation and the quality of these directionally-aligned polymeric fibers.

Keywords: directional alignment into fibers; polyfluorene copolymers; liquid-solid interfaces; polarization spectroscopy; photoconductivity

1. Introduction

Orderly arrangement or directional alignment of organic semiconductor molecules is an effective approach to enhance mobility of charge carriers and is, thus, preferred in efficient organic optoelectronic devices [1–5]. Such directional alignment has been extensively observed in small molecules [6–10]. However, large-scale alignment (>100 µm) behavior has rarely been observed in polymers, although it has been demonstrated in a limited number of polymeric semiconductors [11–15]. Polyfluorene is an efficient light-emitting polymer and its derivatives have been used extensively in light-emitting diodes, transistors, and optically-pumped thin-film lasers [16–19]. Although polyfluorene shows small-scale π-stacking phases in some specially prepared solid films, [20–22] and oriented alignment of poly-9,9-dioctylfluorene (PFO), poly-9,9-dioctylfluorene-co-bethiadisazole (F8BT), and Poly-9,9-dioctylfluorenyl-2,7-diyl-co-bithiophene (F8T2) molecules has been studied on rubbed-polyimide-modified substrates [23], large-scale or long-range alignment has not been reported in the films of these polymers.

In this work, we demonstrate large-scale alignment into defined directions of two derivatives of polyfluorene, poly-9,9-dioctylfluorene-co-bethiadisazole (F8BT) and poly-(9,9-dioctyl-fluorenyl-2,7-diyl)-*co*-(N,N0-diphenyl)-N,N′di(p-butyl-oxy-pheyl)-1,4-diamino-benzene) (PFB) on a patterned substrate.

The direction and the length of the alignment have been controlled by the underneath patterns provided by a one-dimensional dielectric grating. The alignment process was initiated by the surface modification process using an organic solvent of 1,5-pentanediol. Polymer fibers with a length of millimeters, and a diameter ranging from 300 nm to 2 μm, have been produced for both F8BT and PFB, which extended in the same direction as the grating lines. Optical and electrical characterization supply evidence for the directional alignment processes.

2. Experimental Methods

2.1. Materials and Measurement Methods

The polymers of F8BT and PFB were purchased from Sigma-Aldrich and American Dye Source, Inc. (Quebec, QC, Canada), respectively. 1,5-pentanediol and toluene were obtained from the Sinopharm Chemical Reagent Co., Ltd. (Beijing, China). All of these compounds were used as received. A very commonly used positive photoresist S1805 was purchased from Rohm & Haas (Philadelphia, PA, USA) and filtered through a filtration membrane with a pore size of 200 nm before being used. The atomic force microscope (AFM) images were measured using a WiTec Alpha300S system (WITec Wissenschaftliche Instrumente und Technologie Gmbh, Ulm, Germany). Optical microscopic images were acquired using an Olympus BX51 polarization optical microscope (Olympus Corporation, Tokyo, Japan). The photoluminescence (PL) spectra was taken on a time-correlated single-photon counting system (FLS920) from Edinburg Instruments (Livingston, UK). The absorption spectra were measured using an Agilent G1103A UV-VIS spectrometer (Agilent Technologies, Santa Clara, CA, USA). The polarization dependence of the absorption spectra was controlled by placing a polarizer between the sample and the light source. The electrical performance was characterized by an Agilent 4156C semiconductor parameter analyzer (Agilent Technologies, Santa Clara, CA, USA).

2.2. Preparation of the Directionally-Aligned Polymer Fibers

Interference lithography is the first step to produce photoresist (PR) template grating on a silica substrate, as shown in Figure 1a. A 325-nm He-Cd laser (Kimmon Koha Co. Ltd., Tokyo, Japan) was used as the ultraviolet (UV) light source and S-1805 photoresist was used as the photosensitive medium. The PR S1805 is sensitive to UV light at a wavelength shorter than 400 nm and it is hardly dissolved in toluene, but slightly in 1,5-pentanediol, which provided us with a conditional probability to achieve the directional alignment of polymer molecules. The substrate has an area of 15 mm × 15 mm and a thickness of 1 mm. The PR grating employed in this work has a period of about 670 nm. The polymer solution in toluene with a concentration of 15 mg/mL was heated mildly on a hotplate at 60 °C for 20 min before it was spin-coated onto the PR grating at a speed of 2000 rpm for 30 s (Figure 1b). Then, the sample was put into a Petri dish and heated at 80 °C for 20 min on a hotplate, finishing the preparation of the polymer film on the PR grating. After the sample was cooled down in air, 1,5-pentanediol with a volume of about 75 μL was dropped onto the top surface of the sample, as shown in Figure 1c. In the last stage, the sample was annealed at 120 °C for 3.5 h and polymer fibers were grown along the grating lines, as shown in Figure 1d. Similar preparation procedures were adopted for both the F8BT and PFB polymers. The chemical structures of F8BT and PFB are presented in Figure 1e, where PFB exhibits much more amorphous than F8BT.

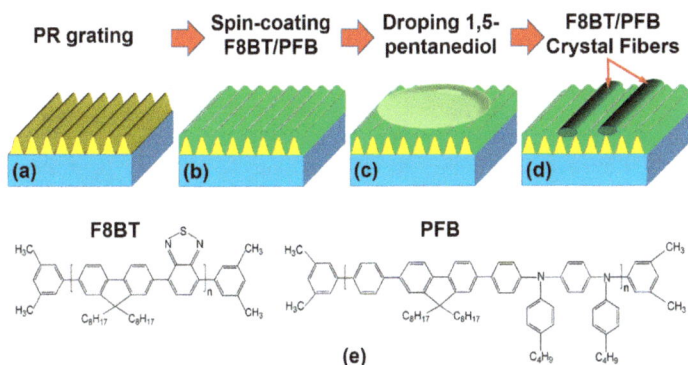

Figure 1. Fabrication procedures for the F8BT fibers. (**a**) The template PR grating with a period of 670 nm on a glass substrate; (**b**) spin-coating of the polymer film; (**c**) drop casting of 1,5-pentanediol onto the top surface of the polymer film; (**d**) growth of the polymeric fibers; and (**e**) chemical structures of F8BT and PFB.

2.3. Device Fabrication and Characterization

Gold (Au) electrodes with a thickness of about 50 nm were deposited by vacuum thermal evaporation through a mask with a channel length of $L = 2.0$ mm and a channel width of $W = 10$ μm at a rate of ca. 0.05 Å per second under a pressure of 10^{-4} Pa. A laser beam at 405 nm with diameter of about 2 mm was used as the excitation in the characterization of the photoconductivity performance.

3. Microscopic Characterization

Figure 2a shows the atomic force microscopic (AFM) image measured on the finished F8BT fibers produced on the template grating structures. From Figure 2a, we cannot only determine the thickness of the directionally-aligned F8BT fibers, but also measure the period of the template grating, which are 2 μm and 676 nm, respectively.

Figure 2b–e show the polarized optical microscopic images of the fibers of the aligned F8BT molecules with the grating lines oriented at 0°, 45°, 90°, and 135°, respectively, with respect to the transmission axis of one of the two orthogonally-arranged polarizers. We can observe bright F8BT fibers only when the grating lines are oriented at 45° and 135° with respect to the transmission axis of the polarizers, which verifies the birefringence or directional alignment properties of F8BT molecules along the direction of the grating lines.

The polymer fibers were produced through the alignment of the F8BT molecules in a direction approximately perpendicular to the fluorene's rigid plane. Such rigid and planar features of the fluorene structures supplied the basic mechanisms for the directional alignment of polyfluorene and its derivatives. Thus, the direction of the template grating lines actually defines the alignment direction. For the observation under a polarization optical microscope equipped with two orthogonally-oriented polarizers, when the alignment direction is parallel or perpendicular to the transmission axis of either polarizer (the fibers are orientated at 0° or 90° with respect to the polarizers), no light can propagate through the system so that a dark field can be observed under the microscope. However, when the fibers are oriented at 45° or 135° with respect to the transmission axes of the polarizers, maximum transmission of light can be observed. Therefore, this polarization- and orientation-dependent performance of optical transmission supplied effective and reliable evidence for the directional alignment nature of the polyfluorene fibers.

We stress further that although the polymeric fibers were grown on template gratings, they are not distributed periodically on the substrate, which can be observed clearly in Figure 2, as well as in the subsequent Figures 3 and 4. Therefore, the color and transmission change with polarization through the fibers in these measurement data definitely did not result from the diffraction processes.

The directional alignment of F8BT molecules the PR grating substrate has been based on the following mechanisms: (1) Slight dissolution of photoresist in 1,5-pentanediol enabled swelling of the PR grating lines and narrowing of the channel width between two adjacent grating lines; (2) The F8BT molecules detached from the solid film and dissolved into 1,5-pentanediol gradually, producing a layer of saturated solution on the tops of the grating-patterned substrate; (3) With evaporation of 1,5-pentanediol, the F8BT molecules solidified out and assembled in the narrowed channels into seeds through π-π stacking, which are the basis for further alignment; (4) Long-range alignment of the F8BT molecules from the seeds along the grating grooves.

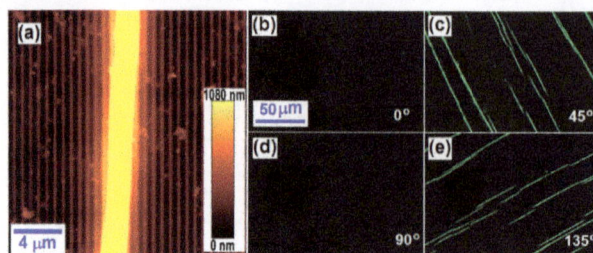

Figure 2. (a) AFM image measured on the F8BT fibers prepared on a photoresist grating template; and (b–e) polarization optical microscopic images of the F8BT fibers.

However, the F8BT fibers shown in Figure 2 are aligned along the direction of the grating lines and have random distributions in their length and separation distance. The quality of the F8BT fibers can be improved by modifying the fabrication process, where continuous single fibers may be selectively achieved by removing the template grating and the low-quality fibers. A series of test experiments enabled optimization of the fabrication parameters, which involve a F8BT/toluene solution concentration of 6 mg/mL, a spin-coating speed of 2200 rpm, an annealing temperature of 120 °C, and an annealing solvent of 1,5-pentanediol. The resultant sample was then immersed in acetone for 20 min to remove the template PR grating. Due to a continuous F8BT film underneath the fibers and above the PR grating, the fibers were not removed as the photoresist was dissolved into acetone. In the last stage, the sample was processed by low-power ultrasonic for about 10 s to remove the F8BT thin film and low-quality short fibers.

Figure 3a–d present the optical polarization microscope images of a single F8BT fiber at an orientation angle of 0°, 45°, 90°, and 135° with respect to the transmission axis of one of the polarizers. Clearly, the bright images of the F8BT fiber with dark backgrounds can be observed only at an orientation angle of 45° and 135°, verifying the directional alignment performance of the F8BT fibers. Figure 3e shows the transmission optical microscope image without polarization performance and Figure 3f shows the fluorescence optical microscope image under UV-light excitation. Strong photoluminescence can be observed from the single F8BT fiber in Figure 3f. It can thus be confirmed from Figure 3e,f that both the template PR grating and the continuous F8BT thin film have been removed completely. Figure 3g,h show the AFM height image and the profile plot of the single F8BT fiber, which has a width of about 350 nm and a height of about 120 nm.

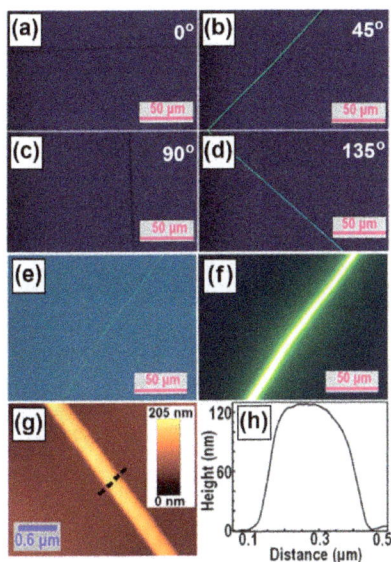

Figure 3. (**a–d**) Polarization optical microscopic images of a single F8BT fiber; (**e,f**) transmission and fluorescence optical microscopic image of the single F8BT fiber; and (**g,h**) AFM height image and plot of the profile of the fiber, respectively.

According to Figure 1e, PFB molecules are much more amorphous than F8BT. However, similar fabrication procedures also applied well to PFB and the directional alignment was achieved on the pattern substrate with the aid of the annealing solvent. Figure 4a–d show the polarization optical microscopic images measured on the fibers of directionally-aligned PFB molecules, corresponding to the orientation of the grating lines at $0°$, $45°$, $90°$, and $135°$, respectively, with respect to one of the two orthogonally mounted polarizers. Clearly, the preparation method for F8BT fibers also applies to PFB. Although the thickness, length, and continuity of the PFB fibers are randomly distributed at different locations, the fibers are aligned very well in the direction of the grating lines.

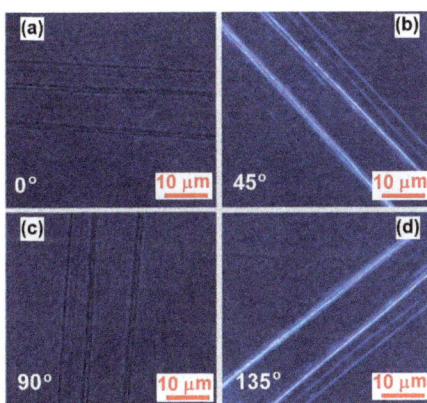

Figure 4. Polarization optical microscopic images measured on the PFB fibers with the grating lines orientated at (**a**) $0°$; (**b**) $45°$; (**c**) $90°$; and (**d**) $135°$ with respect to one of the two orthogonally-arranged polarizers.

4. Spectroscopic Characterization

Figure 5a shows the absorption spectra of the spin-coated F8BT film (black) and the F8BT fibers in Figure 3 aligned along the PR gratings with the light polarized perpendicular (blue) and parallel (red) to the grating lines. The fibers show redshifted absorption from 468 to 513 nm with respect to the continuous film for a parallel polarization. This is commonly observed for π-stacking of organic molecules due to intermolecular coupling along the stacking axis [24–26]. There is also a redshift of about 10 nm between the blue and red spectra, however, this redshift is much smaller than that relative to the spin-coated film, which implies inclination of the molecule plane with respect to the stacking axis and some disorders during the stacking process [27–30].

Figure 5b shows the PL spectra of the spin-coated F8BT film (black) and the fibers for polarization perpendicular (blue) and parallel (red) to the fibers. An obvious broadening of the PL spectrum can be observed for a parallel polarization, implying new structures were produced in the excited states of the fiber that inducing new features in the emission spectrum.

Curve ❸ in Figure 5b is a difference spectrum between the red (❷) and the blue (❶) ones, which is resolved as the emission from the π-stacking phase in the F8BT fibers and peaks at about 580 nm. Therefore, the face-to-face intermolecular coupling between π-stacked F8BT molecules induced a lower-leveled band of the excited states, which is not observable in spin-coated films. Thus, the spectroscopic performances in Figure 5 supplies more solid evidence for the directional alignment of the F8BT molecules.

Figure 5c,d show similar measurements on the PFB samples described in Figure 4. Figure 5c shows the absorption spectra measured on a solid film (black), on the fibers at a parallel (red) and perpendicular (blue) polarization direction. It should be noted that PFB has an absorption mainly in the UV and blue spectral range, which overlaps the absorption by the photoresist. Therefore, although the PFB fibers show obvious red-shifted absorption spectra with respect to the films, in particular for the polarization direction parallel to the fibers, the absorption spectra have been modulated by the substrate absorption. In contrast, the PL spectra are more reliable for the characterization of the spectroscopic properties of the PFB fibers. As shown in Figure 5d, the PFB fibers show a broadened and redshifted PL spectrum for a parallel (❷) polarization with respect to the film (❶). If we assume that the fibers have a similar PL spectrum as the film in the perpendicular polarization, the difference between spectra ❷ and ❶ shows a rough evaluation on the PL spectrum of the π-stacking phase, as shown by spectrum ❸, which peaks at about 480 nm and extends in the blue and green.

Thus, both the optical microscopic characterization in Figures 2–4 and the spectroscopic characterization in Figure 5, which show clear birefringence with strong polarization dependence, have consistently verified the directional alignment nature of the F8BT and PFB molecules in fibers grown along the template grating grooves.

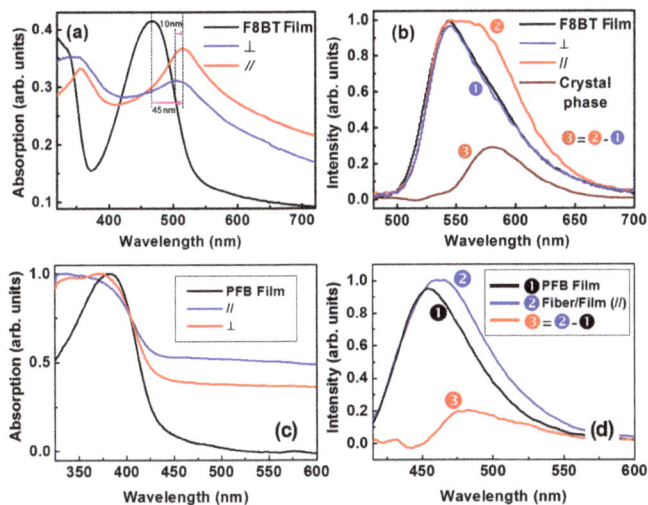

Figure 5. (a) Absorption spectra of the spin-coated F8BT film (black) and the F8BT fibers for light polarized perpendicular (blue) and parallel (red) to the grating lines; (b) The corresponding PL spectra for the sample in (a); (c) Absorption spectra of the spin-coated PFB film (black) and the PFB fibers for light polarized perpendicular (red) and parallel (blue) to the grating lines; and (d) The PL spectra measured on the PFB film (black), PFB fibers at a parallel polarization direction (blue), and the difference between them (red).

5. Photoconductivity Characterization

Figure 6 shows the I-V curves measured on photoconductive device based on a single F8BT fiber in Figure 3 under different excitation power density. The inset of Figure 6 shows the optical microscopic image of the device. Gold was deposited onto the F8BT fiber and functioned as electrodes, leaving a channel as wide as 10 μm. At an applied voltage of 10 V, the dark current of the device is about 170 pA. With the increase of the excitation intensity at 405 nm, the photocurrent of the device was increased from 170 to 230 pA. The performance in Figure 6 implies excellent continuity of the F8BT fiber with good conductivity. However, a slow increase in the photocurrent with increasing the laser power density at 405 nm was observed in Figure 6, this is mainly due to the low absorption by the F8BT fibers at 405 nm, as can be seen in Figure 5a, and the small amount of the fibers with low density within the studied area.

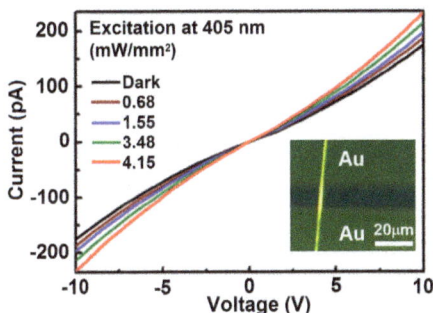

Figure 6. I–V curves measured on the photoconductive device based on a single F8BT fiber under different excitation power density. Inset: optical microscopic image of the device.

6. Conclusions

Directional alignment of F8BT and PFB molecules into large-scale fibers has been achieved by the solvent-annealing in 1,5-pentanediol of the spin-coated F8BT and PFB films on the photoresist grating structures. Microscopic and polarization-dependent spectroscopic characterization verified convincingly the directional π-stacking performance of the F8BT and PFB fibers. Measurements on the photoconductive performance show excellent continuity and conductivity of the single F8BT fibers. This is a first demonstration of millimeter-to-centimeter-scale directional alignment of the polyfluorene molecules, which can be applied in efficient light-emitting or transistor devices based on polyfluorene copolymers. Furthermore, the preparation method for polymeric fibers in this work may be explored to become applicable to other polymeric semiconductors.

Acknowledgments: The authors acknowledge the National Program on Key Basic Research Project of China (2013CB922404) and the National Natural Science Foundation of China (11574015, 11434016), as well as the Beijing Key Lab of Microstructure and Property of Advanced Materials, for the support.

Author Contributions: Xiaolu Pan and Hongwei Li performed the experiments, collected and processed the experimental data, and wrote a draft of the paper; Xinping Zhang designed this research work, supervised the experiments, did the data analysis, and wrote this paper.

Conflicts of Interest: The authors declare no conflict of interest.

References

1. Wang, Y.; Kumashiro, R.; Li, Z.F.; Nouchi, R.; Tanigaki, K. Light Emitting Ambipolar Field-effect Transistors of 2,5-bis(4-biphenyl)bithiophene Single Crystals with Anisotropic Carrier Mobilities. *Appl. Phys. Lett.* **2009**, *95*, 103306. [CrossRef]

2. Singh, M.K.; Kumar, A.; Prakash, R. Self-assembly of Regioregular Poly (3, 3000-didodecylquarterthiophene) in Chloroform and Study of its Junction Properties. *Mater. Sci. Eng. B* **2017**, *217*, 12–17. [CrossRef]

3. Li, X.J.; Xu, Y.X.; Li, F.; Ma, Y.G. Organic Light-emitting Diodes Based on an Ambipolar Single Crystal. *Org. Electron.* **2012**, *13*, 762–766. [CrossRef]

4. Wang, H.; Xie, Z.Q.; Shen, J.C.; Ma, Y.G. Progress on the Optoelectronic Functional Organic Crystals. *Sci. China Ser. B* **2007**, *50*, 433–452. [CrossRef]

5. Shinde, D.B.; Salunke, J.K.; Candeias, N.R.; Tinti, F.; Gazzano, M.; Wadgaonkar, P.P.; Priimagi, A.; Camaioni, N.; Vivo, P. Crystallisation-enhanced Bulk Hole Mobility in Phenothiazine-based Organic Semiconductors. *Sci. Rep.* **2017**, *7*, 46268. [CrossRef]

6. Jang, Y.J.; Lim, B.T.; Yoon, S.B.; Choi, H.J.; Ha, J.U.; Chung, D.S.; Lee, S.G. A Small Molecule Composed of Anthracene and Thienothiophene Devised for High-performance Optoelectronic Applications. *Dyes Pigments* **2015**, *120*, 30–36. [CrossRef]

7. Ozdemir, M.; Choi, D.; Kwon, G.; Zorlu, Y.; Cosut, B.; Kim, H.; Facchetti, A.; Kim, C.; Usta, H. Solution-Processable BODIPY-Based Small Molecules for Semiconducting Microfibers in Organic Thin-Film Transistors. *Appl. Mater. Interfaces* **2016**, *8*, 14077–14087. [CrossRef] [PubMed]

8. Fitzner, R.; Elschner, C.; Weil, M.; Uhrich, C.; Körner, C.; Riede, M.; Leo, K.; Pfeiffer, M.; Reinold, E.; Osteritz, E.M.; et al. Interrelation between Crystal Packing and Small Molecule Organic Solar Cell Performance. *Adv. Mater.* **2012**, *24*, 675–680. [CrossRef]

9. Chou, S.H.; Kang, H.W.; Chang, S.T.; Wu, K.Y.; Bazan, G.C.; Wang, C.L.; Lin, H.L.; Chang, J.H.; Lin, H.W.; Huang, Y.C.; et al. Cofacial Versus Coplanar Arrangement in Centrosymmetric Packing Dimers of Dipolar Small Molecules: Structural Effects on the Crystallization Behaviors and Optoelectronic Characteristics. *Appl. Mater. Interfaces* **2016**, *8*, 18266–18276. [CrossRef]

10. Sun, J.P.; Hendsbee, A.D.; Eftaiha, A.F.; Macaulay, C.; Rutledge, L.R.; Welch, G.C.; Hill, I.G. Phthalimide–thiophene-based Conjugated Organic Dmall Molecules with High Electron Mobility. *J. Mater. Chem. C* **2014**, *2*, 2612–2621. [CrossRef]

11. Sun, Y.; Xiao, G.H.; Lin, Y.; Su, Z.H.; Wang, Q. Self-assembly of Large-scale P3HT Patterns by Confined Evaporation in the Capillary Tube. *RSC Adv.* **2015**, *5*, 20491–20497. [CrossRef]

12. Zhao, K.; Ding, Z.; Xue, L.; Han, Y. Crystallization-Induced Phase Segregation Based on Double-Crystalline Blends of Poly(3-hexylthiophene) and Poly(ethylene glycol)s. *Macromol. Rapid Commun.* **2010**, *31*, 532–538. [CrossRef] [PubMed]

13. Choi, D.; Chang, M.; Reichmanis, E. Controlled Assembly of Poly (3-hexylthiophene): Managing the Disorder to Order Transition on the Nano- through Meso-Scales. *Adv. Funct. Mater.* **2015**, *25*, 920–927. [CrossRef]

14. Yin, Z.; Zheng, Q. Controlled Synthesis and Energy Applications of One-Dimensional Conducting Polymer Nanostructures: An Overview. *Adv. Energy Mater.* **2012**, *2*, 179–218. [CrossRef]

15. Wang, S.; Qu, Y.; Li, S.; Ye, F.; Chen, Z.; Yang, X. Improved Thermal Stability of Polymer Solar Cells by Incorporating Porphyrins. *Adv. Funct. Mater.* **2015**, *25*, 748–757. [CrossRef]

16. Di, D.; Yang, L.; Richter, J.M.; Meraldi, L.; Altamimi, R.M.; Alyamani, A.Y.; Credgington, D.; Musselman, K.P.; MacManus-Driscoll, J.L.; Friend, R.H. Efficient Triplet Exciton Fusion in Molecularly Doped Polymer Light-Emitting Diodes. *Adv. Mater.* **2017**, *29*, 1605987. [CrossRef] [PubMed]

17. Nikolka, M.; Nasrallah, I.; Rose, B.; Ravva, M.K.; Broch, K.; Sadhanala, A.; Harkin, D.; Charmet, J.; Hurhangee, M.; Brown, A.; et al. High Operational and Environmental Stability of High-mobility Conjugated Polymer Field-effect Transistors through the Use of Molecular Additives. *Nat. Mater.* **2017**, *16*, 356–362. [CrossRef] [PubMed]

18. Wang, M.; Zhang, X.P. Ultrafast Injection-locked Amplification in a Thin-film Distributed Feedback Microcavity. *Nanoscale* **2017**, *9*, 2689–2694. [CrossRef] [PubMed]

19. Zhai, T.R.; Xu, Z.Y.; Li, S.T.; Zhang, X.P. Red-green-blue Plasmonic Random Laser. *Opt. Express* **2017**, *25*, 2100–2106. [CrossRef]

20. Liu, L.H.; Wu, K.Q.; Ding, J.Q.; Zhang, B.H.; Xie, Z.Y. Binary Solvent Mixture-induced Crystallization Enhancement for a White Emissive Polyfluorene Copolymer toward Improving Its Electroluminescence. *Polymer* **2013**, *54*, 6236–6241. [CrossRef]

21. Kobayashi, T.; Uda, H.; Nagase, T.; Watanabe, M.; Matsukawa, K.; Naito, H. Correlation between the Crystallization of Polyfluorene and the Surface Free Energy of Substrates. *Thin Solid Films* **2008**, *517*, 1340–1342. [CrossRef]

22. Brinkmann, M. Directional Epitaxial Crystallization and Tentative Crystal Structure of Poly (9,9¢-di-n-octyl-2,7-fluorene). *Macromolecules* **2007**, *40*, 7532–7541. [CrossRef]

23. Whitehead, K.S.; Grell, M.; Bradley, D.D.C.; Inbasekaran, M.; Woo, E.P. Polarized Emission from Liquid Crystal Polymers. *Synth. Met.* **2000**, *111–112*, 181–185. [CrossRef]

24. Hinoue, T.; Shigenoi, Y.; Sugino, M.; Mizobe, Y.; Hisaki, I.; Miyata, M.; Tohnai, N. Regulation of π-Stacked Anthracene Arrangement for Fluorescence Modulation of Organic Solid from Monomer to Excited Oligomer Emission. *Chem. Eur. J.* **2012**, *18*, 4634–4643. [CrossRef] [PubMed]

25. Klein, A.; Hurkes, N.; Kaiser, A.; Wielandt, W. π-Stacking Modulates the Luminescence of [(dppz)Ni(Mes)Br] (dppz = dipyrido[3,2-a:2′,3′-c]phenazine, Mes = 2,4,6-trimethylphenyl). *Zeitschrift für Anorganische und Allgemeine Chemie* **2007**, *633*, 1659–1665. [CrossRef]

26. Kobayashi, K.; Kobayashi, N. Synthesis and Self-Association, Absorption, and Fluorescence Properties of Differentially Functionalized Hexakis(p-substituted phenylethynyl)benzenes. *J. Org. Chem.* **2004**, *69*, 2487–2497. [CrossRef] [PubMed]

27. Dickinson, S.R.; Müller, P.; Tanski, J.M. Crystal Structure of 3-bromo-2-hydroxybenzonitrile. *Acta Cryst.* **2015**, *71*, O523–O524. [CrossRef] [PubMed]

28. Kolnig, O.; Bülrgi, H.B.; Armbruster, T.; Hulliger, J.; Weber, T. A Study in Crystal Engineering: Structure, Crystal Growth, and Physical Properties of a Polar Perhydrotriphenylene Inclusion Compound. *J. Am. Chem. Soc.* **1997**, *119*, 10632–10640. [CrossRef]

29. Funasako, Y.; Mochida, T.; Yoza, K. Order-disorder Phase Transition with Associated Cell-tripling in the (octamethylferrocene)(2,3-dichloro-1,4-naphthoquinone)$_2$ Charge-transfer Complex. *J. Organomet. Chem.* **2012**, *698*, 49–52. [CrossRef]

30. Zhang, X.P.; Sun, B.Q. Organic Crystal Fibers Aligned into Oriented Bundles with Polarized Emission. *J. Phys. Chem. B* **2007**, *111*, 10881–10885. [CrossRef] [PubMed]

polymers

MDPI

Article

Enhancing Stereocomplexation Ability of Polylactide by Coalescing from Its Inclusion Complex with Urea

Ping Liu [1], Xiao-Tong Chen [1] and Hai-Mu Ye [1,2,*]

[1] Department of Materials Science and Engineering, China University of Petroleum, Beijing 102249, China; 18811414503@163.com (P.L.); chenxt_mkrc@163.com (X.-T.C.)
[2] Beijing Key Laboratory of Failure, Corrosion and Protection of Oil/Gas Facilities, China University of Petroleum, Beijing 102249, China
* Correspondence: yehaimu@cup.edu.cn; Tel.: +86-10-8973-3200

Received: 12 October 2017; Accepted: 9 November 2017; Published: 9 November 2017

Abstract: In this study, polylactide/urea complexes were successfully prepared by the electrospinning method, then the host urea component was removed to obtain a coalesced poly(L-lactide) (PLLA)/poly(D-lactide) (PDLA) blend. The crystallization behavior of the coalesced PLLA/PDLA blend (c-PLLA/PDLA) was studied by a differential scanning calorimeter (DSC) and Fourier transform infrared (FTIR) spectroscopy. The c-PLLA/PDLA was found to show better crystallization ability than normal PLLA/PDLA blend (r-PLLA/PDLA). More interestingly, the c-PLLA/PDLA effectively and solely crystallized into stereocomplex crystals during the non-isothermal melt-crystallization process, and the reason was attributed to the equally-distributing state of PLLA and PDLA chains in the PLLA/PDLA/urea complex, which led to good interconnection between PLLA and PDLA chains when the urea frameworks were instantly removed.

Keywords: polylactide; urea; inclusion complex; crystallization; stereocomplex

1. Introduction

Polylactide (PLA) has become one of the most important and commercial biodegradable polymers because it possesses comparable mechanical and thermal properties with polyolefin materials [1]. However, due to the rigid chain structure and relative high glass transition temperature, it shows many drawbacks, such as a low crystallization rate and poor heat resistance; subsequently, the practical applications of PLA are remarkably restricted [2–4]. Thus, various methods have been used to improve the performance of PLA, including chain modification [5,6], blends with other polymers [7–9], and the introduction of efficient nucleating agents [10–13]. In addition to these, stereocomplex crystals formed between two PLA enantiomers, and poly(L-lactide) (PLLA) and poly(D-lactide) (PDLA) have become the most attractive selections [14,15]. The stereocomplex crystals show stronger inter-chain interaction, have a much higher melting point (~225 °C), and display a faster crystallization rate than homocrystallite of either PLLA or PDLA [16–18], resulting in higher mechanical strength and modulus, better heat resistance, etc. [19–21]. Therefore, plenty of research has been carried out. However, the formations of the stereocomplex and homocrystallite are competing during the crystallization process in the PLLA/PDLA blend [22,23]. Especially, the formation of the stereocomplex can be sharply inhibited as the PLA molecular weight increases [24–26]. The intrinsic occupied space leads to the steric repulsion effect among different chains, which goes against the contact between PLLA and PDLA segments, and benefits the formation of homocrystallite. In fact, only high molecular weight PLA can possess useful mechanical properties for thermoplastic applications, so several methods are employed to enhance the formation of the stereocomplex through promoting the mixed degree between PLLA and PDLA segments, such as chain topology regulation [27–29], addition of a compatibility agent [30,31], and the specific design of the process [32].

An inclusion complex formed between guest polymer chains and host small molecular frameworks provides a novel method to isolate polymer chains from each other in small channels [33], which offer us an effective means to obtain polymer blends with better miscibility or polymer materials with less entanglement through removing the host molecules. Until now, Tonelli's group has carried out milestone work in this field [34–40]. For example, they successfully achieved intimately compatible polymer blends from normally immiscible polymers, including a PLLA/poly(ε-caprolactone) (PCL) blend and a nylon 6/nylon 66 blend [35,40]. Additionally, they obtained coalesced PCL by washing a PCL/urea inclusion complex and found that the crystallization ability and mechanical properties of PCL are significantly improved [36,39]. Recently, based on the similar method, we have successfully prepared the extended-chain crystals of poly(butylene succinate) (PBS) under atmosphere [41]. As to PLA, Howe et al. had proven that PLLA and urea could form an inclusion complex [42,43]. This, it is wondered whether PDLA chains can be adopted equally as PLLA chains in urea frameworks? Furthermore, can the complex-coalescence method be used to promote the compatibility between PLLA and PDLA and the formation of PLA stereocomplex? In this study the PLLA/urea, PDLA/urea, and PLLA/PDLA/urea complexes were prepared by electrospinning and the crystallization behaviors of the coalesced PLLA/PDLA blend were studied in detail.

2. Materials and Methods

2.1. Materials

PLLA with M_w of 2.20×10^5 and 9.15×10^5 g/mol and PDLA with M_w of 2.20×10^5 and 9.23×10^5 g/mol were purchased from Ji'nan Daigang Biological Engineering Company (Ji'nan, China). Urea (AR grade) was obtained from Shanghai Aladdin Industrial Inc. (Shanghai, China). All reagents were used without further purification.

2.2. Preparation of the PLA/Urea Complex and Coalesced PLA

The solution for electrospinning was prepared by dissolving PLA and urea in hexafluoroisopropanol (HFIP) with a PLA concentration fixed at 2.5 wt %. The electrospinning process was optimized as following condition: a DC voltage of 30 kV, a collector-to-needle tip distance of 18 cm, and an inner needle diameter of 0.6 mm. The as-electrospun species was dried in a vacuum oven at room temperature for two days before sealing. The weight ratio of PLA/urea was optimally selected as 1:7 based on the maximization of the experimental melting enthalpy of the complex (as seen in Figure S1).

The coalesced PLA was obtained by washing the PLA/urea complex with sufficient deionized water three times, and methanol once, to completely remove the urea component, and then drying in vacuum at 45 °C for three days.

2.3. Characterizations

Non-isothermal crystallization and melting behaviors of the samples were performed on a differential scanning calorimeter (DSC, 204 F1, NETZSCH, Berlin, Germany) equipped with an intercooler as cooling system under an argon atmosphere; the heating and cooling rates were set as 10 °C/min. Fourier transformation infrared (FTIR) spectra were recorded on a Hyperion spectrometer (Bruker, Karlsruhe, Germany) by signal averaging over 32 scans in the wavenumber range of 4000~400 cm^{-1}; the spectrometer was equipped with a hot stage (THMS-600, Linkam, Surrey, UK) for temperature-resolution FTIR measurement.

3. Results

3.1. Characterization of the PLA/Urea Complex

DSC measurement was performed to determine the formation of inclusion complex between PLLA ($M_w = 2.20 \times 10^5$ g/mol) and urea. As shown in Figure 1A, the as-spun product displayed a single melting point at 137.3 °C, which was different from those of urea and PLLA at 134.0 and 174.8 °C (the first melting peak), respectively. The new melting point, which was consistent with previous reports on other polymer/urea complexes, indicated the successful preparation of the PLLA/urea inclusion complex [42]. More interestingly, inclusion complexes between poly(R-3-hydroxybutyrate) (PHB) and urea, and between polypropylene and urea, showed almost the same melting point (136.8 and 138.0 °C) as PLLA/urea complex [44,45], so the melting point at around 137 °C might be a common phenomenon for the polymer/urea inclusion complexes when polymer chains contain pendant methyl groups. Figure 1B shows the FTIR spectra of in the regions from 3500 to 3300 cm^{-1} and 1820 to 1400 cm^{-1} obtained for urea, the PLLA/urea complex, and PLLA. Obviously, PLLA/urea showed quite a different FTIR spectrum from either neat urea or neat PLLA, further confirming the formation of the PLLA/urea complex. The strong N–H stretching vibration bands split and shifted from 3443 and 3347 to 3455, 3439, and 3344 cm^{-1}, respectively; the C=O stretching vibration band red-shifted from 1681 to 1691 cm^{-1}; the N–H bending vibration bands shifted from 1625 and 1605 to 1634 and 1602 cm^{-1}; and the N–C–N antisymmetric stretching band shifted from 1465 to 1468 cm^{-1} after the urea molecules had been complexed with PLLA from its traditional tetragonal modification. As for the PLLA component in the complex, it exhibited a single C=O stretching vibration band at 1757 cm^{-1}, which was quite different from the neat crystalline PLLA that had two C=O stretching vibration bands at 1755 and 1749 cm^{-1}. The absence of 1749 cm^{-1} in the PLLA/urea complex revealed that PLLA chains in the complex were in an amorphous state, and the blue-shift of 2 cm^{-1}, from 1755 to 1757 cm^{-1}, might be due to the isolated and confined effect of the urea frameworks. The blue-shift of the C=O stretching vibration of PLLA induced by confinement had also been observed in other systems [10].

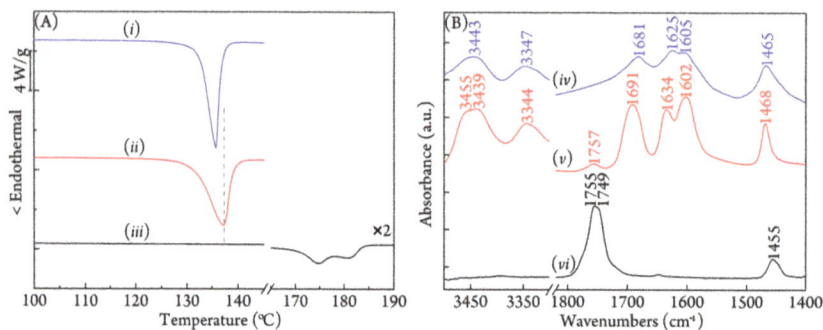

Figure 1. (**A**) DSC curves of (*i*) urea, (*ii*) as-prepared PLLA/urea complex and (*iii*) PLLA at a heating rate of 10 °C/min; and (**B**) the FTIR spectra of (*iv*) urea, (*v*) as-prepared PLLA/urea complex, and (*vi*) PLLA.

PDLA (2.2 × 10^5 g/mol) had been also used to produce a PDLA/urea complex, and the DSC curve in Figure S2 confirms the successful preparation of PDLA/urea complex. Thus, it was expected that the guest PLLA and PDLA could be equally treated by the host urea frameworks, and equal amounts of high molecular weight PLLA (9.15 × 10^5 g/mol) and PDLA (9.23 × 10^5 g/mol) were employed to prepare a ternary complex with urea. Figure 2A shows that the ternary complex had a melting point of 138.3 °C; Figure 2B affirms that the ternary complex exhibited the similar FTIR spectrum as the PLLA/urea complex. Thus, the PLLA/PDLA/urea complex adopted the same structure as the PLLA/urea complex.

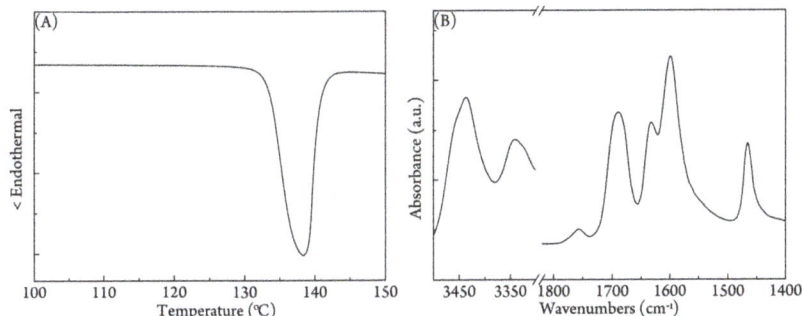

Figure 2. (**A**) DSC heating curve and (**B**) FTIR spectrum of the PLLA/PDLA/urea complex.

3.2. Crystallization Behavior of Coalesced PLLA/PDLA Blend

Compared with the referential PLLA/PDLA (*r*-PLLA/PDLA, prepared by the same electrospinning process without urea), the coalesced PLLA/PDLA (*c*-PLLA/PDLA) from the PLLA/PDLA/urea complex showed significant enhancement of melt-crystallization during the cooling process in Figure 3A, displaying a crystallization peak at 130.6 °C. The *r*-PLLA/PDLA did not show crystallization under the same cooling setting. During the subsequent heating (seen in Figure 3B), the *c*-PLLA/PDLA presented a single melting point at 214.9 °C, besides the normal glass transition, which demonstrated that only the stereocomplex formed between PLLA and PDLA forms during the previous cooling process. Nevertheless, a rather complicated DSC curve, which contained one exothermal peak at 109.2 °C and two endothermal peaks at 177.5 and 212.2 °C after the glass transition appeared for *r*-PLLA/PDLA; the three peaks corresponded to the cold crystallization, melt of homocrystallites of PLA, and melt of the stereocomplex of PLA, respectively. On the basis of the enthalpies of peaks at 177.5 and 212.2 °C, it is clear that PLA chains predominately cold-crystallized to homocrystals even during the heating process, which had been observed in other high molecular weight PLLA/PDLA blends [24–32]. Therefore, the combining processes of the PLA/urea complexation and coalescent greatly promoted the stereocomplexation ability between PLLA and PDLA. Due to the equal accommodation of PLA chains in urea frameworks, PLLA and PDLA chains would pack densely and realize an even distribution in statistics when the urea molecules were instantaneously removed by washing, which helped promote the inter-contact between PLLA and PDLA and then accelerated the stereocomplexation efficiency.

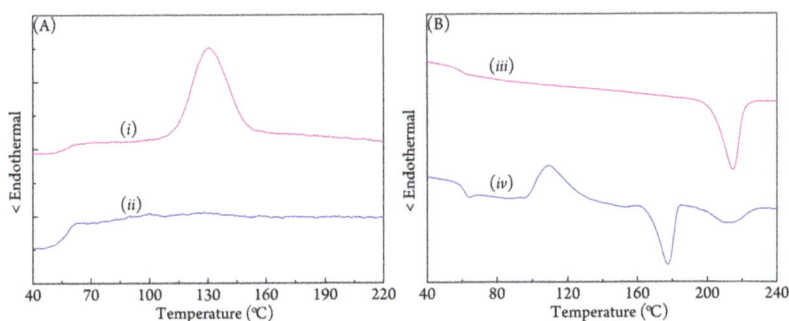

Figure 3. (**A**) Non-isothermal crystallization DSC curves of (*i*) coalesced PLLA/PDLA blend (*c*-PLLA/PDLA) and (*ii*) the referential PLLA/PDLA blend (*r*-PLLA/PDLA) from 240 °C; and (**B**) the subsequent DSC heating curves of (*iii*) *c*-PLLA/PDLA and (*iv*) *r*-PLLA/PDLA. Both the cooling and heating rates are 10 °C/min.

3.3. FTIR Study on Stereocomplexation Ability

To understand more about the enhancing crystallization ability for *c*-PLLA/PDLA, FTIR spectra were employed to characterize the melt-quenched sample and melting process. Figure 4 shows the FTIR spectra of *c*-PLLA/PDLA and *r*-PLLA/PDLA in a wavenumber range from 980 to 860 cm^{-1}, in which a characteristic band around 909 cm^{-1} responds to the vibration mode of the 3_1 helical conformation in the stereocomplex [46]. The *r*-PLLA/PDLA displayed a very weak absorption band at 909 cm^{-1} when melt-quenched from 240 °C and no absorption band at 920 cm^{-1}, which is the characteristic band of homocrystallite [47], indicating that the PLA chains were almost amorphous and only a small amount transformed to stereocomplex crystals. However, a rather obvious absorption band at 909 cm^{-1} appeared in the melt-quenched *c*-PLLA/PDLA specimen, meaning that the stereo-crystallization occurred remarkably.

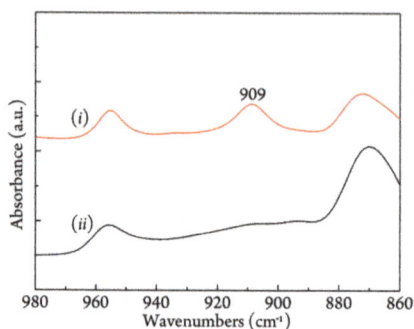

Figure 4. FTIR spectra of melt-quenched (*i*) *c*-PLLA/PDLA and (*ii*) *r*-PLLA/PDLA in the range from 980 to 860 cm^{-1}.

The subsequent temperature-dependent FTIR spectra of *c*-PLLA/PDLA are presented in Figure 5A. The characteristic absorption band of the 3_1 helical conformation in the stereocomplex at 909 cm^{-1} remained almost constant until 190 °C, then weakened at higher temperatures (i.e., 210 and 215 °C), and finally disappeared at 240 °C. For comparison, the temperature-dependent FTIR spectra of *r*-PLLA/PDLA were collected and are lined in Figure 5B. When heated to 100 °C, a characteristic band of homocrystallites rose at 920 cm^{-1}, while the intensity of the 909 cm^{-1} band was still rather weak. The intensity of 920 cm^{-1} band further increased until 150 °C, then decreased, and finally disappeared at 180 °C, confirming the endothermal peak in Figure 4B originated from the cold-crystallization of homocrystallites. The further heating process after melting of homocrystals led to the formation of some sterocomplex (i.e., 180 and 210 °C), but the amount was small; and all crystallites melted at 240 °C. Both the *c*-PLLA/PDLA and the *r*-PLLA/PDLA crystallites melted and the 3_1 helical conformations disappeared completely at 240 °C, so the remarkable promotion of stereocomplexation in *c*-PLLA/PDLA should be due to the better inter-contacting state between PLLA and PDLA chains, which was usually rather poor in high molecular weight PLLA/PDLA blends [24–26]. The isolated and extended state of PLA chains in the complex would facilitate the interconnection between PLLA and PDLA during coalescing process, leading to enhancing the formation of the stereocomplex.

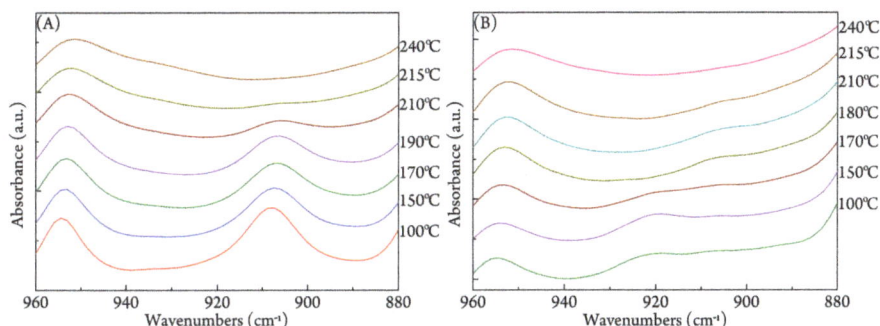

Figure 5. The temperature-dependent FTIR spectra of (**A**) *c*-PLLA/PDLA and (**B**) *r*-PLLA/PDLA during the heating process.

4. Conclusions

In this research, a complexation-coalescent method was employed to enhance the stereocomplexation ability of PLA. The crystallization ability of high molecular weight *c*-PLLA/PDLA was found to be stronger than *r*-PLLA/PDLA. Furthermore, the *c*-PLLA/PDLA solely formed stereocomplex crystals during the non-isothermal cooling process, and the reason was ascribed to the equal distribution of PLLA and PDLA in the PLLA/PDLA/urea complex, which led to good interconnection between PLLA and PDLA chains when the urea frameworks were instantly removed.

Supplementary Materials: The following are available online at http://www.mdpi.com/2073-4360/9/11/592/s1; Figure S1: Dependencies of melting points and enthalpy of the complex on the PLA mass fraction; Figure S2: DSC heating curve of the PDLA/urea complex.

Acknowledgments: This work was supported by National Natural Science Foundation of China (NSFC, grant No. 21674128) and China University of Petroleum, Beijing.

Author Contributions: Hai-Mu Ye conceived and designed the experiments; Ping Liu and Xiao-Tong Chen performed the experiments; Ping Liu and Hai-Mu Ye analyzed the data; Ping Liu and Hai-Mu Ye wrote the paper.

Conflicts of Interest: The authors declare no conflict of interest.

References

1. Chen, G.Q.; Patel, M.K. Plastics Derived from Biological Sources: Present and Future: A Technical and Environmental Review. *Chem. Rev.* **2012**, *112*, 2082–2099. [CrossRef] [PubMed]
2. Cicero, J.A.; Dorgan, J.R.; Garrett, J.; Runt, J.; Lin, J.S. Effects of molecular architecture on two-step, melt-spun poly(lactic acid) fibers. *J. Appl. Polym. Sci.* **2002**, *86*, 2839–2846. [CrossRef]
3. Drieskens, M.; Peeters, R.; Mullens, J.; Lemstra, P.J.; Hristova-Bogaerds, D.G. Structure versus properties relationship of poly(lactic acid). I. Effect of crystallinity on barrier properties. *J. Polym. Sci. Part B Polym. Phys.* **2009**, *47*, 2247–2258. [CrossRef]
4. Wang, L.; Wang, Y.N.; Huang, Z.G.; Weng, Y.X. Heat resistance, crystallization behavior, and mechanical properties of polylactide/nucleating agent composites. *Mater. Des.* **2015**, *66*, 7–15. [CrossRef]
5. Courgneau, C.; Domenek, S.; Lebossé, R.; Guinault, A.; Avérous, L.; Ducruet, V. Effect of crystallization on barrier properties of formulated polylactide. *Polym. Int.* **2012**, *61*, 180–189. [CrossRef]
6. Loiola, L.M.D.; Más, B.A.; Duek, E.A.R.; Felisberti, M.I. Amphiphilic multiblock copolymers of PLLA, PEO and PPO blocks: Synthesis, properties and cell affinity. *Eur. Polym. J.* **2015**, *68*, 618–629. [CrossRef]
7. Lu, J.; Qiu, Z.; Yang, W. Fully biodegradable blends of poly(L-lactide) and poly(ethylene succinate): Miscibility, crystallization, and mechanical properties. *Polymer* **2007**, *48*, 4196–4204. [CrossRef]
8. Pan, P.; Shan, G.; Bao, Y. Enhanced nucleation and crystallization of poly(L-lactic acid) by immiscible blending with poly(vinylidene fluoride). *Ind. Eng. Chem. Res.* **2014**, *53*, 3148–3156. [CrossRef]

9. Zhao, Y.; Qiu, Z. Effect of poly(vinyl alcohol) as an efficient crystallization-assisting agent on the enhanced crystallization rate of biodegradable poly(L-lactide). *RSC Adv.* **2015**, *5*, 49216–49223. [CrossRef]

10. Ye, H.M.; Hou, K.; Zhou, Q. Improve the thermal and mechanical properties of poly(L-lactide) by forming nanocomposites with pristine vermiculite. *Chin. J. Polym. Sci.* **2016**, *34*, 1–12. [CrossRef]

11. Bai, H.W.; Huang, C.M.; Xiu, H.; Zhang, Q.; Fu, Q. Enhancing mechanical performance of polylactide by tailoring crystal morphology and lamellae orientation with the aid of nucleating agent. *Polymer* **2014**, *55*, 6924–6934. [CrossRef]

12. Barrau, S.; Vanmansart, C.; Moreau, M.; Addad, A.; Stoclet, G.; Lefebvre, J.M.; Seguela, R. Crystallization behavior of carbon nanotube-polylactide nanocomposites. *Macromolecules* **2011**, *44*, 6496–6502. [CrossRef]

13. Zhao, L.; Liu, X.; Zhang, R.; He, H.F.; Jin, T.; Zhang, J. Unique morphology in polylactide/graphene oxide nanocomposites. *J. Macromol. Sci. Phys.* **2015**, *54*, 45–57. [CrossRef]

14. Ikada, Y.; Jamshidi, K.; Tsuji, H.; Hyon, S.H. Stereocomplex formation between enantiomeric poly(lactides). *Macromolecules* **1987**, *20*, 904–906. [CrossRef]

15. Tsuji, H. Poly(lactide) stereocomplexes: Formation, structure, properties, degradation, and applications. *Macromol. Biosci.* **2005**, *5*, 569–597. [CrossRef] [PubMed]

16. Schmidt, S.C.; Hillmyer, M.A. Polylactide stereocomplex crystallites as nucleating agents for isotactic polylactide. *J. Polym. Sci. Part B Polym. Phys.* **2001**, *39*, 300–313. [CrossRef]

17. Tsuji, H.; Tezuka, Y. Stereocomplex formation between enantiomeric poly(lactic acid)s. 12. Spherulite growth of low-molecular-weight poly(lactic acid)s from the melt. *Biomacromolecules* **2004**, *5*, 1181–1186. [CrossRef] [PubMed]

18. Tsuji, H.; Takai, H.; Saha, S.K. Isothermal and non-isothermal crystallization behavior of poly(L-lactic acid): Effects of stereocomplex as nucleating agent. *Polymer* **2006**, *47*, 3826–3837. [CrossRef]

19. Tsuji, H.; Ikada, Y. Stereocomplex formation between enantiomeric poly(lactic acid)s. XI. Mechanical properties and morphology of solution-cast films. *Polymer* **1999**, *40*, 6699–6708. [CrossRef]

20. Ma, P.M.; Shen, T.F.; Xu, P.W.; Dong, W.F.; Lemstra, P.J.; Chen, M.Q. Superior performance of fully biobased poly(lactide) via stereocomplexation-induced phase separation: Structure versus property. *ACS Sustain. Chem. Eng.* **2015**, *3*, 1470–1478. [CrossRef]

21. Tsuji, H.; Fukui, I. Enhanced thermal stability of poly(lactide)s in the melt by enantiomeric polymer blending. *Polymer* **2003**, *44*, 2891–2896. [CrossRef]

22. Tsuji, H.; Hyon, S.H.; Ikada, Y. Stereocomplex formation between enantiomeric poly(lactic acid)s. 3. Calorimetric studies on blend films cast from dilute solution. *Macromolecules* **1991**, *24*, 5651–5656. [CrossRef]

23. Tsuji, H.; Ikada, Y. Stereocomplex formation between enantiomeric poly(lactic acids). 9. Stereocomplexation from the melt. *Macromolecules* **1993**, *26*, 6918–6926. [CrossRef]

24. Pan, P.J.; Han, L.L.; Bao, J.N.; Xie, Q.; Shan, G.R.; Bao, Y.Z. Competitive stereocomplexation, homocrystallization, and polymorphic crystalline transition in poly(L-lactic acid)/poly(D-lactic acid) racemic blends: Molecular weight effects. *J. Phys. Chem. B* **2015**, *119*, 6462–6470. [CrossRef] [PubMed]

25. Han, L.L.; Pan, P.J.; Shan, G.R.; Bao, Y.Z. Stereocomplex crystallization of high-molecular-weight poly(L-lactic acid)/poly(D-lactic acid) racemic blends promoted by a selective nucleator. *Polymer* **2015**, *63*, 144–153. [CrossRef]

26. Tsuji, H.; Tashiro, K.; Bouapao, L.; Hanesaka, M. Synchronous and separate homo-crystallization of enantiomeric poly(L-lactic acid)/poly(D lactic acid) blends. *Polymer* **2012**, *53*, 747–754. [CrossRef]

27. Han, L.L.; Shan, G.R.; Bao, Y.Z.; Pan, P.J. Exclusive stereocomplex crystallization of linear and multiarm star-shaped high-molecular-weight stereo diblock poly(lactic acid)s. *J. Phys. Chem. B* **2015**, *119*, 14270–14279. [CrossRef] [PubMed]

28. Han, L.L.; Yu, C.T.; Zhou, J.; Shan, G.R.; Bao, Y.Z.; Yun, X.Y.; Dong, T.; Pan, P.J. Enantiomeric blends of high-molecular-weight poly(lactic acid)/poly(ethylene glycol) triblock copolymers: Enhanced stereocomplexation and thermomechanical properties. *Polymer* **2016**, *103*, 376–386. [CrossRef]

29. Fukushima, K.; Kimura, Y. An efficient solid-state polycondensation method for synthesizing stereocomplexed poly(lactic acid)s with high molecular weight. *J. Polym. Sci. Part A Polym. Chem.* **2008**, *46*, 3714–3722. [CrossRef]

30. Li, S.H.; Woo, E.M. Kinetic Analysis on Effect of Poly(4-vinyl phenol) on Complex-Forming Blends of Poly(L-lactide) and Poly (D-lactide). *Polym. J. Tokyo Jpn.* **2009**, *41*, 374–382. [CrossRef]

31. Pan, P.J.; Bao, J.N.; Han, L.L.; Xie, Q.; Shan, G.R.; Bao, Y.Z. Stereocomplexation of high-molecular-weight enantiomeric poly(lactic acid)s enhanced by miscible polymer blending with hydrogen bond interactions. *Polymer* **2016**, *98*, 80–87. [CrossRef]

32. Bao, R.Y.; Yang, W.; Jiang, W.R.; Liu, Z.Y.; Xie, B.H.; Yang, M.B.; Fu, Q. Stereocomplex formation of high-molecular-weight polylactide: A low temperature approach. *Polymer* **2012**, *53*, 5449–5454. [CrossRef]

33. Lu, J.; Mirau, P.A.; Tonelli, A.E. Chain conformations and dynamics of crystalline polymers as observed in their inclusion compounds by solid-state NMR. *Prog. Polym. Sci.* **2002**, *27*, 357–401. [CrossRef]

34. Gurarslan, A.; Joijode, A.S.; Tonelli, A.E. Polymers Coalesced from Their Cyclodextrin Inclusion Complexes: What Can They Tell Us about the Morphology of Melt-Crystallized Polymers? *J. Polym. Sci. Part B Polym. Phys.* **2012**, *50*, 813–823. [CrossRef]

35. Wei, M.; Shuai, X.; Tonelli, A.E. Melting and crystallization behaviors of biodegradable polymers enzymatically coalesced from their cyclodextrin inclusion complexes. *Biomacromolecules* **2003**, *4*, 783–792. [CrossRef] [PubMed]

36. Williamson, B.R.; Krishnaswamy, R.; Tonelli, A.E. Physical properties of poly(ε-caprolactone) coalesced from its α-cyclodextrin inclusion compound. *Polymer* **2011**, *52*, 4517–4527. [CrossRef]

37. Gurarslan, A.; Shen, J.; Tonelli, A.E. Behavior of poly(ε-caprolactone)s (PCLs) coalesced from their stoichiometric urea inclusion compounds and their use as nucleants for crystallizing PCL melts: Dependence on PCL molecular weights. *Macromolecules* **2012**, *45*, 2835–2840. [CrossRef]

38. Wei, M.; Davis, W.; Urban, B.; Song, Y.; Porbeni, F.E.; Wang, X.; White, J.L.; Balik, C.M.; Rusa, C.C.; Fox, J.; Tonelli, A.E. Manipulation of Nylon-6 Crystal Structures with Its α-Cyclodextrin Inclusion Complex. *Macromolecules* **2002**, *35*, 8039–8044. [CrossRef]

39. Gurarslan, A.; Caydamli, Y.; Shen, J.L.; Tse, S.; Yetukuri, M.; Tonelli, A.E. Coalesced poly(ε-caprolactone) fibers are stronger. *Biomacromolecules* **2015**, *16*, 890–893. [CrossRef] [PubMed]

40. Wei, M.; Shin, I.D.; Urban, B.; Tonelli, A.E. Partial miscibility in a nylon-6/nylon-66 blend coalesced from their common α-cyclodextrin inclusion complex. *J. Polym. Sci. Part B Polym. Phys.* **2004**, *42*, 1369–1378. [CrossRef]

41. Ye, H.M.; Chen, X.T.; Liu, P.; Wu, S.Y.; Jiang, Z.Y.; Xiong, B.J.; Xu, J. Preparation of Poly(butylene succinate) Crystals with Exceptionally High Melting Point and Crystallinity from Its Inclusion Complex. *Macromolecules* **2017**, *50*, 5425–5433. [CrossRef]

42. Howe, C.; Vasanthan, N.; MacClamrock, C.; Sanker, S.; Skin, I.D.; Simonsen, I.K.; Tonelli, A.E. Inclusion compound formed between poly(L-lactic acid) and urea. *Macromolecules* **1994**, *27*, 7433–7436. [CrossRef]

43. Howe, C.; Sankar, S.; Tonelli, A.E. ^{13}C n.m.r observation of poly (L-lactide) in the narrow channels of its inclusion compound with urea. *Polymer* **1993**, *34*, 2674–2676. [CrossRef]

44. Eaton, P.; Vasanthan, N.; Shin, I.D.; Tonelli, A.E. Formation and Characterization of Polypropylene-Urea Inclusion Compounds. *Macromolecules* **1996**, *29*, 2531–2536. [CrossRef]

45. Ravindran, P.; Vasanthan, N. Formation of Poly(3-hydroxybutyrate) (PHB) Inclusion Compound with Urea and Unusual Crystallization Behavior of Coalesced PHB. *Macromolecules* **2015**, *48*, 3080–3087. [CrossRef]

46. Okihara, T.; Tsuji, M.; Kawaguchi, A.; Katayama, K.I.; Tsuji, H.; Hyon, S.H.; Ikada, Y. Crystal structure of stereocomplex of poly(L-lactide) and poly(D-lactide). *J. Macromol. Sci. Part B Phys.* **1991**, *30*, 119–140. [CrossRef]

47. Kister, G.; Cassanas, G.; Vert, M.; Pauvert, B.; Térol, A. Vibrational analysis of poly(L-lactic acid). *J. Raman Spectrosc.* **1995**, *26*, 307–311. [CrossRef]

Article

pH-Responsive Host–Guest Complexation in Pillar[6]arene-Containing Polyelectrolyte Multilayer Films

Henning Nicolas [1], Bin Yuan [2], Jiangfei Xu [2], Xi Zhang [2] and Monika Schönhoff [1,*]

[1] University of Muenster, Institute of Physical Chemistry, Corrensstraße 28/30, 48149 Münster, Germany; h_nico01@uni-muenster.de

[2] Key Laboratory of Organic Optoelectronics & Molecular Engineering, Department of Chemistry, Tsinghua University, Beijing 100084, China; shibingriji@163.com (B.Y.); xujf@mail.tsinghua.edu.cn (J.X.); xi@mail.tsinghua.edu.cn (X.Z.)

* Correspondence: schonho@uni-muenster.de; Tel.: +49-251-83-23419

Received: 28 November 2017; Accepted: 13 December 2017; Published: 16 December 2017

Abstract: A water-soluble, anionic pillar[6]arene derivative (WP6) is applied as monomeric building block for the layer-by-layer self-assembly of thin polyelectrolyte multilayer films, and its pH-dependent host–guest properties are employed for the reversible binding and release of a methylviologen guest molecule. The alternating assembly of anionic WP6 and cationic diazo resin (DAR) is monitored in-situ by a dissipative quartz crystal microbalance (QCM-D). In solution, the formation of a stoichiometric inclusion complex of WP6 and cationic methylviologen (MV) as guest molecule is investigated by isothermal titration calorimetry and UV-vis spectroscopy, respectively, and attributed to electrostatic interactions as primary driving force of the host–guest complexation. Exposure of WP6-containing multilayers to MV solution reveals a significant decrease of the resonance frequency, confirming MV binding. Subsequent release is achieved by pH lowering, decreasing the host–guest interactions. The dissociation of the host–guest complex, release of the guest from the film, as well as full reversibility of the binding event are identified by QCM-D. In addition, UV-vis data quantify the surface coverage of the guest molecule in the film after loading and release, respectively. These findings establish the pH-responsiveness of WP6 as a novel external stimulus for the reversible guest molecule recognition in thin films.

Keywords: host-guest chemistry; layer-by-layer self-assembly; pillar[6]arene; pH-responsiveness

1. Introduction

The unique host–guest chemistry of macrocyclic molecules has experienced tremendous attention in the past years and promoted intensive scientific efforts in the fields of surface chemistry, polymer research, and material science. A variety of host–guest inclusion complexes, supramolecular polymers, and self-assembled nanostructures based on the stimuli-responsive properties of host molecules as the key recognition component were developed [1–4]. In continuation of a series of well-established hosts such as cyclodextrins [4–9] or cucurbiturils [10–15], pillar[n]arenes (P[n]) are rather novel representatives as they were only recently reported by Ogoshi et al. in 2008 [16–19]. The Lewis-acid catalyzed condensation of 1,4-dimethoxybenzene or hydroquinone units, respectively, yielded pillar[5]arene derivatives bridged by methylene groups in their 2- and 5-positions, therefore revealing a macrocyclic structure with a highly symmetrical shape. Since then, this novel type of host has attracted great attention on account of the ease of chemical modification, resulting from the reactive substituents located at the upper and lower rim, respectively. Due to the selective addressability of the reactive groups, a series of homologues (n = 5–15) bearing a large variety of properties are available to date, although the synthesis of higher homologues still suffers from low yields [20,21]. The π-electron-rich cavity combined with the opportunity of structure-tailoring motivated scientists to investigate the potential of P[*n*]s in the field of host–guest chemistry. The molecular recognition of P[*n* = 5, 6] towards a series of guest molecules has been reported, showing that cooperative host–guest-interactions such as electrostatics, hydrophobic interactions, charge transfer interactions, hydrogen bonding, and π-π-stacking are responsible for driving the complexation [22,23]. In addition, the introduction of stimuli-responsiveness on either the guest or the host enabled the fabrication of P[*n*]-guest based complexes, receptors, and drug delivery systems whose assembly could reversibly be switched by tuning either a single or multiple stimuli such as pH, light irradiation, or redox conditions [24–31].

Surface molecular imprinted polyelectrolyte multilayer films (SmiLbL films) are based on the layer-by-layer self-assembly of oppositely charged polyelectrolytes [32–35] and have opened an established class of thin and structured nanomaterial that bears imprinted binding sites for the reversible binding and release of small organic molecules [36]. In such polymer films, the binding sites are located comparatively close to a surface, therefore solving the problem of long diffusion pathways and buried binding sites known from macroscopic molecular imprinted materials [37,38]. Shi et al. prepared the first SmiLbL film in 2007 by employing a macromolecular complex first fabricated from both (Poly)acrylic acid and a porphyrin derivative, and then used it as building block for the LbL preparation of a polyelectrolyte multilayer film [39]. This initial SmiLbL approach was based on a pH-tunable electrostatic interaction to yield quantitative and reversible uptake and release [40,41]. Similar concepts aim at an optimization of selectivity by employing more specific interactions for molecular recognition, such as disulfide [42] or amide bonds [43] charge transfer interactions [44] and hydrogen bonds [45,46].

However, since the soft and flexible nature of such films allows swelling and structural deformation of the imprinted binding sites, their performance suffered in terms of reversibility and binding efficiency. For this reason, we recently applied a molecular host, cucurbit[8]uril (CB[8]), as artificial binding site within a polyelectrolyte multilayer film in order to take advantage of the intrinsic structural rigidity. It was found that, although swelling of the film was still observed, CB[8] was indeed unaffected from the film dynamics, i.e., binding site deformation was successfully avoided. Therefore, the host–guest chemistry of CB[8] could be implemented in polyelectrolyte multilayer films and be able to perform the reversible binding and release of a guest molecule controlled by either redox-conditions [47] or light irradiation [48]. A quite high binding efficiency over several binding and release cycles was observed [49]. Furthermore, the initial imprinting step, realized via pre-complexation during preparation, was no longer required as a result of the ready-made character of CB[8] as artificial binding site. Similarly, cyclodextrin hosts have more recently been employed in LbL films to achieve reversible, triggered guest uptake and release [50–52].

More recently, we have also employed a monomeric and water-soluble pillar[6]arene derivative (WP6) as negatively charged building block for the layer-by-layer self-assembly of a multilayer film, which bears anionic layers solely consisting of host molecules [53]. A proof of principle of a host–guest inclusion complex formation was provided by exposing the host-containing multilayer films to a viologen guest molecule. After complexation, release was achieved by washing off the guest with the help of a solvent change, thus demonstrating the full reversibility of the binding event in the film. In this way, the pre-association of the host with a polyelectrolyte, as it was previously required in order to incorporate CB[8] into the multilayer films, was no longer required and therefore the volume of inoperative polymer material in such novel WP6-containing films was drastically reduced.

In the present work, we have developed this approach further, as we introduce a new trigger to this optimized release system and report on the pH value as an external stimulus that enables precise control over the alternating binding and release processes of a viologen guest in WP6-containing multilayer films.

We engaged the pH sensitivity of the carboxylated rims of WP6 (see Figure 1) in order to manipulate the electrostatic interactions between host and guest, and therefore control the binding by de-/protonation, respectively. For this purpose, the pH-dependent host–guest complexation was first investigated in solution by employing UV-vis and isothermal titration calorimetry (ITC) experiments. Then, a dissipative quartz crystal microbalance (QCM-D) was employed in order to monitor in-situ the LbL self-assembly of multilayer films (DAR/WP6)$_{30}$ obtained from the alternating deposition of both the photosensitive polycation diazo resin (DAR) and WP6 (see Figure 1, left). The multilayer films were then crosslinked by photo-induced formation of covalent ester bonds between the decomposed azido groups of DAR and carboxylate groups of WP6 [54]. Subsequently, the pH-dependence of the host–guest complex formation is investigated by first loading the guest molecule into the film at basic conditions and then releasing it by protonation of the host molecule as a result of pH lowering; see schematic illustration in Figure 2. In addition, we prepared the same multilayer films on UV-permeable quartz slides and addressed the quantification of the guest molecule surface coverage (Γ_{MV}) by employing UV-vis spectroscopy.

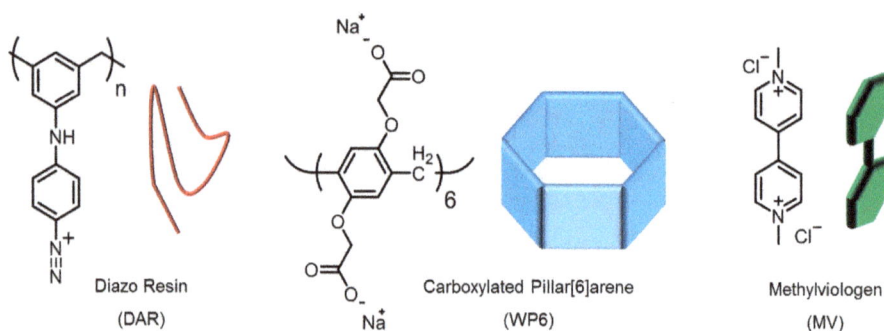

Figure 1. Chemical structures of the building blocks diazo resin polymer (DAR) and carboxylated pillar[6]arene (WP6), as well as the guest molecule methylviologen dichloride (MV).

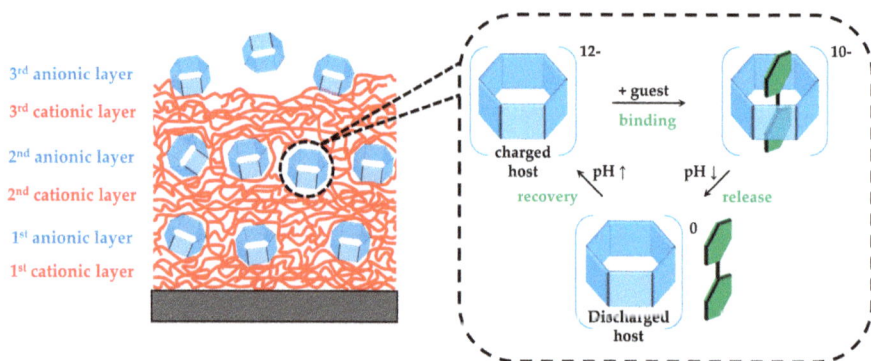

Figure 2. Sketch of the multilayers formed of cationic diazo resin polymer (DAR) and anionic carboxylated pillar[6]arene (WP6), indicating the mechanism of binding and release of the guest molecule methylviologen dichloride (MV). Net charges of the complex are ideal values, neglecting the charge loss upon photo-crosslinking.

2. Materials and Methods

2.1. Materials

Sodium acetate, sodium dihydrogen phosphate, and disodium hydrogen phosphate were obtained from Merck (Kenilworth, NJ, USA). Acetic acid was purchased from VWR International (Radnor, PA,

USA). Methylviologen dichloride hydrate (98%) and diluted solutions of NH$_3$ (25%) and H$_2$O$_2$ (30%) were provided by Sigma Aldrich. DAR and WP6 (see structures in Figure 1) were synthesized as described before [53]. All compounds were used as received.

2.2. Solutions

All compounds were prepared in buffer solutions of different pH values. Phosphate buffer (pH = 7.8) and acetate buffer (pH = 4.8 or 3.6) were prepared in ultrapure water (Merck Millipore, Darmstadt, Germany, resistivity > 18 MΩ·cm) with a concentration of both 10 mmol L^{-1} and 50 mmol L^{-1}, respectively. Solutions of MV (1 mmol L^{-1}) and WP6 (0.1 mmol L^{-1}) were prepared in both phosphate buffer (50 mmol L^{-1}) and acetate buffer (50 mmol L^{-1}), and used for ITC experiments. For multilayer film preparation, DAR (0.1 mmol L^{-1}) and WP6 (0.1 mmol L^{-1}) solutions, respectively, were prepared in 10 mM phosphate buffer (pH 7.8). Solutions for UV-vis experiments were diluted from such stock solutions.

2.3. Characterization of the MV–WP6 Complex

The thermodynamic parameters of the host–guest complexation between MV (1 mmol L^{-1}) and WP6 (0.1 mmol L^{-1}) were investigated at different pH values (pH = 7.8 or 4.8) by isothermal titration calorimetry. The solutions were degassed for 15 min and the experiments were then performed on a Nano ITC (TA Instruments, New Castle, DE, USA). MV solution (250 µL) was titrated into WP6 solution (953 µL). The successively titrated volume was 10 µL and injected in intervals of 5 min. The obtained data was evaluated by using the software "NanoAnalyze" (Version 3.3.0, TA Instruments, New Castle, DE, USA, 2005/2014).

The host–guest complexation was further investigated by UV-vis spectroscopy (UV-2550 photometer, Shimadzu, Kyoto, Japan). The absorbance spectrum of MV (0.05 mmol L^{-1}, in 10 mM phosphate buffer, pH 7.8) was recorded by using pure solvent as reference. Then, a stoichiometric amount of WP6 was added to the MV solution in order to fabricate a 1:1 inclusion complex. The absorbance spectrum of the mixture was taken and referenced to an equimolar solution of WP6. In this way, solely the contribution of MV after complexation was recorded.

2.4. Substrate Cleaning

Gold-coated quartz sensors (QSX-301, Q-sense, Biolin Scientific, Gothenburg, Sweden) with a resonance frequency of f_0 = 4.95 MHz were used to study both the multilayer film formation and the subsequent guest molecule binding and release processes. Before use, the sensors were heated in RCA solution (NH$_3$/H$_2$O$_2$/H$_2$O = 1:1:5) for 20 min at 70 °C. UV-permeable quartz slides (Hellma Analytics, Mülheim, Germany) were employed as substrates for the preparation of dip-coated multilayer films and previously cleaned using the same procedure.

2.5. In Situ Preparation of Multilayer Films and Monitoring by QCM-D

Layer-by-layer self-assembled multilayer films (DAR/WP6)$_{30}$/DAR were prepared by alternating deposition of the building blocks DAR and WP6. The preparation was performed in flow cells allowing in situ dissipative quartz crystal microbalance (QCM-D) experiments in a Q-Sense E4 apparatus (Biolin Scientific, Gothenburg, Sweden). All experimental work involving the use of DAR was carried out in a dark lab in order to avoid the light-induced decomposition of the photosensitive diazo groups. Tubes and flow cell were initially filled with phosphate buffer (pH 7.8) and equilibrated for at least 20 min. The flow rate was 0.1 mL min^{-1} and kept constant throughout all QCM-D experiments. The resonance frequency f_0 as well as its overtones were then determined several times until a stable baseline (Δf < ± 1 Hz within a period of 10 min) was detected. A programmable autosampler was used to control the alternating flow of solutions and phosphate buffer. DAR solution overflows the sensor for 20 min to form an initial cationic layer. Its completion was indicated by a plateau of the deviation from the baseline, Δf. Subsequently, phosphate buffer was used for 20 min to remove loosely attached polyelectrolytes from the surface. Then, the WP6 solution was flown over the sensor for 10 min in order to adsorb an anionic layer. Finally, the film was overflowed by phosphate buffer again for 10 min. By following this cyclic procedure, (DAR/WP6)$_{30}$/DAR was fabricated. After preparation, the sensor was irradiated by visible light (λ = 440 nm) for 5 min in order to photo-crosslink the multilayer film.

2.6. Dip-Coating of Multilayer Films

In order to enable UV-vis investigations of the multilayer films, (DAR/ WP6)$_{30}$/DAR films were prepared via dip-coating by a programmable dipping robot (DR-1, Riegler and Kirstein, Berlin, Germany). Cleaned quartz slides were alternatingly immersed into the respective solutions by the robot, thereby using the same sequence as well as identical adsorption times as for the QCM-D experiments. After each layer deposition, the samples were washed in three individual buffer solutions for 2 min, respectively. The multilayer films were then characterized by UV-vis before and after photo-crosslinking. As reference, an uncoated quartz slide was used.

2.7. Guest Molecule Binding and Release

After photo-crosslinking, the coated QCM-D sensor was remounted and phosphate buffer (pH 7.8) was exposed to the sample for 2 h in order to equilibrate the multilayer film with respect to the content of hydration water. Again, resonance frequency and overtones were determined in terms of a stable baseline as described above. The multilayer films were then exposed to a solution of MV (1 mmol L^{-1}, pH 7.8) for 3 h. This was followed by flowing acetate buffer (pH 3.6) over the sensor for 1 h in order to induce the guest molecule release. Finally, the acidic buffer was replaced by the phosphate buffer (pH 7.8) to recreate basic conditions.

3. Results

3.1. pH-Tunable Host–Guest Chemistry in Aqueous Solution

The host–guest complexation of WP6 and MV is observed by UV-vis spectroscopy. Spectra of free MV (green line in Figure 3) and MV complexed with WP6 (blue line) show a distinct difference, which imply the interaction of the guest with the host. The stoichiometric addition of WP6 to an MV solution significantly reduces the intensity of the absorbance band of the guest at 257 nm and therefore indicates intermolecular interactions between both components. Also, this absorbance maximum is bathochromically shifted by about 6 nm. Furthermore, an additional absorbance band can be observed at around 310 nm. Since WP6 alone bears an absorbance maximum at 292 nm (see Supplementary Figure S1 in the supporting information), the new band might be the result of changed absorbance properties of the host due to the intermolecular interactions with the guest molecule. Additionally, a very weak charge transfer band is detected in the regime between 400 and 550 nm [22]. The latter was even more clearly identified by measuring the complex solution at an increased concentration (see Supplementary Figure S2), thus a charge transfer interaction between the electron-rich cavity and the electron-poor guest molecule is indicated.

We further employed ITC experiments in order to identify the thermodynamic properties of the host–guest complexation. MV was first titrated into a solution of WP6 at pH 7.8 and a binding constant of k_a = (1.53 $\pm$ 0.39) 10^6 M^{-1} was extracted from a single site binding model (see Table 1 and Supplementary Figure S3 for raw data). The determined stoichiometry of n = 0.92 $\pm$ 0.07 further confirmed that a 1:1 complex was formed, as expected [22]. We then repeated the ITC experiments at pH 4.8 in order to figure out if acidic conditions are able to prevent the host–guest complexation due to the protonation of WP6. ITC data still indicate the formation of a 1:1 complex, however, the binding constant was lowered by about one order of magnitude (see Table 1 and Supplementary Figure S4). We therefore concluded that WP6 is not fully protonated at pH 4.8, and still allows the formation of the inclusion complex due to electrostatic interactions. A reduced number of dissociated carboxyl groups on WP6 at pH 4.8 might also lead to a lower solvent contribution to the entropy and act as the reason for the significantly reduced binding entropy, again indicating a weakened complexation. We intended to perform the experiments at an even lower pH of 3.6, however, WP6 was not soluble at all at this conditions, although the concentration required for ITC is very low. From this observation, we concluded that the charge of the host is sufficiently reduced by such a pH of 3.6 [30]. Thus, the experiments in solutions established pH-dependent conditions suitable for uptake and release of the guest from WP6.

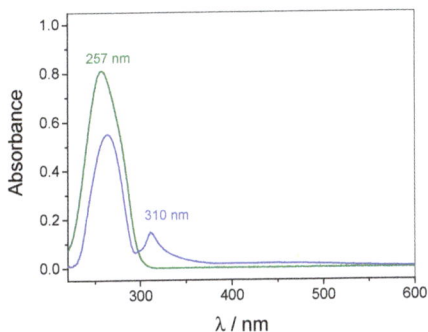

Figure 3. UV-vis spectra of MV solution (50 μmol L^{-1}) against water as reference (green line) and MV solution containing an equimolar ratio of WP6 against WP6 solution as reference (blue line).

Table 1. Thermodynamic parameters obtained from ITC (isothermal titration calorimetry) data fitted by a single site binding model for the titration of MV solution (1 mM) into WP6 solution (0.1 mM) at different pH values.

pH	k_a/M^{-1}	n	$\Delta H/kJ\ mol^{-1}$	$\Delta S/J\ mol^{-1}\ K^{-1}$
7.8	$(1.53 \pm 0.39) \times 10^6$	0.92 ± 0.07	-18.2 ± 1.0	57.2 ± 1.6
4.8	$(1.64 \pm 0.42) \times 10^5$	1.01 ± 0.02	-17.5 ± 1.0	40.9 ± 5.4

3.2. In Situ QCM-D Study of Multilayer Formation

A typical time-dependent change of the resonance frequency, Δf, was obtained for the alternating layer-by-layer deposition of DAR and WP6 (see Figure 4). Please note that for clarity, only the fifth overtone of the resonance frequency is shown, the entire ensemble of overtones can be seen in Supplementary Figure S5. For the deposition of each layer, a decrease of the resonance frequency was monitored, thus the formation of a multilayer film (DAR/WP6)$_{30}$/DAR was demonstrated.

From the insert in Figure 4, it can further be seen that the adsorption of DAR causes a significantly stronger decrease of Δf in comparison to WP6. In order to analyze the growth behavior of the multilayer film, the frequency changes detected per individual layer, $-\Delta(\Delta f)$, were evaluated and given in dependence of the bilayer number n in Figure 5. It is obvious that the mass increment per DAR layer can be separated into three individual regimes (see red squares in Figure 5). Initially, the DAR adsorption causes Δf changes per individual DAR layer $-\Delta(\Delta f_{DAR})$ of 8 Hz to 12 Hz ($n = 1$–10). Then, continuously thickened layers are monitored (13 Hz $\leq -\Delta(\Delta f_{DAR}) \leq$ 59 Hz for $n = 10$–25). Finally, clearly decreasing $-\Delta(\Delta f)$ changes are indicated at the end of the preparation (58 Hz $\geq -\Delta(\Delta f_{DAR}) \geq$ 25 Hz).

Polymers 2017, 9, 719

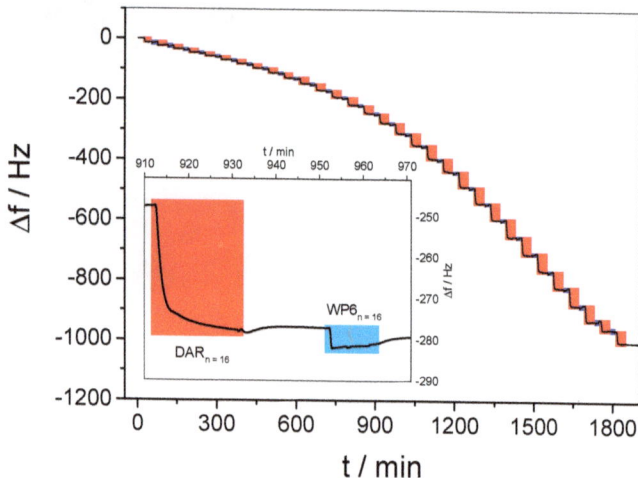

Figure 4. Development of Δf for the Layer-by-Layer (LbL) preparation of a multilayer film (DAR/WP6)$_{30}$/DAR by alternating deposition of DAR (highlighted in red) and WP6 (highlighted in blue), respectively. A close-up of the deposition of a typical bilayer is shown as insert.

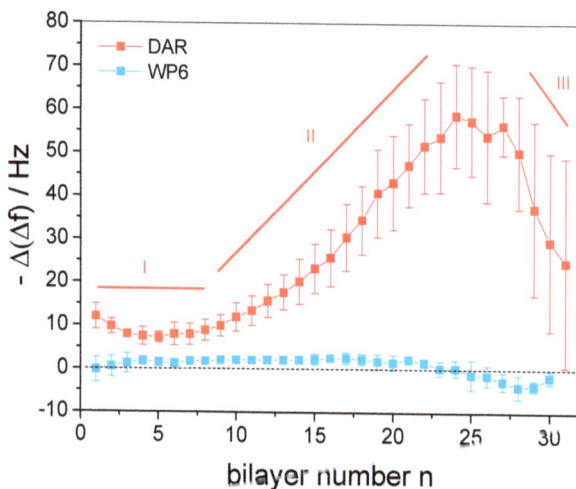

Figure 5. Changes of the resonance frequency for the individual adsorption of a DAR ($-\Delta(\Delta f_{DAR})$, red squares) and a WP6 ($-\Delta(\Delta f_{WP6})$, blue squares) layer during preparation of multilayer films (DAR/WP6)$_{30}$/DAR.

As a potential reason for the different growth regimes, the basic experimental conditions must be taken into account. In order to keep the WP6 permanently deprotonated, all employed sample solutions during film preparation were adjusted to pH 7.8, although DAR is sensitive to basic conditions [55]. To identify the potential decomposition of DAR in basic solution, a reference experiment monitored the time-dependent development of the DAR absorbance (see Supplementary Figure S6). A continuous decrease of the main DAR absorption band was detected, indicating the slow decomposition of the azido functions and giving a plausible explanation for the different growth regimes. This decomposition yields the same spectral changes as the crosslinking, compare

Supplementary Figures S6 and S7. Since the cationic diazo groups are essential for the electrostatically driven layer-by-layer assembly of the multilayer film, their slow decomposition causes the need of increasing amounts of DAR in order to accomplish the charge overcompensation of the previous WP6 layer. The mass increment per layer is supported by a softening of the film, as shown by the strongly increasing dissipation values in Supplementary Figure S8. Such additional mass, increasing with layer number, might be represented by $-\Delta(\Delta f_{DAR})$ of the second growth regime. After several hours, the remaining charge density of DAR might be too weak to overcompensate WP6 and therefore leads to thinner DAR layers, resulting in the third growth regime.

In contrast, $-\Delta(\Delta f)$ values observed for the layer formation of WP6 do not suggest different growth regimes (see blue squares in Figure 5). The $-\Delta(\Delta f_{WP6})$ is in the range from -4 Hz to $+2$ Hz for all n, thus indicating a similar content of the host for each anionic layer. Such small values and the occurrence of negative values might, on the one hand, be due to changes of layer hydration upon WP6 adsorption, as the surface coverage of hydration water in the films is included in the detected frequency change. On the other hand, one would expect a much stronger frequency decrease as a consequence of the deposition of a full WP6 monolayer. The obtained data therefore indicate that the WP6 molecules are not deposited as a closely packed monolayer, but are rather located with large intermolecular distances. This might be due to the high charge of WP6, requiring only a low mass coverage for charge overcompensation. On the other hand, strong mutual electrostatic repulsion of WP6 molecules might play a role since it was recently shown that strongly charged gold nanoparticles yield only sub-monolayer coverage in LbL assembly [56].

In addition, the low WP6 amount questions the role of electrostatic interactions regarding the LbL assembly of the multilayer film. A control experiment shed some light into this uncertainty, wherein DAR was initially adsorbed on a QCM sensor, resulting in a typical Δf_{DAR} value of -12 Hz (see Supplementary Figure S9). After washing, DAR was again offered to the surface in order to examine whether the omitted deposition of WP6 influences the multilayer growth. Indeed, only a very slight mass deposition was detected after such procedure was performed twenty times in total (compare Supplementary Figure S9). This reasons the conclusion that the low amounts of deposited WP6 observed during film preparation is fundamentally essential for a regular buildup, i.e., the electrostatic interactions between both building blocks were identified as the driving force of the LbL assembly.

3.3. In Situ Investigation of Guest Binding and Release

After photo-crosslinking the multilayer film, the film was exposed to a basic solution of MV (pH 7.8) in order to achieve binding of the guest into the WP6 layers. In QCM-D, a significant decrease of the resonance frequency over a period of 3 h was observed (see Figure 6, highlighted in green), indicating that the guest molecule is incorporated into the multilayer film. The main mass increment occurs within only a few minutes, while a subsequent, lower mass increment rate was detected over several hours, until a plateau was reached, thus indicating the complete saturation of the film.

We note here that the carboxylic groups of WP6 are partly turned into ester bonds due to the crosslinking with DAR, however, as the uptake and release experiments show, a sufficient fraction of carboxylic groups remains unreacted and can thus act as a negatively charged site enabling MV uptake.

In order to release MV, the loaded multilayer film was exposed to acetate buffer (pH 3.6) for 1 h to reduce the attractive electrostatic interaction between host and guest. As a result, the corresponding response of the Δf development reveals a rapid increase that clearly confirms the release of mass from the multilayer film (see Figure 6, highlighted in yellow). However, the second plateau clearly exceeds $\Delta f = 0$, therefore indicating that the released amount of mass was significantly higher than the previously incorporated mass. Since QCM-D detects mass changes of all components involved, the occurrence of additional molecular processes besides guest molecule binding and release must be considered as a contribution to the enhanced mass release. The identification of such processes will be discussed later.

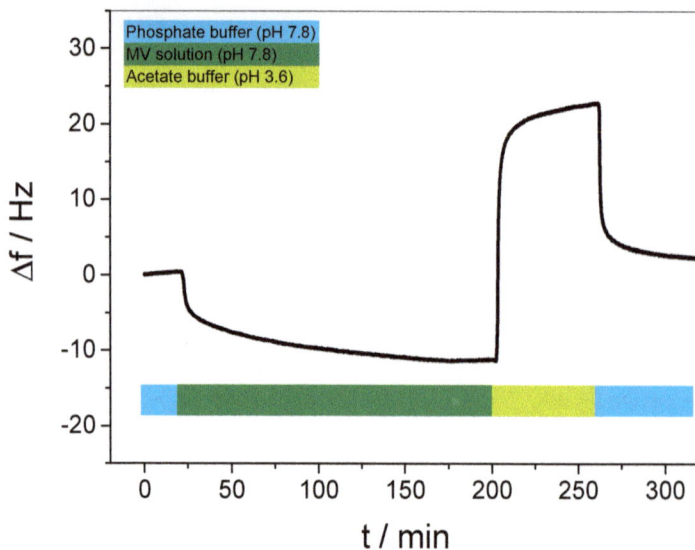

Figure 6. Resonance frequency difference, Δf, upon exposure of a $(DAR/WP6)_{30}/DAR$ multilayer film to MV solution (pH 7.8, green) and acetic buffer (pH 3.6) as the releasing agent (yellow). The blue bars represent the exposure of the film to phosphate buffer (pH 7.8).

The subsequent reestablishment of the basic conditions by increasing the pH to 7.8 sheds some light into such additional molecular processes. In Figure 6, highlighted in blue, a clear decrease of Δf indicated an increase of mass of the multilayer film, which ends in a third plateau at $\Delta f \approx +2$ Hz, indicating that the multilayer film has returned to the original state before guest binding, thus a complete removal of MV can be concluded. For verification, the pH-dependent binding and release was performed several times by repeating the cycle (see Supplementary Figure S10), revealing a similar Δf development for each cycle. Binding and release data, extracted from the plateau values and averaged over four individual samples, is shown in Figure 7. Please note that the given values for the release (yellow bars in Figure 7) were calculated by subtracting the plateau value of guest saturation from the plateau value after reestablishing the basic conditions (blue highlighted in Figure 6). The data clearly demonstrate that the binding and release of MV in WP6-containing multilayer films is pH-responsive and full reversible.

It was already mentioned that the loss of mass resulting from the acidic treatment (Figure 6, highlighted in yellow) indicates the simultaneous occurrence of both the release of the guest and other molecular processes in the multilayer film. In order to identify such secondary processes, we employed acidic conditions to a multilayer film after the guest was fully released (see Figure 8).

It is obvious that the alternating pH already causes strong and fully reversible changes in the resonance frequency. This is most likely due to the fact that the pH-induced protonation/deprotonation of WP6 in the film results in a disturbance of the charge equilibrium. A film with a net charge is self-repelling, requiring a larger mass of hydration water, leading to swelling. Such reversible, pH-induced LbL film swelling effects have been observed earlier in a simple pH-controlled electrostatic release system [40], and pH-controlled changes of the charge equilibrium were even used to control ion uptake in multilayer films [57]. In the present case, it is interesting to note that swelling occurs in basic, and de-swelling in acidic environment. Thus, it can be concluded that the as-prepared multilayer bears a negative excess charge, which is reduced upon protonation.

In order to identify the role of additional contributions, a second control experiment with an alternating buffer change was performed on a clean, uncoated QCM-D sensor (see Supplementary

Figure S11). Although even here a pH-dependent increase and decrease was found, the order of magnitude of the Δf change (ca. ±2 Hz) was much smaller than observed in the same experiment performed on the multilayer film. Such minor effect might be caused by slight differences between both liquids in density and viscosity, respectively [58]. Summarizing all these contributions, the significant Δf increase observed in Figure 6 displays the simultaneous occurrence of three different processes, i.e., the release of the guest, de-swelling of the multilayer film, and a negligible solvent effect.

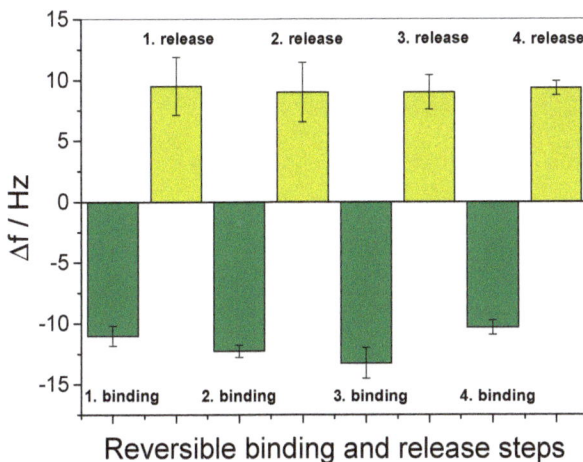

Figure 7. Dissipative quartz crystal microbalance (QCM-D) experiments of binding and release of MV, averaged over four individual (DAR/WP6)$_{30}$/DAR films (green bar: Δf change upon guest molecule binding; yellow bar: Δf change detected after guest molecule release by pH lowering (7.8 to 3.6) and subsequent re-equilibration of the multilayer film back to pH 7.8. Yellow bars thus compensate for hydration changes and include guest molecule mass changes only.

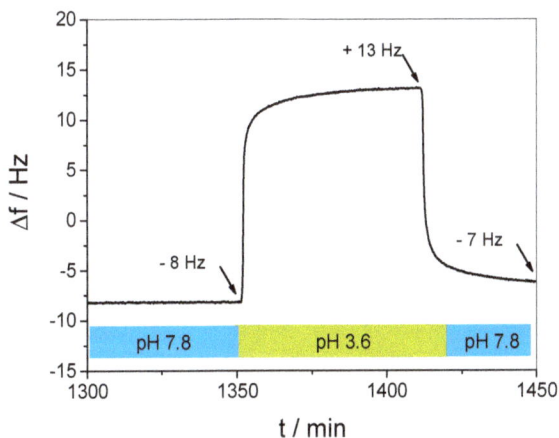

Figure 8. Δf change of a (DAR/WP6)$_{30}$/DAR multilayer film resulting from alternatingly changing the pH from 7.8 (blue) to 3.6 (yellow) without presence of a guest molecule.

Finally, we decided to deeper investigate the role of electrostatic interactions concerning the binding and release of the guest molecule. Therefore, we adjusted the unloaded multilayer film to acidic conditions to protonate WP6 (see Figure 9, highlighted in yellow), which yields the expected de-swelling. Subsequently, the film was exposed to MV solution (pH 3.6) in order to attempt binding to protonated WP6. As seen in Figure 9 (highlighted in orange), no significant change of Δf was monitored. This observation clearly demonstrates that the guest can only be bound with WP6 in the deprotonated state, therefore proving that the electrostatic interactions between host and guest is the main driving force for the host–guest complexation. In addition, the pH-responsiveness of the host–guest complexation can be concluded.

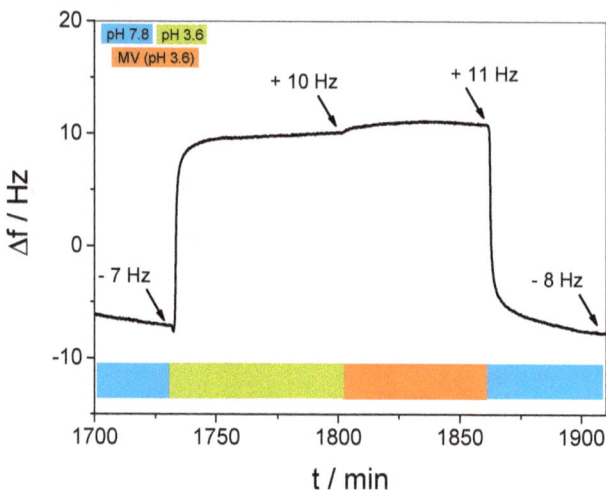

Figure 9. Δf change of the control experiment employing an acidic guest molecule solution (orange) to the multilayer films bearing WP6 in a fully protonated state (yellow).

3.4. Quantification of Guest Surface Coverage by UV-Vis Spectra

The building blocks DAR and WP6 as well as the guest comprise a strong electronic absorbance in the UV-vis regime, respectively (see Supplementary Figures S1 and S7). Thus, we prepared dip-coated multilayer films (DAR/ WP6)$_{30}$/DAR on quartz slides in order to determine the surface coverage of MV in the films by UV-vis spectroscopy. After preparation, the film spectrum on the quartz slides before crosslinking (see Supplementary Figure S12) agrees with that of DAR in solution (compare to Supplementary Figure S7), as expected, and therefore indicates a successful LbL assembly, as already described in our previous work [53].

The absorbance of DAR as the cationic building block of the film is clearly visible, having its maximum at 381 nm before photo-crosslinking (see Supplementary Figure S12, black line). In contrast, the absorbance of WP6, expected at ca. 290 nm (see Supplementary Figure S1), can only be weakly identified in Supplementary Figure S12. This observation corresponds to the formation of rather thin WP6 layers in the film, as indicated by the QCM-D data described above. The subsequent light irradiation decreases the DAR absorption band and therefore indicates a successful photo-crosslinking.

The WP6-containing films were saturated with MV in order to detect the surface coverage (Γ_{MV}) of the incorporated guest molecule. Γ_{MV} was determined by measuring the absorption spectra of MV-saturated films referenced to an identical, but unloaded multilayer film (see Figure 10). Assuming that the incorporation of MV into the film does not influence the absorption properties of the components, the significant absorbance at 265 nm can be attributed to the incorporated MV and thus used in order to quantify the surface coverage Γ_{MV} with the help of a calibration line taken from MV solutions of different concentrations.

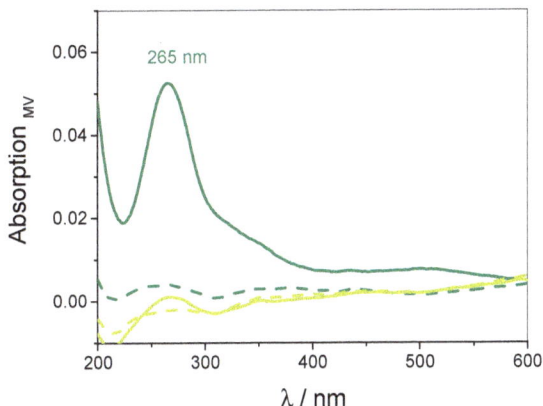

Figure 10. UV-vis spectra of MV after binding into a multilayer film at pH 7.8 (green continuous line) and subsequent release by pH 3.6 (green dotted line). The yellow spectra show analogue binding (continuous line) and release (dotted line) experiments performed at pH 3.6.

Following this procedure over several samples, each undergoing ten pH-stimulated binding and release cycles, averaged data of Γ_{MV} was obtained and is given in Figure 11 (green squares). Γ_{MV} within a saturated film amounts to about 0.1 nmol cm^{-2} per layer. Please note that the decrease of the MV absorbance due to the complexation with WP6 is not regarded in the calculation of the surface coverage (compare Figure 3), therefore the corrected values are supposed to be slightly higher as indicated by Figure 11.

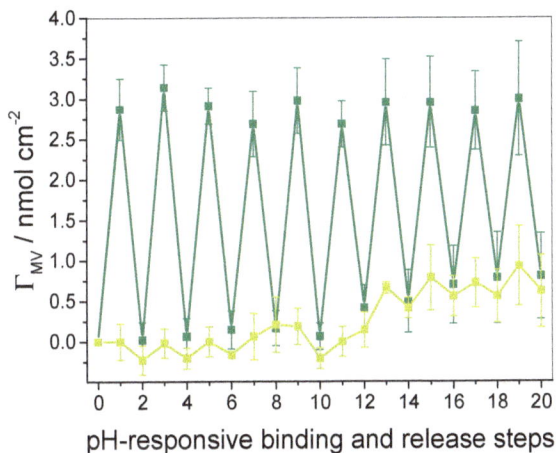

Figure 11. MV surface coverage per total film during ten binding and release cycles of MV into WP6-containing multilayer films. Γ_{MV} is given in dependence of the pH (green squares: binding at pH 7.8, release at pH 3.6; yellow squares: binding and release conducted at pH 3.6). Data are averaged over several samples.

Within cycles one to five, identical amounts of MV were first bound and subsequently released from the film, thus reversibility as well as pH responsiveness of the binding event is again demonstrated. It is noteworthy that pH lowering allows the complete removal of the guest from the film without leaving any residuals. In conclusion, the WP6-containing multilayer films do not provide other sufficiently strong attractive interactions towards the guest, except of the desired host–guest interactions. This fact demonstrates a rare and outstanding property of these multilayer films compared to previous approaches, which were found to suffer from a decomposition of the binding sites, and moderate to enhanced unspecific binding of the guest due to secondary interactions [42,43,49,59].

The essential role of electrostatics was proven by a control experiment, wherein a series of identical films was exposed to an acidic MV solution. We once more conducted the binding and release procedure, but observed no incorporation of the guest into the films, as expected (see Figure 11, yellow squares). Therefore, such control experiments clearly identify the driving force of the host–guest complexation in WP6-containing multilayer films, in compliance with the data obtained from QCM-D experiments. However, for both experimental series, we noted a slight enrichment of the guest in the films after the fifth cycle (see green and yellow squares for steps 10 to 20 in Figure 11, respectively). This might be due to the multiple swelling and de-swelling events within the film as a consequence of the alternating pH changes, which might have caused structural rearrangements of the multilayer film and therefore lead to a low amount of irreversibly incorporated guest molecule.

From the surface coverage of the guest molecule, conclusions can be drawn about the surface coverage of WP6: Assuming that the stoichiometry of the formed complex in the film is identical to that in solution, i.e., all WP6 cavities are accessible to MV binding, the surface coverage of WP6 is identical to that of MV, meaning that the average host coverage is 0.1 nmol cm^{-2} per bilayer. Taking our previous approaches employing cucurbituril into account [48,49,59,60], the CB[8]-based layer systems revealed a host coverage of 0.5 nmol cm^{-2} per bilayer, while classical SmiLbL films also yielded a binding site coverage of 0.1 nmol cm^{-2} per bilayer [40]. A potential explanation for the reduced host coverage of WP6 as compared to CB[8] might be the fact that WP6 was deposited as single molecule and without pre-association to a polyelectrolyte in order to reduce the content of non-binding and "inactive" material. Thus, the lower surface coverage of WP6 can probably be reasoned with the formation of rather thin anionic layers, as it was detected by QCM-D, due to the monomeric character of the building block and its high negative charge in the deprotonated state, which can yield low coverages due to electrostatic repulsion of neighboring host molecules.

4. Conclusions

In conclusion, we have proceeded in developing WP6-based polyelectrolyte multilayer release systems, and established the pH value as novel external stimulus for the control of the reversible complexation of a guest molecule. For this purpose, carboxylated WP6 was employed as negatively charged monomeric building block in the LbL assembly and then used as artificial binding site in the film. In this way, no second polymeric building block is required and thus the fraction of inactive polyelectrolyte material was reduced in comparison to previous host-containing films. The novel pH stimulus is based on the pH-dependent de-/protonation of the carboxylated rims of WP6. Since electrostatic attraction was determined as the main driving force of the host–guest complexation, the bound guest molecule could be easily released by a simple pH decrease, which leads to the protonated state of the WP6 and reduces the attractive host–guest interaction. Full reversibility of the pH-responsive binding and release over at least ten cycles, accompanied by a pH-dependent de/-swelling of the multilayer film, was clearly demonstrated. It is thus shown that the carboxylate groups can simultaneously accomplish three different function in the LbL release systems, as they (i) serve as charged groups to provide LbL assembly and (ii) partially react with DAR to accomplish crosslinking, and finally, (iii), the remaining fraction of unreacted carboxylic groups is sufficient to control guest binding via pH changes. Thus, we have therefore contributed to the promising field of host-containing polyelectrolyte multilayer films and future reports on further pillar[*n*]arene-containing multilayer films are anticipated due to the extensive modifiability of the host structure.

Supplementary Materials: The following are available online at www.mdpi.com/2073-4360/9/12/719/s1, Figure S1: Absorbance spectra of carboxylated pillar[6]arene (WP6, 50 µM in 10 mM phosphate buffer); Figure S2: Absorbance spectra of the MV@WP6 complex (c = 250 µmol L^{-1}) in 10 mM phosphate buffer (pH 7.8) showing

Polymers **2017**, *9*, 719

the charge transfer band in the regime between 380 and 640 nm; Figure S3: Heating rate (upper plot) and binding isotherm (lower plot) of an isothermal titration calorimetry ITC experiment wherein a methylviologen dichloride (MV) solution (1 mmol L^{-1} in phosphate buffer, pH 7.8) was titrated into a WP6 solution (0.1 mmol L^{-1} in phosphate buffer, pH 7.8); Figure S4: Heating rate (upper plot) and binding isotherm (lower plot) of an ITC experiment wherein an MV solution (1 mmol L^{-1} in acetate buffer, pH 4.8) was titrated into a WP6 solution (0.1 mmol L^{-1} in acetate buffer, pH 4.8); Figure S5: Time-dependent change of the resonance frequency, Δf, for the nth overtone (n = 3, 5, 7, 9, 11) of the Layer-by-Layer assembly of a multilayer film (DAR/WP6)30/DAR; Figure S6: Absorbance spectra of (left plot, continuous line) a fresh DAR solution (50 µmol L^{-1}, in 10 mM phosphate buffer, pH 7.8) before preparation and after 48h (left plot, dotted line). The chemical decomposition of DAR was followed by using the steady wavelength mode. Therefore, the fresh DAR solution was used in order to monitor the absorbance band of the diazo groups at 372 nm for 45h (right plot), giving the decrease in the absorbance value at a fixed wavelength with time; Figure S7: Absorbance spectra of DAR (50 µmol L^{-1}) in 10 mM phosphate buffer (pH 7.8) after preparation (continuous line) and after 5 min VIS light irradiation (dotted line); Figure S8: Simultaneously detected change of dissipation (ΔD, red curve) during the LbL assembly of the multilayer film (DAR/WP6)30/DAR; Figure S9: Dissipative quartz crystal microbalance (QCM-D) control experiment: consecutive adsorption of 20 individual DAR layers without participation of WP6 as anionic building block; Figure S10: Four consecutive binding and release cycles of MV into a WP6-containing multilayer film (DAR/WP6)30/DAR. The binding of MV is represented by the green background, while the release performed by acetic buffer (pH 3.6) is highlighted in yellow. Blue displays the use of phosphate buffer (pH 7.8); Figure S11: Alternating change between phosphate buffer (pH 7.8) and acetic buffer (pH 3.8) on an uncoated QCM-D sensor. The procedure was repeated several times; Figure S12: UV-vis absorbance spectra of a multilayer film (DAR/WP6)30/DAR before (black line) and after photo-crosslinking (grey line).

Acknowledgments: We are grateful to the Deutsche Forschungsgemeinschaft (DFG) and the National Natural Science Foundation of China (NFSC) for funding this work as a part of the Sino-German collaborative research center TRR61 (project B11).

Author Contributions: The compounds DAR and WP6 were synthesized and characterized by Bin Yuan and Jangfei Xu. Both were advised by Xi Zhang. Henning Nicolas performed the measurements on LbL assembly of the multilayer film as well as on the subsequent characterization of reversible guest molecule binding and release. He was supervised by Monika Schönhoff, including for the preparation of the manuscript. Monika Schönhoff and Xi Zhang designed and planned the project within the frame of the Sino-German collaboration "TRR 61".

Conflicts of Interest: The authors declare no conflict of interest.

References

1. Masson, E.; Ling, X.; Joseph, R.; Kyeremeh-Mensah, L.; Lu, X. Cucurbituril chemistry: A tale of supramolecular success. *RSC Adv.* **2012**, *2*, 1213–1247. [CrossRef]
2. Barrow, S.J.; Kasera, S.; Rowland, M.J.; Del Barrio, J.; Scherman, O.A. Cucurbituril-based molecular recognition. *Chem. Rev.* **2015**, *115*, 12320–12406. [CrossRef] [PubMed]
3. Yu, G.; Jie, K.; Huang, F. Supramolecular amphiphiles based on host-guest molecular recognition motifs. *Chem. Rev.* **2015**, *115*, 7240–7303. [CrossRef] [PubMed]
4. Del Valle, E.M.M. Cyclodextrins and their uses: A review. *Process Biochem.* **2004**, *39*, 1033–1046. [CrossRef]
5. Wang, H.; Shao, N.M.; Qiao, S.N.; Cheng, Y.Y. Host-guest chemistry of dendrimer-cyclodextrin conjugates: Selective encapsulations of guests within dendrimer or cyclodextrin cavities revealed by NOE NMR techniques. *J. Phys. Chem. B* **2012**, *116*, 11217–11224. [CrossRef] [PubMed]
6. Zhao, J.; Zhang, Y.M.; Sun, H.L.; Chang, X.Y.; Liu, Y. Multistimuli-responsive supramolecular assembly of cucurbituril/cyclodextrin pairs with an azobenzene-containing bispyridinium guest. *Chemistry* **2014**, *20*, 15108–15115. [CrossRef] [PubMed]
7. Isenbügel, K.; Gehrke, Y.; Ritter, H. Photo-switchable behavior of azobenzene-dye-modified silica nanoparticles and their assembly with cyclodextrin derivatives. *Macromol. Chem. Phys.* **2012**, *213*, 227–233. [CrossRef]
8. Karakhanov, E.A.; Maximov, A.L. Molecular imprinting technique for the design of cyclodextrin based materials and their application in catalysis. *Curr. Org. Chem.* **2010**, *14*, 1284–1295. [CrossRef]
9. Ravoo, B.J.; Darcy, R. Cyclodextrin bilayer vesicles. *Angew. Chem. Int. Ed.* **2000**, *39*, 4324–4326. [CrossRef]
10. Behrend, R.; Meyer, E.; Rusche, F.I. Ueber condensationsproducte aus glycoluril und formaldehyd. *Eur. J. Org. Chem.* **1905**, *339*, 1–37. [CrossRef]
11. Freeman, W.A.; Mock, W.L.; Shih, N.Y. Cucurbituril. *J. Am. Chem. Soc.* **1981**, *103*, 7367–7368. [CrossRef]
12. Mock, W.L.; Shih, N.Y. Structure and selectivity in host-guest complexes of cucurbituril. *J. Org. Chem.* **1986**, *51*, 4440–4446. [CrossRef]

13. Isaacs, L. Stimuli responsive systems constructed using cucurbit[*N*]uril-type molecular containers. *Acc. Chem. Res.* **2014**, *47*, 2052–2062. [CrossRef] [PubMed]

14. Appel, E.A.; Biedermann, F.; Rauwald, U.; Jones, S.T.; Zayed, J.M.; Scherman, O.A. Supramolecular cross-linked networks via host-guest complexation with cucurbit[8]uril. *J. Am. Chem. Soc.* **2010**, *132*, 14251–14260. [CrossRef] [PubMed]

15. Rauwald, U.; Scherman, O.A. Supramolecular block copolymers with cucurbit[8]uril in water. *Angew. Chem. Int. Ed.* **2008**, *47*, 3950–3953. [CrossRef] [PubMed]

16. Ogoshi, T.; Yamagishi, T. Pillararenes: Versatile synthetic receptors for supramolecular chemistry. *Eur. J. Org. Chem.* **2013**, *2013*, 2961–2975. [CrossRef]

17. Ogoshi, T.; Kanai, S.; Fujinami, S.; Yamagishi, T.A.; Nakamoto, Y. Para-bridged symmetrical pillar[5]arenes: Their lewis acid catalyzed synthesis and host-guest property. *J. Am. Chem. Soc.* **2008**, *130*, 5022–5023. [CrossRef] [PubMed]

18. Cao, D.R.; Meier, H. Synthesis of pillar[6]arenes and their host-guest complexes. *Synthesis-Stuttgart* **2015**, *47*, 1041–1056. [CrossRef]

19. Ogoshi, T.; Yamagishi, T. New synthetic host pillararenes: Their synthesis and application to supramolecular materials. *Bull. Chem. Soc. Jpn.* **2013**, *86*, 312–332. [CrossRef]

20. Strutt, N.L.; Zhang, H.C.; Schneebeli, S.T.; Stoddart, J.F. Functionalizing pillar[n]arenes. *Acc. Chem. Res.* **2014**, *47*, 2631–2642. [CrossRef] [PubMed]

21. Ogoshi, T.; Ueshima, N.; Sakakibara, F.; Yamagishi, T.; Haino, T. Conversion from pillar[5]arene to pillar[6-15]arenes by ring expansion and encapsulation of c-60 by pillar[n]arenes with nanosize cavities. *Org. Lett.* **2014**, *16*, 2896–2899. [CrossRef] [PubMed]

22. Ogoshi, T.; Hashizume, M.; Yamagishi, T.A.; Nakamoto, Y. Synthesis, conformational and host-guest properties of water-soluble pillar[5]arene. *Chem. Commun.* **2010**, *46*, 3708–3710. [CrossRef] [PubMed]

23. Tan, L.L.; Yang, Y.W. Molecular recognition and self-assembly of pillarenes. *J. Incl. Phenom. Macrocycl. Chem.* **2015**, *81*, 13–33. [CrossRef]

24. Hu, X.Y.; Jia, K.K.; Cao, Y.; Li, Y.; Qin, S.; Zhou, F.; Lin, C.; Zhang, D.M.; Wang, L.Y. Dual photo- and ph-responsive supramolecular nanocarriers based on water-soluble pillar[6]arene and different azobenzene derivatives for intracellular anticancer drug delivery. *Chem. Eur. J.* **2015**, *21*, 1208–1220. [CrossRef] [PubMed]

25. Zhou, Q.Z.; Jiang, H.J.; Chen, R.; Qiu, F.L.; Dai, G.L.; Han, D.M. A triply-responsive pillar[6]arene-based supramolecular amphiphile for tunable formation of vesicles and controlled release. *Chem. Commun.* **2014**, *50*, 10658–10660. [CrossRef] [PubMed]

26. Cao, Y.; Hu, X.Y.; Li, Y.; Zou, X.C.; Xiong, S.H.; Lin, C.; Shen, Y.Z.; Wang, L.Y. Multistimuli-responsive supramolecular vesicles based on water-soluble pillar[6]arene and saint complexation for controllable drug release. *J. Am. Chem. Soc.* **2014**, *136*, 10762–10769. [CrossRef] [PubMed]

27. Yu, G.C.; Zhou, X.R.; Zhang, Z.B.; Han, C.Y.; Mao, Z.W.; Gao, C.Y.; Huang, F.H. Pillar[6]arene/paraquat molecular recognition in water: High binding strength, ph-responsiveness, and application in controllable self-assembly, controlled release, and treatment of paraquat poisoning. *J. Am. Chem. Soc.* **2012**, *134*, 19489–19497. [CrossRef] [PubMed]

28. Wang, P.; Yao, Y.; Xue, M. A novel fluorescent probe for detecting paraquat and cyanide in water based on pillar[5]arene/10-methylacridinium iodide molecular recognition. *Chem. Commun.* **2014**, *50*, 5064–5067. [CrossRef] [PubMed]

29. Ogoshi, T.; Kida, K.; Yamagishi, T. Photoreversible switching of the lower critical solution temperature in a photoresponsive host-guest system of pillar[6]arene with triethylene oxide substituents and an azobenzene derivative. *J. Am. Chem. Soc.* **2012**, *134*, 20146–20150. [CrossRef] [PubMed]

30. Li, Z.T.; Yang, J.; Yu, G.C.; He, J.M.; Abliz, Z.; Huang, F.H. Synthesis of a water-soluble pillar[9]arene and its ph-responsive binding to paraquat. *Chem. Commun.* **2014**, *50*, 2841–2843. [CrossRef] [PubMed]

31. Chen, H.Q.; Jia, X.S.; Li, C.J. A pillar[6]arene-[2]pseudorotaxane based ph-sensitive molecular switch. *Chin. J. Chem.* **2015**, *33*, 343–345. [CrossRef]

32. Decher, G.; Hong, J.D.; Schmitt, J. Buildup of ultrathin multilayer films by a self-assembly process: 3. Consecutively alternating adsorption of anionic and cationic polyelectrolytes on charged surfaces. *Thin Solid Films* **1992**, *210–211*, 831–835. [CrossRef]

33. Decher, G. Fuzzy nanoassemblies: Toward layered polymeric multicomposites. *Science* **1997**, *277*, 1232–1237. [CrossRef]

34. Jaber, J.A.; Schlenoff, J. Recent developments in the properties and applications of polyelectrolyte multilayers. *Curr. Opin. Colloid Interface Sci.* **2006**, *11*, 324–329. [CrossRef]

35. Klitzing, R.V. Internal structure of polyelectrolyte multilayer assemblies. *Phys. Chem. Chem. Phys.* **2006**, *8*, 5012–5033. [CrossRef] [PubMed]

36. Xu, H.P.; Schönhoff, M.; Zhang, X. Unconventional layer-by-layer assembly: Surface molecular imprinting and its applications. *Small* **2012**, *8*, 517–523. [CrossRef] [PubMed]

37. Wulff, G. Molecular imprinting in cross-linked materials with the aid of molecular templates—A way towards artificial antibodies. *Angew. Chem. Int. Ed.* **1995**, *34*, 1812–1832. [CrossRef]

38. Wulff, G. Molecular imprinting in crosslinked polymers—The role of the binding sites. *Mol. Cryst. Liq. Cryst.* **1996**, *276*, 1–6. [CrossRef]

39. Shi, F.; Liu, Z.; Wu, G.L.; Zhang, M.; Chen, H.; Wang, Z.Q.; Zhang, X.; Willner, I. Surface imprinting in layer-by-layer nanostructured films. *Adv. Funct. Mater.* **2007**, *17*, 1821–1827. [CrossRef]

40. Gauczinski, J.; Liu, Z.; Zhang, X.; Schönhoff, M. Mechanism of surface molecular imprinting in polyelectrolyte multilayers. *Langmuir* **2010**, *26*, 10122–10128. [CrossRef] [PubMed]

41. Zhou, Y.; Cheng, M.; Zhu, X.; Zhang, Y.; An, Q.; Shi, F. A facile method to prepare molecularly imprinted layer-by-layer nanostructured multilayers using postinfiltration and a subsequent photo-cross-linking strategy. *ACS Appl. Mater. Interfaces* **2013**, *5*, 8308–8313. [CrossRef] [PubMed]

42. Niu, J.; Shi, F.; Liu, Z.; Wang, Z.Q.; Zhang, X. Reversible disulfide cross-linking in layer-by-layer films: Preassembly enhanced loading and ph/reductant dually controllable release. *Langmuir* **2007**, *23*, 6377–6384. [CrossRef] [PubMed]

43. Guan, G.; Liu, R.; Wu, M.; Li, Z.; Liu, B.; Wang, Z.; Gao, D.; Zhang, Z. Protein-building molecular recognition sites by layer-by-layer molecular imprinting on colloidal particles. *Analyst* **2009**, *134*, 1880–1886. [CrossRef] [PubMed]

44. Zhang, J.W.; Liu, Y.L.; Wu, G.L.; Schönhoff, M.; Zhang, X. Bolaform supramolecular amphiphiles as a novel concept for the buildup of surface-imprinted films. *Langmuir* **2011**, *27*, 10370–10375. [CrossRef] [PubMed]

45. Niu, J.; Liu, Z.; Fu, L.; Shi, F.; Ma, H.; Ozaki, Y.; Zhang, X. Surface-imprinted nanostructured layer-by-layer film for molecular recognition of theophylline derivatives. *Langmuir* **2008**, *24*, 11988–11994. [PubMed]

46. Gauczinski, J.; Liu, Z.H.; Zhang, X.; Schönhoff, M. Surface molecular imprinting in layer-by-layer films on silica particles. *Langmuir* **2012**, *28*, 4267–4273. [CrossRef] [PubMed]

47. Zhang, J.W.; Liu, Y.L.; Yuan, B.; Wang, Z.Q.; Schönhoff, M.; Zhang, X. Multilayer films with nanocontainers: Redox-controlled reversible encapsulation of guest molecules. *Chem. Eur. J.* **2012**, *18*, 14968–14973. [CrossRef] [PubMed]

48. Nicolas, H.; Yuan, B.; Zhang, X.; Schönhoff, M. Cucurbit[8]uril-containing multilayer films for the photocontrolled binding and release of a guest molecule. *Langmuir* **2016**, *32*, 2410–2418. [CrossRef] [PubMed]

49. Nicolas, H.; Yuan, B.; Zhang, J.; Zhang, X.; Schönhoff, M. Cucurbit[8]uril as nanocontainer in a polyelectrolyte multilayer film: A quantitative and kinetic study of guest uptake. *Langmuir* **2015**, *31*, 10734–10742. [CrossRef] [PubMed]

50. Bian, Q.; Jin, M.M.; Chen, S.; Xu, L.P.; Wang, S.T.; Wang, G.J. Visible-light-responsive polymeric multilayers for trapping and release of cargoes via host-guest interactions. *Polym. Chem.* **2017**, *8*, 5525–5532. [CrossRef]

51. Xuan, H.Y.; Ren, J.Y.; Zhang, J.H.; Ge, L.Q. Novel highly-flexible, acid-resistant and self-healing host-guest transparent multilayer films. *Appl. Surf. Sci.* **2017**, *411*, 303–314. [CrossRef]

52. Xu, G.; Pranantyo, D.; Xu, L.Q.; Neoh, K.G.; Kang, E.T.; Teo, S.L.M. Antifouling, antimicrobial, and antibiocorrosion multilayer coatings assembled by layer-by-layer deposition. Involving host-guest interaction. *Ind. Eng. Chem. Res.* **2016**, *55*, 10906–10915. [CrossRef]

53. Yuan, B.; Xu, J.F.; Sun, C.L.; Nicolas, H.; Schönhoff, M.; Yang, Q.Z.; Zhang, X. Pillar[6]arene containing multilayer films: Reversible uptake and release of guest molecules with methyl viologen moieties. *ACS Appl. Mater. Interfaces* **2016**, *8*, 3679–3685. [CrossRef] [PubMed]

54. Sun, J.Q.; Wu, T.; Liu, F.; Wang, Z.Q.; Zhang, X.; Shen, J.C. Covalently attached multilayer assemblies by sequential adsorption of polycationic diazo-resins and polyanionic poly(acrylic acid). *Langmuir* **2000**, *16*, 4620–4624. [CrossRef]

55. Morgan, G.T.; Alcock, M. The colour and constitution of diazonium salts. Part I. *J. Chem. Soc.* **1909**, *95*, 1319–1329. [CrossRef]

56. Ostendorf, A.; Cramer, C.; Decher, G.; Schönhoff, M. Humidity-tunable electronic conductivity of polyelectrolyte multilayers containing gold nanoparticles. *J. Phys. Chem. C* **2015**, *119*, 9543–9549. [CrossRef]

57. Parveen, N.; Schönhoff, M. Quantifying and controlling the cation uptake upon hydrated ionic liquid-induced swelling of polyelectrolyte multilayers. *Soft Matter* **2017**, *13*, 1988–1997. [CrossRef] [PubMed]

58. Nomura, T.; Okuhara, M. Frequency-shifts of piezoelectric quartz crystals immersed in organic liquids. *Anal. Chim. Acta* **1982**, *142*, 281–284. [CrossRef]

59. Nicolas, H.; Yuan, B.; Zhang, J.W.; Zhang, X.; Schönhoff, M. Correction to "Cucurbit[8]uril as nanocontainer in a polyelectrolyte multilayer film: A quantitative and kinetic study of guest uptake". *Langmuir* **2017**, *33*, 4879. [CrossRef] [PubMed]

60. Nicolas, H.; Yuan, B.; Zhang, X.; Schönhoff, M. Correction to "Cucurbit[8]uril-containing multilayer films for the photocontrolled binding and release of a guest molecule". *Langmuir* **2017**, *33*, 5098. [CrossRef] [PubMed]

polymers

MDPI

Article

Antibacterial Films Made of Ionic Complexes of Poly(γ-glutamic acid) and Ethyl Lauroyl Arginate

Ana Gamarra-Montes [1], Beatriz Missagia [2], Jordi Morató [2] and Sebastián Muñoz-Guerra [1,*]

[1] Departament d'Enginyeria Química, Universitat Politècnica de Catalunya, ETSEIB, Diagonal 647, 08028 Barcelona, Spain; anagamarramontes@gmail.com
[2] Health and Environmental Microbiology Lab & UNESCO Chair on Sustainability, Universitat Politècnica de Catalunya, ESEIAAT, Edifici Gaia, Pg. Ernest Lluch/Rambla Sant Nebridi, 08222 Terrassa, Spain; beatrizmissagia@gmail.com (B.M.); jordi.morato@upc.edu (J.M.)
* Correspondence: sebastian.munoz@upc.edu; Tel.: +34-93401-6680

Received: 23 November 2017; Accepted: 20 December 2017; Published: 24 December 2017

Abstract: The biocide agent LAE (ethyl $^\alpha$N-lauroyl L-arginate chloride) was coupled with poly(γ-glutamic acid) (PGGA) to form stable ionic complexes with LAE:PGGA ratios of 1 and 0.5. The nanostructure adopted by these complexes and its response to thermal changes were examined in detail by Differential scanning calorimetry (DSC) and X-ray diffraction (XRD) using synchrotron radiation in real time. A layered biphasic structure with LAE filling the space between the polypeptidic sheets was adopted in these complexes. The complexes were stable up to above 250 °C, non-water soluble, and were able to form consistent transparent films. The release of LAE from the complexes upon incubation in aqueous buffer was examined and found to depend on both pH and complex composition. The antibacterial activity of films made of these complexes against Gram-positive (*L. monocytogenes* and *S. aureus*) and Gram-negative (*E. coli* and *S. enterica*) bacteria was preliminary evaluated and was found to be very high against the formers and only moderate against the later. The bactericide activity displayed by the LAE·PGGA complexes was directly related with the amount of LAE that was released from the film to the environment.

Keywords: ionic polyglutamic acid complexes; biocide polyglutamic acid; comb-like polyglutamic acid complexes; ethyl lauroyl arginate; antibacterial polymer complexes

1. Introduction

Food safety is today an issue of major concern that is receiving great social and technological attention. It has been estimated that as much as 30% of people in industrialized countries suffer yearly from a food borne disease, and that in 2000, at least two millions of people died from diarrheal diseases worldwide, the major proportion being attributable to microbial contamination of food and water [1]. There are more than 200 of active agents causing gastrointestinal illnesses, about 60% of which being due to infection by food borne bacterial pathogens. *Salmonella* spp., *Campylobacter* spp. and *Escherichia coli* are the bacteria traditionally attracting major attention [2], but in the last decades concerns have included not only an increasing number of additional pathogens as *L. monocytogenes* but also the expansion of modified traditional strains displaying antimicrobial resistance [3]. The use of bactericides, both of synthetic and natural origin, constitutes today the most common practice applied to prevent food spoilage, so the demand for these compounds has increased considerably in these last years [4,5]. Methods followed for impregnating the targeted food with the antimicrobial agent include blending in bulk, surface treatment, and controlled delivery from active films used either for wrapping or coating the food [6,7].

The incorporation of antimicrobials into polymeric films in contact with food to be gradually released during shelf-life has unquestionable advantages over those procedures in which the active compound is directly loaded into or onto the food. (a) Deactivation of the antimicrobial by the food components is largely prevented, and (b) a higher effectiveness in the inhibition of pathogens growing on the food surface, which is the most common way of food contamination, may be achieved. As a result, smaller amounts of active compounds will be required by the film activation approach to reach satisfactory outcomes. This is a very remarkable benefit, since additive minimization constitutes a major challenge today for food quality and safety [8,9].

In this paper we wish to report on a new antibacterial polymeric system based on an ionic polymer complex made of poly(γ-glutamic acid) (PGGA) and a guanidinium-based compound (LAE). PGGA is an emerging biopolymer that is edible and biodegradable, and that has an enormous potential as biomaterial [10]. PGGA is generated by bacterial fermentation of a wide variety of substrates and it is produced at industrial scale to be used as a food complement, in healthcare and for water treatment, among others. As it is much expected for a polycarboxylic compound, PGGA is highly hygroscopic, and a number of modifications, consisting mainly of esterification and amidation of the carboxylic side groups, has been reported with the purpose of making the polymer higher water-resistant [11]. The innocuity of PGGA makes it an excellent candidate for designing antimicrobial polymeric materials for food packaging. On the other hand the polyanionic nature of this biopolymer makes it very suitable for the efficient loading of organocationic compounds by ionic coupling. In fact, ionic complexes of PGGA with both alkylammonium [12–14] and alkylphosphonium [15] soaps have been reported, and the capacity of the later to display biocide activity has been demonstrated. Furthermore, the capacity of PGGA to inhibit, by itself, the growth of some pathogenic bacteria has been also announced [16]. Nevertheless the references on the application of PGGA in food packaging are very scarce in the accessible literature [17].

LAE (ethyl $^\alpha$N-lauroyl L-arginate) is one of the most potent food preservative agents that is known today, which displays a broad spectrum of activity against food-borne bacteria [18,19]. The high biocide activity of LAE has been attributed to its capacity for altering the metabolic processes of microorganisms without causing cellular lysis [20]. LAE has been assessed to be nontoxic, since after consumption, it is rapidly metabolized to naturally occurring amino acids, among which arginine and ornithine appear to be majority [21,22]. The Food and Drug Administration (FDA) has classified LAE as a GRAS (Generally Recognized as Safe) food preservative at concentrations up to 200 ppm. Antibacterial films containing LAE were firstly prepared from synthetic polymers of common use in packaging such as PP, EVA, and EVOH [19,23,24]. In these last years, efforts has been redirected towards the development of systems made of either bio-based polymers as PLA [25], biopolymers as chitosan [26,27], and others [28,29], which are able to be biodegraded, and even fit to be eaten.

Organocationic compounds are extensively used as bactericides in a diversity of applications, but their utilization in active films is severely limited by the difficulty in achieving suitable mixing with polymers that are commonly used for packaging. Coupling the organocation with anionic polymers is a useful approach that allows for designing active films with the desired stability and releasing properties. The ionic interaction of LAE with anionic polysaccharides has been examined to evaluate the influence that these compounds may have on its biocide activity when they are used as food ingredients [30–32]. However, to our knowledge, no study addressed assessing the potential of ionic LAE complexes as active films has been described so far. In this work, LAE has been coupled with the polyanionic PGGA to obtain ionic stable complexes (LAE·PGGA) with antibacterial properties. Firstly, the LAE·PGGA complexes are extensively characterized by physical-chemical methods (Fourier Transform Infrartd (FTIR), Nuclear Magnetic Resonance (NMR), Thermogravimetric analysis (TGA), Differential scanning calorimetry (DSC), X-ray diffraction (XRD), and polarizing optical microscope (POM) to establish their chemical and supramolecular structure. Then, the dissociation of the complexes into their components upon incubation at different pH is examined. Finally the antibacterial properties of the complexes against Gram-positive bacteria (*Listeria monocytogenes* and *Staphylococcus aureus*) and Gram-negative bacteria (*Salmonella enterica* and *Escherichia coli*) are preliminary estimated in order to evaluate their potential for food preserving and packaging applications.

2. Experimental Section

2.1. Materials

The sodium salt of poly (γ-glutamic acid) (PGGA-Na) sample that was used in this work was kindly supplied by Dr. Kubota of Meiji. Co. (Tokyo, Japan). It was obtained by biosynthesis with a weight-average molecular weight of ~300,000 Da and a D:L enantiomeric ratio of 59:41. Ethyl $^\alpha$N-lauroyl L-arginate chloride (LAE) was a sample gifted by Vedeqsa (LAMIRSA Group, Terrassa, Barcelona, Spain).

2.2. Measurements

FTIR spectra (Perkin Elmer, Waltham, MA, USA) within the 4000–600 nm range were recorded from powder samples on a Perkin Elmer Frontier equipment provided with an ATR accessory. ^{1}H and ^{13}C NMR spectra were recorded on a Bruker AMX-300 NMR instrument (Billerica, MA, USA) operating at 300.1 and 75.5 MHz, respectively. Samples were dissolved in deuterated methanol and (tetramethylsilane) TMS was used as internal reference. 128 (Free induction decay) FIDs for ^{1}H NMR spectra were recorded with 2.3 µs pulse width, 3.4 s acquisition time, 20 s relaxation delay, and 4.9 KHz spectral width. For ^{13}C NMR spectra, 1000 to 10,000 FIDs were recorded using pulse and spectral widths of 4.3 µs and 18 KHz, respectively. TGA was performed at a heating rate of 10 °C·min^{-1} from 30 to 600 °C under an inert atmosphere on a Mettler-Toledo (Zurich, Switzerland) TGA/DSC 1 Star System thermobalance. Sample weights of 10–15 mg were used for this analysis. DSC was carried out on a Perkin-Elmer (Waltham, MA, USA) DSC 8000 instrument that was calibrated with indium and zinc. Heating-cooling cycles at a rate of 10 °C·min^{-1} under a nitrogen atmosphere within the temperature range of –30 to 200 °C were applied for the analysis of sample weights of about 2–5 mg. X-ray diffraction studies were performed using X-ray synchrotron radiation at the BL11 beamline (NCD, Non-Crystalline Diffraction, Cerdanyola del Vallès, España) of ALBA synchrotron in Cerdanyola del Vallès, Barcelona. Simultaneous small angle region (SAXS) and wide-angle region (WAXS) were taken in real time from powder samples subjected to heating-cooling cycles at a rate of 10 °C·min^{-1}. The radiation energy employed corresponded to a 0.1 nm wavelength, and spectra were calibrated with silver behenate (AgBh) and Cr_2O_3 for small and wide angle diffraction, respectively. Optical microscopy was carried out on an Olympus BX51 POM (Allentown, PA, USA), which was outfitted with a digital camera. For observation, samples were prepared as films casted from 5% (*w/w*) methanol solutions and were placed in a Linkam THMS-600 (Tadworth, UK) hot stage provided with a nitrogen gas circulating system.

2.3. Complexes Formation and Film Preparation

The methodology originally used by Ponomarenko et al. [33] for the preparation of ionic complexes from poly(α-amino acids) and ionic surfactants were applied in this work. This methodology with some slight modifications has been used previously by us for the synthesis of ionic complexes made from either PGGA [12–14] or polyuronic acids [34,35], and quaternary ammonium salts bearing linear alkyl chains with 12–22 carbon atoms. The procedure is essentially as follows: A solution of LAE hydrochloride in water was slowly poured into a solution of PGGA-Na in water under stirring at a temperature around 35 °C. The formed complex precipitated as a white powder after several hours of standing. The precipitate was recovered by centrifugation, and repeatedly washed with water and dried under vacuum for at least 48 h. Complexes were prepared from mixtures containing both 1:1 and 1:2 molar ratios of LAE to PGGA (LAE·PGGA-1 and LAE·PGGA-0.5).

LAE·PGGA films were prepared by casting from a dilute solution of 400 mg of complex in methanol on 3 × 3 cm^2 Petri plates. After drying at room temperature for 24 h and applying vacuum for 24 h further, consistent films were formed and cut in 1 × 1 cm^2 squares. The average films thickness measured using a Mitutoyo micrometer (Osaka, Japan) was 100 ± 2 µm.

2.4. Complex Dissociation and Antibacterial Activity

The dissociation rate of the LAE·PGGA complexes taking place upon incubation in aqueous medium was followed by measuring the absorbance at 220 nm of the released compounds as a function of time. Assays were carried out by placing 1×1 cm^2 squares of LAE·PGGA films into cellulose dialysis tubes (2000 Da cut-off) that were immersed in 20 mL of buffer solutions at pH = 9.2, 7.4, 5.5, and 4.5 at 25 °C, and were left under mild stirring for one week.

The antibacterial activity of LAE·PGGA complexes was tested in vitro against both Gram-negative and Gram-positive bacteria in liquid culture media over time. Bacteria for this study were selected for their widespread occurrence and well-known ability to cause food-borne diseases by uncontrolled ingestion [2]. Cultures of *E. coli* NCTC 9001 isolated from human urine cystitis, *S. enterica* CECT 4594 from septicemic liver from bovine, *L. monocytogenes* ATCC 19115 from human, and *S. aureus* ATCC 6538 isolated from human lesion were obtained from the National Collection of Type Cultures (NCTC, Public Health England, Porton Down Salisbury, UK), the Spanish Type Culture Collection (CECT, Valencia, Spain), and the American Type Culture Collection (ATCC, Manassas, VA, USA), respectively.

The organisms were stored at -20 °C in tryptic soy broth (TSB; Merck, Darmstadt, Germany) containing 50% (v/v) glycerol until needed. To activate them, a loopful of each bacterium was streaked on tryptic soy agar (TSA; Difco Laboratories, Livonia, MI, USA) petri dishes. After 24 h at 37 °C, a single colony of each strain was picked and suspended into 10 mL tubes of TSB pH 7 and incubated at 37 °C for 24 h to obtain early stationary phase cells (optical density of 0.9 at 600 nm). The cultures were then further inoculated (100 µL) into fresh TSB and were incubated at 37 °C for 18 h to reach the exponential phase (optical density of 0.2 at 600 nm). At this stage, 100 µL of TSB containing 10^5 CFU/mL and approximately 1 cm^2 of each film (PGGA, LAE·PGGA, LAE·PGGA-0.5, and the control) were placed into sterile tubes with 10 mL of fresh TSB and were incubated at 37 °C. For quantification, 100 µL aliquots were removed from the suspension at selected periods of time (2, 8, 24 and 168 h) and plated on petri dishes with 15 mL of TSA culture medium. Serial dilutions were performed with peptone water (1% v/v) depending on the turbidity produced. Controls without films (blank) and with unmodified PGGA films (negative controls, NC) were also tested and experiments were performed in triplicates. All of the films were sterilized before using by UV light for 15 minutes. Data are represented as logarithm of colony forming units (LogCFU). Formula for logarithm reduction value (LRV) and percentage reduction calculations are shown below [36],

$$\text{Log reduction value} = \log_{10} (A/B)$$

$$\text{Percentage reduction} = [(A-B)/A] \cdot 100$$

where A is the number of viable bacteria in the negative control and B is the number of viable bacteria after treatment with either LAE·PGGA-1 or LAE·PGGA-0.5.

3. Results and Discussion

3.1. Synthesis of Complexes

The synthesis of the complexes of PGGA and LAE did not entail any special difficulty since they were spontaneously formed upon mixing the aqueous solutions of LAE and PGGA-Na (Scheme 1). Ionic coupling of the guanidinium cation of LAE and the carboxylate anion of PGGA resulted in non-water soluble stable complexes that precipitated from the aqueous solution upon standing. Two LAE:PGGA ratios, i.e., 1:1 and 1:2, were used with the purpose of evaluating the effect of composition on properties. The complexes were recovered by centrifugation in the form of white powders in 50–70% yields. Conditions that were used in these experiments and results attained are given in Table 1.

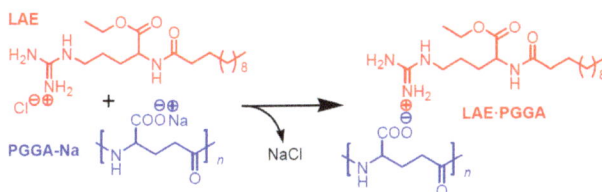

Scheme 1. Coupling reaction leading to ionic ionic stable complexes ((ethyl $^{\alpha}$N-lauroyl L-arginate chloride) LAE·PGGA (poly(γ-glutamic acid)) complexes.

Table 1. Results for the preparation of LAE·PGGA complexes.

Complex	LAE:PGGA [a]	Mixing Conditions		Yield (%)	Color	Composition [d]
		c (M) [b]	*T* (°C) [c]			
LAE·PGGA-1	1.0:1.0	0.01	35	70	white	0.9:1.0
LAE·PGGA-0.5	0.5:1.0	0.01	35	57	white	0.5:1.0

[a] Molar ratio of LAE to PGGA used for coupling; [b] Concentration of the solutions mixed to form the complex; [c] Minimum temperature at which LAE is soluble in water at the used concentration; [d] Molar ratio of LAE to PGGA in the complex determined by ^{1}H NMR.

3.2. Chemical Characterization

The presence of the two components in the LAE·PGGA complexes was evidenced by FTIR, as it is shown in Figure 1. The characteristic absorptions of both LAE and PGGA are present in the spectra of the complexes with the expected relative transmittance values.

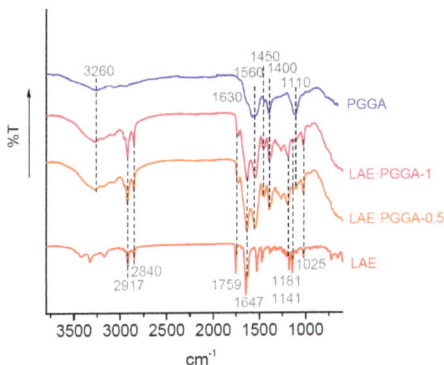

Figure 1. Compared FTIR spectra of LAE, PGGA, and LAE·PGGA complexes.

The ^{1}H NMR analysis of the complexes ascertained the presence of the two components and provided an accurate quantification of their composition. The spectrum recorded from LAE·PGGA-1 is depicted in Figure 2, with indication of peak assignments. The area ratio of the signal arising from the γ-methylene of PGGA to the area of the two partially overlapped signals, including the 3-11 methylenes of the lauroyl chain of LAE revealed that the actual composition of the complexes LAE·PGGA-1 and LAE·PGGA-0.5 were 0.9:1 and 0.5:1, respectively, which are values that are very close to those expected from the relative amounts of the two components that were used for their preparation. The ^{1}H NMR spectrum of LAE·PGGA-0.5, as well as the ^{13}C NMR of the two complexes are shown in the Supporting Information (SI) file associate to this paper (Figures SI-1 and SI-2 in the supplementary materails).

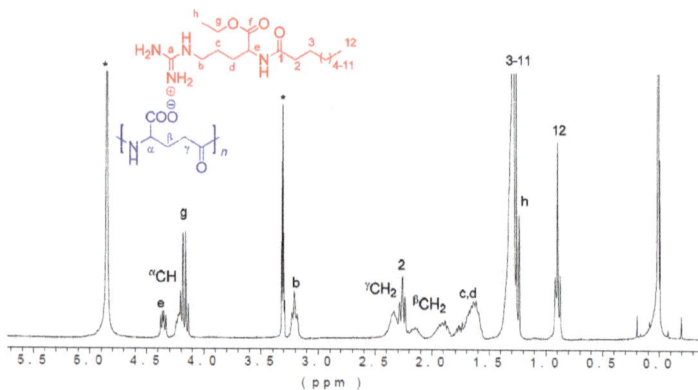

Figure 2. ^{1}H NMR spectra of LAE·PGGA-1 recorded at 25 °C in MeOD. *Asterisked signals are those arising from water and non-deuterated solvent.

3.3. Thermal Properties and Structure

The thermal decomposition of the LAE·PGGA complexes was examined by TGA under inert atmosphere, and the possible thermal transitions were explored by DSC. The TGA traces recorded for the two complexes as well as their respective derivative curves are compared with that of LAE in Figure 3, and decomposition temperatures and remaining weights measured on these traces are given in Table 2.

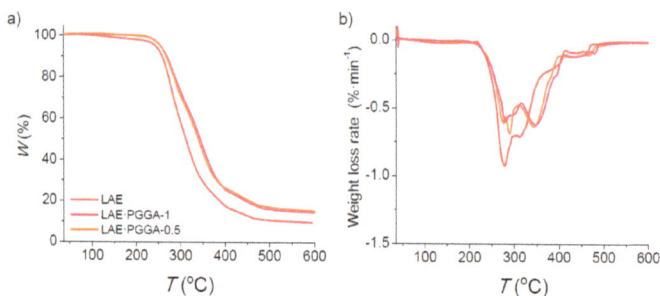

Figure 3. Thermogravimetric analysis (TGA) traces (**a**) and derivative curves (**b**) of LAE and its ionic complexes with PGGA.

It was found that decomposition of the complexes started about 10 °C above that of LAE and that the whole process happened through two stages in the three cases. Temperatures at which the first decomposition step took place at maximum rate were not very dissimilar for the three samples, whereas values that were observed for the second step were much higher for the complexes than for LAE. The thermal decomposition of ionic complexes of PGGA with trimethylalkylammonium surfactants (nATMA·PGGA) has been studied by us in some detail [37]. It was then reported that nATMA·PGGA started to decompose around 200 °C by the decoupling of the complex, and that decomposition proceeded along two steps at rate temperatures within the 270–280 °C and 320–380 °C ranges, respectively. Given the resemblance of the overall thermal decomposition patterns of LAE·PGGA and nATMA·PGGA, a similar mechanism may be assumed to occur in both types of complexes, in spite that decomposition of LAE should be expected to be much more intricate than that of alkylammonium compounds.

Table 2. Thermal parameters of LAE·PGGA complexes.

Sample	$^{\circ}T_d$ [a] (°C)	$^{max}T_d$ [a] (°C)	W [a] (%)	T_m (°C)
LAE	245	275/311	27/10	62
LAE·PGGA-1	255	280/349	68/15	-
LAE·PGGA-0.5	253	273/342	63/16	-

[a] $^{\circ}T_d$ and $^{max}T_d$ are onset and maximum rate decomposition temperatures, respectively; W is the remaining weight at the end of each decomposition process; [b] T_m is the melting temperature recorded by DSC.

3.4. Supramolecular Structure and Thermal Transitions

The DSC traces recorded at heating from LAE and the complexes in the 0–200 °C range are depicted in Figure 4. A well-defined sharp endothermal peak was observed for LAE at 62 °C, which doubtlessly arises from melting. In fact, the XRD of LAE in the WAXS produced multiple discrete scattering in the 0.3–0.5 nm range characteristic of crystalline organic material (Figure 5a). Furthermore, the examination of this sample under the polarizing optical microscope (POM), while heating revealed the presence of a typical crystalline texture that disappeared at temperatures nearly above 60 °C (see Figure SI-3 in the SI file in the supplementary materails). On the contrary, the DSC traces of both LAE·PGGA-1 and LAE·PGGA-0.5 did not show below 100 °C any sign that was indicative of crystallinity. Accordingly, no definite diffraction peak, but a broad peak centered on 0.45 nm characteristic of disordered material, was the only scattering detected in the WAXS of the complexes, which is taken as indicative that the LAE counterpart must be in the amorphous state.

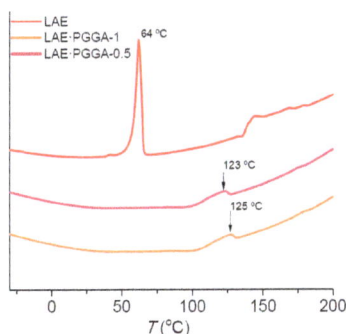

Figure 4. Differential scanning calorimetry (DSC) traces of LAE and its ionic complexes with PGGA.

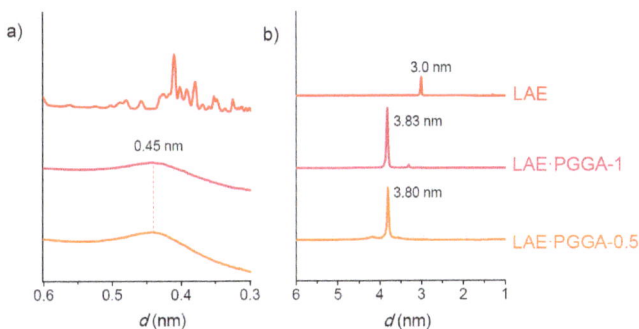

Figure 5. XRD profiles recorded from LAE and its complexes with PGGA in the wide-angle region (WAXS) (**a**) and small angle regions (SAXS) (**b**) at room temperature.

Inspection of the SAXS of the XRD patterns revealed in every case the presence of one sharp diffraction peak corresponding to a spacing of 3.0 nm for LAE and of ~3.8 nm for the complexes (Figure 5b). The presence of a peak in the ~3–4 nm range of SAXS is a distinctive characteristic of the ionic complexes made of PGGA and tetraalkylammonium surfactants bearing long linear alkyl chains that are arranged in a biphasic structure made of alternating polypeptidic and paraffinic layers [12,13]. According to such antecedents, the ~3.8 nm spacing observed for both LAE·PGGA-1 and LAE·PGGA-0.5 can be interpreted as arising from the periodicity of the biphasic layered arrangement that was adopted by these complexes, although the LAE moiety is in the non-crystallized state. The long spacing displayed by LAE·PGGA (~3.8 nm) is consistent with that observed for LAE (3.0 nm), since the space occupied by the PGGA layer has to be added in the complex. It is also in agreement with the long spacing reported for 12ATMA·PGGA (3.1 nm) [13] provided that the LAE non-alkyl moiety is much bulkier than the trimethylammonium group of 12ATMA.

In order to have a deeper insight into the structure of the LAE·PGGA complexes, an XRD study at variable temperature was carried out using synchrotron radiation and the spacing data measured in this study are listed in Table 3. Both WAXS and SAXS traces were simultaneously registered at real time from samples while heated or cooled over the 10–120 °C range at a rate of 10 °C·min^{-1}. The evolution of the SAXS and WAXS profiles recorded for LAE is shown in Figure 6a,a'. The scattering that was initially present in both regions was retained until heating up to 60 °C to completely disappear at higher temperatures in full agreement with what was observed by DSC. No changes were detected after cooling (Figure SI-4, SI file in the Supplementary Materials), confirming the incapacity of LAE to crystallize from the melt, such as was evidenced before by both DSC and POM.

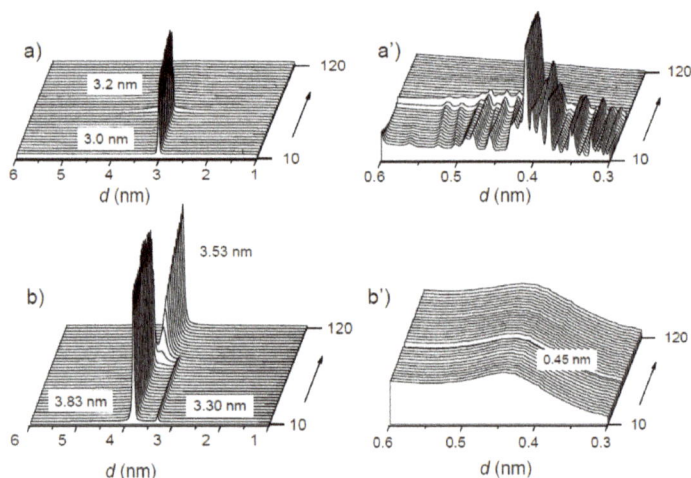

Figure 6. Evolution of the SAXS and WAXS profiles of LAE (**a,a'**) and LAE·PGGA-1 (**b,b'**) at heating from 10 to 120 °C.

The results obtained in the thermal XRD study of LAE·PGGA-1 are shown in Figure 6b,b'. In this case, the SAXS peak at 3.83 nm was kept almost invariable over the whole temperature interval, indicating that the layered arrangement present in the complex was essentially retained at the applied temperatures. In the WAXS region, the broad peak that was observed at 0.45 nm was unaffected by heating as it should be expected for a disordered scattering. It should be noted, however, that a small jump of the 3.83 peak down to 3.53 nm was observed around 60 °C. The occurrence of small jumping in the SAXS peaks in the 30–60 °C range is a frequently observed fact in the heating of comb-like ionic complexes of PGGA. Jumping may be either upwards or downwards, and it is invariably attributed to the occurrence of small rearrangements that take place in the layered structure upon melting of the alkyl chain [38].

The jump observed here for LAE·PGGA-1 cannot be explained in the same manner as for *n*ATMA·PGGA complexes, since the alkyl chain of LAE is not crystallized. It may be speculated however that some spatial rearrangement could occur in the molecular assembly of the LAE nanophase involving a light shortening of the interlayer distance of the complex. In this regard, it is interesting to notice that no jump was observed for LAE·PGA-0.5 (Figure SI-5 in SI file in the supplementary materials) where the low concentration of LAE may be insufficient to adopt a close continuous packing in this phase. POM observations of LAE·PGGA-1 subjected to heating revealed an initial typical liquid-crystalline texture at room temperature that slightly changed above 60 °C to fully disappear when the temperature reached the proximities of 125–130 °C (Figure SI-6 in SI file in the supplementary materials). This behavior is in agreement with DSC results that showed the presence of a small peak at 123 °C characteristic of a liquid crystal-isotropic phase transition.

Table 3. X-ray diffraction data of LAE·PGGA complexes.

Sample	SAXS			WAXS		
	$L_0^{10\,°C}$	$L_0^{120\,°C}$	$L_0^{10\,°C}$	$d^{10\,°C}$	$d^{120\,°C}$	$d^{10\,°C}$
LAE	3.0	-	-	Multiple	-	-
LAE·PGGA-1	3.8	3.5	3.5	0.45	0.45	0.45
LAE·PGGA-0.5	3.8	3.5	3.5	0.45	0.45	0.45

L_0: interlamellar distance (layered structure window); d: interplanar distances arising from Bragg spacings

3.5. LAE Release and Antibacterial Properties

Although it has been reported that PGGA is a moderate microbiocide [16], it is the LAE counterpart of the LAE·PGGA complexes that is expected to play the main biocide activity in these systems. Accordingly, the biocide activity of the LAE·PGGA films in aqueous medium will be largely determined by the LAE concentration that is attained in the environment upon dissociation of the complex. To substantiate this hypothesis the accumulative amount of LAE that is released from the LAE·PGGA films to the incubation medium at 25 °C was estimated by measuring the absorbance at 220 nm of the supernatant solution as a function of time. The results obtained from these assays for both LAE·PGGA-1 and LAE·PGGA-0.5 at different pH ranging from ~4.5 to ~9.5 are shown in Figure 7a,b As it could be logically expected, the general trend is that the amount of LAE present in the buffer increased exponentially with time to finally reach a more or less constant concentration. The influence of pH on the delivery of LAE is clearly illustrated in the bar graphics shown in Figure 7c,d. In these plots, both the amount of LAE that is present in the incubation medium and the weight lost by the film is compared for the two complexes after one week of incubation at the different assayed pH. It is clearly seen that LAE is liberated much faster at basic pH and that the minimum release rate happens at pH 5.5, a result that may be explained by taking into account the pKa of the two complex components, i.e., PGGA and LAE. It is also evidenced that the liberated amount of LAE is higher in LAE·PGGA-1 than in LAE·PGGA-0.5, which is much according to expectations, whereas the weight that is lost by the latter is significantly greater, a difference that is more ostensible at pH 9.2. This apparent conflict may be rationalized by taken into account the partial hydrolysis that it is probable undergone by PGGA upon incubation. The hydrolytic degradation of PGGA is a well-known process that is favored at higher pH [39]. This process is expected to happen more extensively in the case of LAE·PGGA-0.5 due to the higher accessibility of the PGGA backbone to water in this complex.

The antibacterial activity of LAE·PGGA complexes against both Gram-negative (*S. enterica* and *E. coli*) and Gram-positive bacteria (*L. monocytogenes* and *S. aureus*) as a function of time was evaluated by the liquid medium method. Single colony of each strain was suspended into the buffer placed in essay tubes, and the incubation effects were followed both visually and by measuring the optical density. The turbidity appreciated by visual appearance of the supernatant was a preliminary indication of how the bacterial growing is affected by the presence of the complex (Figure SI-7 in the SI file in the supplementary materials). The results that were obtained by spectroscopic measurement are graphically depicted in Figure 8, together with those obtained for neat PGGA (negative control NC), and the blank. Average values that were obtained from triplicate counting and calculations are presented. Numerical data of these results expressed as Log(CFU) (logarithm of colony forming units), as well as their corresponding logarithmic reduction values (LRV) and reduction percentages (PR) calculated by means of the expressions given above are given in Table 4.

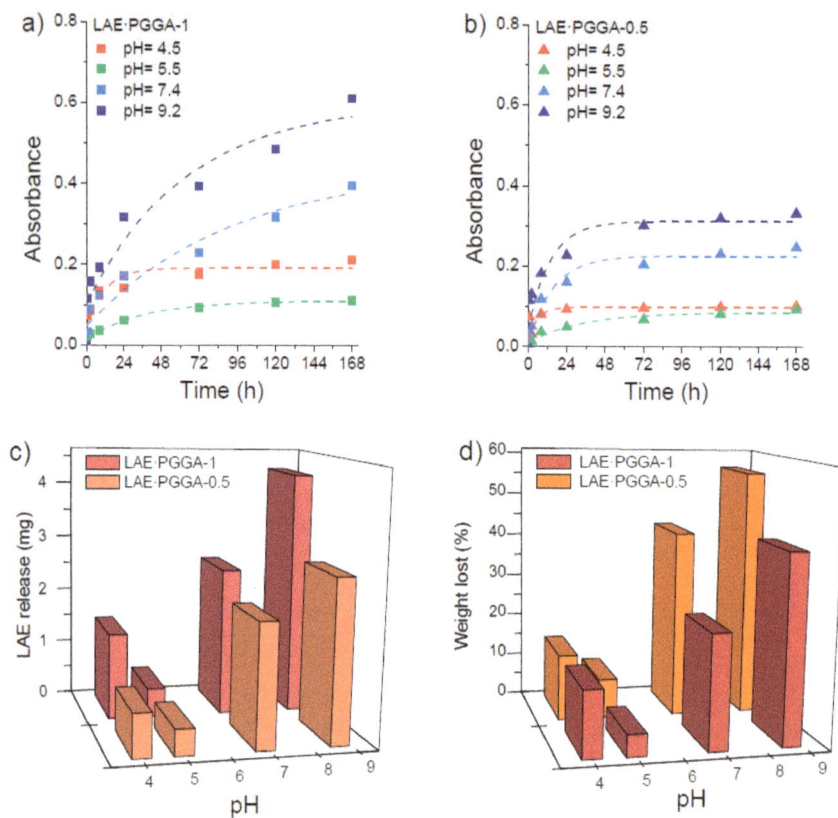

Figure 7. Dissociation of LAE·PGGA-1 (**a**) and LAE·PGGA-0.5 (**b**) complexes in aqueous buffer at the indicated pH. In (**c,d**) the amount of released LAE and the film weight loss are respectively compared for the two complexes after seven days of incubation at the assayed pH.

An overall inspection of the results obtained in the biocide activity assays leads us to conclude that, in agreement with the amount of LAE that is delivered in each case, the bactericide effect is in general more pronounced for LAE·PGGA-1 than for LAE·PGGA-0.5. The higher bacterial growth reduction capacity observed for the complex containing more LAE is according to previous observations that were made on antimicrobial films based on LAE loaded chitosan [27].

The two complexes displayed a similar great biocide activity against Gram-positive bacteria, reaching an almost total growth inhibition for both *L. monocytogenes* and *S. aureus* after seven days of incubation, and >99.99% of reduction after 24 h. On the contrary, both of the complexes were less effective against Gram-negative bacteria (*S. enterica* and *E. coli*), which is agreement with observations made by other authors on the antibacterial activity of LAE [27,40]. The LAE·PGGA-0.5 films achieved only 2.3% reduction of *S. enterica* after 24 h and 70.9% after seven days, although the double concentrated complex reached 99.6% reduction of the same bacteria after 24 h. Since Gram-negative organisms have a greater defense system due to the outer lipopolysaccharide coat that surrounds the cell wall, the diffusion of hydrophobic compounds is expected to be much more hindered than in Gram-positive species [41].

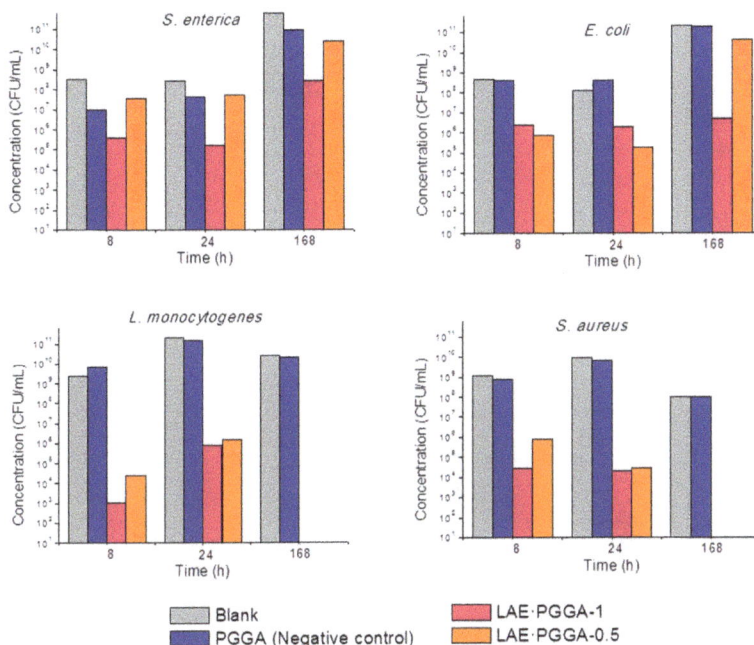

Figure 8. Antibacterial activity of LAE·PGGA-1 and LAE·PGGA-0.5 films against Gram-negative and Gram-positive bacteria expressed as concentration of colony forming units (CFU·mL^{-1}).

Table 4. Bactericide activity of films made of LAE·PGGA complexes against Gram-negative and Gram-positive bacteria after 1 and 7 days of incubation at pH 7.4 and 37 °C.

		S. enterica		*E. coli*		*L. monocytogenes*		*S. aureus*	
		24 h	7 d	24 h	7 d	24 h	7 d	24 h	7 d
PGGA	Log(CFU) [a]	7.6	10.6	8.6	11.3	11.2	10.30	9.8	8.0
LAE·PGGA-1	Log(CFU)	5.2	8.4	6.3	6.7	3.9	total	4.3	0.3
	LRV [b]	2.4	2.5	2.3	4.6	7.3		5.5	7.7
	PR (%) [c]	99.6	99.7	99.5	>99.9	>99.9	100	>99.9	>99.9
LAE·PGGA-0.5	Log(CFU)	7.6	10.4	5.3	10.6	4.2	total	4.5	total
	LRV	0.01	0.5	3.3	0.6	7.0		5.4	
	PR (%)	2.3	70.9	99.9	76.8	>99.9	100	>99.9	100

[a] Antibacterial activity expressed as logarithm of colony forming units; [b] Antibacterial activity expressed as log reduction value; [c] Percentage of reduction.

Since our results were obtained from one single experiment and only one strain was used for each bacterium, they should be taken as a first evaluation of the bactericide potential of these complexes. Further studies, including different incubation conditions (pH, temperature, etc.), additional microorganisms, and a statistical analysis of data collected from replicated experiments,

Polymers **2018**, *10*, 21

will be required to attain a more definite appraisal of the biocide capacity of the LAE·PGGA system. Furthermore, the possibility that a VBNC (viable but non-culturable) state is adopted by the bacteria upon the action of the bactericide cannot be completely discarded since it has been described for some of the pathogens studied in this work [42]. In this case, however, such a situation is highly improbable since LAE is a biocide in use for long time, and, to our knowledge, the occurrence of dormant bacteria after the action of this agent has not been reported.

4. Conclusions

The widely known antimicrobial agent LAE (ethyl $^\alpha N$-lauroyl L-arginate) was ionically coupled with poly(γ-glutamic acid) (PGGA) to generate host-guest ionic complexes containing either stoichiometric molar amounts of the two components or an half amount of LAE. The complexes may be used to prepare consistent films that are non-water soluble and are thermally stable. Both complexes adopt a layered biphasic structure similar to that described for similar ionic complexes of PGGA made of alkyltrimethylammonium surfactants. Although LAE is a crystalline compound, the LAE·PGGA complexes are essentially amorphous with the lauroyl chains staying in a disordered state. LAE is released from the complexes upon incubation in aqueous buffer at a rate that is depending on pH and in less degree on the LAE/PGGA ratio. Films of these complexes displayed antibacterial activity against both Gram positive and Gram-negative strains, although the biocide effect on the former was much more ostensible. The biocide effect is motivated by both the LAE released to the aqueous environment and the fraction bounded to the complex. Since the two components that integrate the complexes are edible, these complexes offer interesting potential as antibacterial materials to be used in both food additives or packaging.

Supplementary Materials: The following are available online at www.mdpi.com/2073-4360/10/1/21/s1.

Acknowledgments: This work received financial support from MCINN (Spain) with Grant MAT2012-38044-C03 and MAT-2016-77345-CO3-03. Portions of this research were carried out at the BL11 line of ALBA synchrotron (Cerdanyola del Vallès, Barcelona, Spain) with the invaluable support of Christina Kamma-Lorger. Thanks to Marta Fernández García from CSIC (Madrid) for her valuable comments on the methodology followed for biocide activity estimation. Authors acknowledge the LAE sample kindly gifted by Vedeqsa (LAMIRSA GROUP, Terrassa, Barcelona, Spain). Thanks also to the MICINN for the Ph.D. grant awarded to Ana Gamarra Montes.

Author Contributions: Ana Gamarra-Montes designed and performed most of the experiments, and analyzed the results; Beatriz Missagia and Ana Gamarra-Montes designed and performed the bactericide assays; Jordi Morató supervised and discussed the microbiology related part. Sebastián Muñoz-Guerra conceived and supervised the research, and wrote the paper.

Conflicts of Interest: The authors declare no conflict of interest.

References

1. Mead, P.S.; Slutsker, L.; Dietz, V.; McCaig, L.F.; Bresee, J.S.; Shapiro, C.; Griffin, P.M.; Tauxe, R.V. Food-related illness and death in the United States. *Emerg. Infect. Dis.* **1999**, *5*, 607–625. [CrossRef] [PubMed]

2. Newell, D.G.; Koopmans, M.; Verhoef, L.; Duizer, E.; Aidara-Kane, A.; Sprong, H.; Opsteegh, M.; Langelaar, M.; Threlall, J.; Scheutz, F.; et al. Food-borne diseases. The challenges of 20 years ago still persist while new ones continue to emerge. *Int. J. Food Microbiol.* **2010**, *139*, S3–S15. [CrossRef] [PubMed]

3. Lai, E.P.C.; Iqbal, Z.; Avis, T.J. Combating Antimicrobial Resistance in Foodborne Microorganisms. *J. Food Prot.* **2016**, *79*, 321–336. [CrossRef] [PubMed]

4. Dobrucka, R. Antimicrobial packaging with natural compounds—A review. *LogForum* **2016**, *12*, 193–202. [CrossRef]

5. Fu, Y.; Sarkar, P.; Bhunia, A.K.; Yao, Y. Delivery systems of antimicrobial compounds to food. *Trends Food Sci. Technol.* **2016**, *54*, 165–177. [CrossRef]

6. Appendini, P.; Hotchkiss, J.H. Review of antimicrobial food packaging. *Innov. Food Sci. Emerg. Technol.* **2002**, *3*, 113–126. [CrossRef]

7. Muñoz-Bonilla, A.; Fernández-García, M. Polymeric materials with antimicrobial activity. *Prog. Polym. Sci.* **2012**, *37*, 281–339. [CrossRef]

8. Malhotra, B.; Keshwani, A.; Kharkwal, H. Antimicrobial food packaging: Potential and pitfalls. *Front. Microbiol.* **2015**, *6*, 611. [CrossRef] [PubMed]
9. Erickson, M.C.; Doyle, M.P. The challenges of eliminating or substituting antimicrobial preservatives in foods. *Ann. Rev. Food Sci. Technol.* **2017**, *8*, 371–390. [CrossRef] [PubMed]
10. Bajaj, I.; Singhal, R. Poly (glutamic acid)—An emerging biopolymer of commercial interest. *Bioresour. Technol.* **2011**, *102*, 5551–5561. [CrossRef] [PubMed]
11. Muñoz-Guerra, S.; García-Álvarez, M.; Portilla-Arias, J.A. Chemical modification of microbial poly(γ-glutamic acid). *J. Renew. Mater.* **2013**, *1*, 42–60. [CrossRef]
12. Pérez-Camero, G.; García-Álvarez, M.; Martínez de Ilarduya, A.; Fernández, C.; Campos, L.; Muñoz-Guerra, S. Comb-like complexes of bacterial poly(γ,D-glutamic acid) and cationic surfactants. *Biomacromolecules* **2004**, *5*, 144–152. [CrossRef] [PubMed]
13. García-Álvarez, M.; Álvarez, J.; Alla, A.; Martínez de Ilarduya, A.; Herranz, C.; Muñoz-Guerra, S. Comb-like ionic complexes of cationic surfactants with bacterial poly(γ-glutamic acid) of racemic composition. *Macromol. Biosci.* **2005**, *5*, 30–38. [CrossRef] [PubMed]
14. Portilla-Arias, J.A.; García-Alvarez, M.; Martínez de Ilarduya, A.; Muñoz-Guerra, S. Ionic complexes of biosynthetic poly(malic acid) and poly(glutamic acid) as prospective drug-delivery systems. *Macromol. Biosci.* **2007**, *7*, 897–906. [CrossRef] [PubMed]
15. Gamarra, A.; Martínez de Ilarduya, A.; Vives, M.; Morató, J.; Muñoz-Guerra, S. Ionic complexes of poly(γ-glutamic acid) with alkyltrimethylphosphonium surfactants. *Polymer* **2017**, *116*, 43–54. [CrossRef]
16. Lee, N.; Go, T.; Lee, S.; Jeong, S.; Park, G.; Hong, C.; Son, H. In vitro evaluation of new functional properties of poly-γ-glutamic acid) produced by *Bacillus subtilis* D7. *Saudi J. Biol. Sci.* **2014**, *21*, 153–158. [CrossRef] [PubMed]
17. Siracusa, V.; Rocculi, P.; Romani, S.; Marco, D. Biodegradable polymers for food packaging: A review. *Trends Food Sci. Technol.* **2008**, *19*, 634–643. [CrossRef]
18. Becerril, R.; Manso, S.; Nerín, C.; Gómez-Lus, R. Antimicrobial activity of Lauroyl Arginate Ethyl (LAE), against selected food-borne bacteria. *Food Control* **2013**, *32*, 404–408. [CrossRef]
19. Otero, V.; Becerril, R.; Santos, J.A.; Rodríguez-Calleja, J.M.; Nerín, C.; García-López, M.L. Evaluation of two antimicrobial packaging films against Escherichia coli O157:H7 strains in vitro and during storage of a Spanish ripened sheep cheese. *Food Control* **2014**, *42*, 296–302. [CrossRef]
20. Rodríguez, E.; Seguer, J.; Rocabayera, X.; Manresa, A. Cellular effects of monohydrochloride of L-arginine, N-α-lauroyl ethyl ester (LAE) on exposure to *Salmonella typhimurium* and *Staphylococcus aureus*. *J. Appl. Microbiol.* **2004**, *96*, 903–912. [CrossRef] [PubMed]
21. Ruckman, S.A.; Rocabayera, X.; Borzelleca, J.F.; Sandusky, C.B. Toxicological and metabolic investigations of the safety of N-α-lauroyl-L-arginine ethyl ester monohydrochloride (LAE). *Food Chem. Toxicol.* **2004**, *42*, 245–259. [CrossRef] [PubMed]
22. Hawkins, D.R.; Rocabayera, X.; Ruckman, S.; Segret, R.; Shaw, D. Metabolism and pharmacokinetics of ethyl N-α-lauroyl-L-arginate hydrochloride in human volunteers. *Food Chem. Toxicol.* **2009**, *47*, 2711–2715. [CrossRef] [PubMed]
23. Muriel-Galet, V.; Cran, M.J.; Bigger, S.W.; Hernández-Muñoz, P.; Gavara, R. Antioxidant and antimicrobial properties of ethylene vinyl alcohol copolymer films based on the release of oregano essential oil and green tea extract components. *J. Food Eng.* **2015**, *149*, 9–16. [CrossRef]
24. Muriel-Galet, V.; López-Carballo, G.; Gavara, R.; Hernández-Muñoz, P. Antimicrobial food packaging film based on the release of LAE from EVOH. *Int. J. Food Microbiol.* **2012**, *157*, 239–244. [CrossRef] [PubMed]
25. Guo, M.; Jin, T.Z.; Yang, R. Antimicrobial polylactic acid packaging films against *Listeria* and *Salmonella* in culture medium and on ready-to-eat meat. *Food Bioprocess Technol.* **2014**, *7*, 3293–3307. [CrossRef]
26. Ma, Q.; Zhang, Y.; Zhong, Q. Physical and antimicrobial properties of chitosan films incorporated with lauric arginate, cinnamon oil, and ethylenediaminetetraacetate. *LWT Food Sci. Technol.* **2016**, *65*, 173–179. [CrossRef]
27. Higueras, L.; López-Carballo, G.; Hernández-Muñoz, P.; Gavara, R.; Rollini, M. Development of a novel antimicrobial film based on chitosan with LAE (ethyl-N$^\alpha$-dodecanoyl-L-arginate) and its application to fresh chicken. *Int. J. Food Microbiol.* **2013**, *165*, 339–345. [CrossRef] [PubMed]
28. Pattanayaiying, R.; H-Kittikun, A.; Cutter, C.N. Optimization of formulations for pullulan films containing lauric arginate and nisin Z. *LWT Food Sci. Technol.* **2015**, *63*, 1110–1120. [CrossRef]

29. Kashiri, M.; Cerisuelo, J.P.; Domínguez, I.; López-Carballo, G.; Hernández-Muñoz, P.; Gavara, R. Novel antimicrobial zein film for controlled release of lauroyl arginate (LAE). *Food Hydrocoll.* **2016**, *61*, 547–554. [CrossRef]

30. Asker, D.; Weiss, J.; McClements, D.J. Analysis of the Interactions of a Cationic Surfactant (Lauric Arginate) with an anionic biopolymer (Pectin): Isothermal Titration Calorimetry, Light Scattering, and Microelectrophoresis. *Langmuir* **2009**, *25*, 117–122. [CrossRef] [PubMed]

31. Asker, D.; Weiss, J.; McClements, D.J. Formation and Stabilization of antimicrobial delivery systems based on electrostatic complexes of cationic-non-ionic mixed micelles and anionic polysaccharides. *J. Agric. Food Chem.* **2011**, *59*, 1041–1049. [CrossRef] [PubMed]

32. Loeffler, M.; McClements, D.J.; McLandsborough, L.; Terjung, N.; Chang, Y.; Weiss, J. Electrostatic interactions of cationic lauric arginate with anionic polysaccharides affect antimicrobial activity against spoilage yeasts. *J. Appl. Microbiol.* **2014**, *117*, 28–39. [CrossRef] [PubMed]

33. Ponomarenko, E.A.; Waddon, A.J.; Tirrell, D.A.; Macknight, W.J. Structure and properties of stoichiometric complexes formed by sodium poly(α,L-glutamate) and oppositely charged surfactants. *Langmuir* **1996**, *12*, 2169–2172. [CrossRef]

34. Tolentino, A.; Alla, A.; Martínez de Ilarduya, A.; Muñoz-Guerra, S. Comb-like ionic complexes of pectinic and alginic acids with alkyltrimethylammonium surfactants. *Carbohyd. Polym.* **2011**, *86*, 484–490. [CrossRef]

35. Tolentino, A.; Alla, A.; Martínez de Ilarduya, A.; Muñoz-Guerra, S. Comb-like ionic complexes of hyaluronic acid with alkyltrimethylammonium surfactants. *Carbohyd. Polym.* **2013**, *92*, 691–696. [CrossRef] [PubMed]

36. Durán, N.; Marcato, P.D.; De Souza, G.I.H.; Alves, O.L.; Espósito, E. Antibacterial effect of silver nanoparticles produced by fungal process on textile fabrics and their effluent treatment. *J. Biomed. Nanotechnol.* **2007**, *3*, 203–208. [CrossRef]

37. Portilla-Arias, J.A.; García-Álvarez, M.; Martínez de Ilarduya, A.; Muñoz-Guerra, S. Thermal decomposition of microbial poly(γ-glutamic acid) and poly(γ-glutamate)s. *Polym. Degrad. Stab.* **2007**, *92*, 1916–1924. [CrossRef]

38. Tolentino, A.; Leon, S.; Alla, A.; Martínez de Ilarduya, A.; Muñoz-Guerra, S. Comblike ionic complexes of poly(γ-glutamic acid) and alkanoylcholines derived from fatty acids. *Macromolecules* **2013**, *46*, 1607–1617. [CrossRef]

39. Kubota, H.; Nambu, Y.; Endo, T. Alkaline hydrolysis of poly(γ-glutamic acid) produced by microorganism. *J. Polym. Sci. Part A Polym. Chem.* **1996**, *34*, 1347–1351. [CrossRef]

40. Ma, Q.; Davidson, P.M.; Zhong, Q. Antimicrobial properties of lauric arginate alone or in combination with essential oils in tryptic soy broth and 2% reduced fat milk. *Int. J. Food Microbiol.* **2013**, *166*, 77–84. [CrossRef] [PubMed]

41. Vaara, M. Agents that increase the permeability of the outer membrane. *Microbiol. Rev.* **1992**, *56*, 395–411. [PubMed]

42. Oliver, J.D. The viable but nonculturable state in bacteria. *J. Microbiol.* **2005**, *43*, 93–100. [PubMed]

polymers

MDPI

Article

Effect of Dendrigraft Generation on the Interaction between Anionic Polyelectrolytes and Dendrigraft Poly(L-Lysine)

Feriel Meriem Lounis, Joseph Chamieh, Laurent Leclercq, Philippe Gonzalez, Jean-Christophe Rossi and Hervé Cottet *

IBMM, Université de Montpellier, CNRS, ENSCM, 34095 Montpellier, France; lounisferiel@hotmail.fr (F.M.L.); Joseph.chamieh@umontpellier.fr (J.C.); laurent.leclercq@umontpellier.fr (L.L.); Philippe.gonzalez@univ-montp2.fr (P.G.); jean-christophe.rossi@umontpellier.fr (J.-C.R.)
* Correspondence: herve.cottet@umontpellier.fr; Tel.: +33-4-6714-3427

Received: 23 November 2017; Accepted: 28 December 2017; Published: 4 January 2018

Abstract: In this present work, three generations of dendrigraft poly(L-Lysine) (DGL) were studied regarding their ability to interact with linear poly (acrylamide-*co*-2-acrylamido-2-methyl-1-propanesulfonate) (PAMAMPS) of different chemical charge densities (30% and 100%). Frontal analysis continuous capillary electrophoresis (FACCE) was successfully applied to determine binding constants and binding stoichiometries. The effect of DGL generation on the interaction was evaluated for the first three generations (G2, G3, and G4) at different ionic strengths, and the effect of ligand topology (linear PLL vs. dendrigraft DGL) on binding parameters was evaluated. An increase of the biding site constants accompanied with a decrease of the DGL-PAMAMPS (n:1) stoichiometry was observed for increasing DGL generation. The logarithm of the global binding constants decreased linearly with the logarithm of the ionic strength. This double logarithmic representation allowed determining the extent of counter-ions released from the association of DGL molecules onto one PAMAMPS chain that was compared to the total entropic reservoir constituted by the total number of condensed counter-ions before the association.

Keywords: polyelectrolyte complexes; dendrimers; frontal analysis continuous capillary electrophoresis; counter-ion release; binding constants; ionic strength dependence

---

1. Introduction

Dendrimers are nano-sized, radially symmetrical molecules with well-defined and monodisperse structure consisting of tree-like arms or branches [1]. Due to their exceptional architecture, dendrimers have found various applications in supramolecular chemistry, particularly in host–guest reactions and self-assembly processes. They constitute very promising candidates in many biomedical applications because of their possibility to perform controlled and specified drug delivery [2–5], their use in anticancer therapy [6–8] and imaging diagnostic analysis [9,10]. Dendrimers can also be used as solubility enhancers [11–14], for layer-by-layer deposition [15] and catalysis [16–18]. Complexes containing dendrimers and linear polyelectrolytes or (bio)macromolecules have attracted great attention [19–23]. Some experimental and theoretical investigations were interested in determining size and structural properties of linear-dendritic polyelectrolyte complexes [22,24–30]. However, little is known about the influence of ramification on the thermodynamic binding parameters (stoichiometry, binding constant, enthalpy, entropy) of such polyelectrolyte complexes (PEC).

Giri et al. [19] studied the binding of human serum albumin (HSA) and poly(amidoamine) (PAMAM) dendrimers. Binding constants gradually increased with dendrimer generation (from 1.67×10^5 M^{-1} for G0 to 5.42×10^6 M^{-1} for G6) followed by a slight decrease for G8 dendrimer

$(3.3 \times 10^6 \text{ M}^{-1})$. This study showed that binding constants depended on the chemical structure of the core and the terminal group of dendrimers. Furthermore, DNA-PAMAM dendrimer complex stability and binding constant were found to increase with dendrimer generation [31]. Kabanov et al. [28] showed that the complexes of poly(propylene imine) dendrimers with DNA or synthetic linear polyanions containing equal amounts of cationic and anionic groups were stoichiometric and insoluble in water. Water-soluble non-stoichiometric complexes were obtained when dendrimers were introduced in default in the complex.

Dendrigraft poly-L-lysine (DGL) are dendritic synthetic cationic polypeptides synthesized by successive polycondensation of *N*-trifluoroacetyl-L-lysine-*N*-carboxyanhydride in water. Compared to dendrimers, DGL have a linear core (and not point core) and more flexible structures [32]. In previous investigations, frontal analysis continuous capillary electrophoresis (FACCE) was found to be a straightforward method to study interactions between dendrigraft poly(L-Lysine) (DGL) (G3) and oppositely charged biomolecules such as adenosine monophosphate (AMP), adenosine triphosphate (ATP) ligands [21], and human serum albumin (HSA) [33,34]. HSA-G3 interactions studies under physiological conditions, demonstrated that HSA had two cooperative binding sites with G3 with the following successive constants $K_1 = 31.2 \times 10^3 \text{ M}^{-1}$ and $K_2 = 30.6 \times 10^3 \text{ M}^{-1}$. Increasing DGL generation (G1 to G5) led to an increase of the binding constant accompanied with a decrease of the HSA:DGL (1:*n*) stoichiometry and a decrease of the cooperativity with dendrimer generation [33].

In the present work, we propose to study the effect of DGL generation and the influence of the polylysine topology by comparison with a linear poly-L-Lysine (PLL) on polyelectrolyte complexes (PEC) formation. The interaction between DGL (G2–G4) (or linear PLL) and statistical copolymers of acrylamide and 2-acrylamido-2-methyl-1-propanesulfonate (PAMAMPS) with chemical charge densities of 30% and 100%, were investigated by FACCE. The DGL-PAMAMPS interactions and PLL-PAMAMPS interactions are discussed in term of stoichiometry, binding constants, and amount of released counter-ions during the complex formation. This study represents a quantitative investigation of how the ionic strength, the chemical charge density and the topology (linear vs. dendritic) of polyelectrolytes influence the thermodynamic binding parameters when PEC are formed. This works also brings new experimental data about oppositely charged macromolecules, including the experimental estimation of condensed and released counter-ions, which constitutes a topic of interest from a theoretical point of view [35,36].

2. Materials and Methods

2.1. Chemicals

Random copolymers of acrylamide and 2-acrylamido-2-methyl-1-propanesulfonate (PAMAMPS) with chemical charge densities *f* of 30% and 100% were synthesised by free radical copolymerization as described by McCormick et al. [37]. The details of the synthesis are reported elsewhere [38] and briefly described in the Supplementary. DP_w of PAMAMPS 30% (respectively 100%) were 3689 (respectively 4166) as obtained by Size-Exclusion Chromatography Coupled with Multi-Angle Laser Light Scattering (SEC-MALLS) and published elsewhere [38]. Molar mass distributions obtained by SEC-MALLS and charge density (or chemical composition) distributions obtained by capillary electrophoresis are available in supporting information of a previously published study [38]. Average chemical composition was also confirmed by ^{1}H NMR [38]. Poly-L-Lysine (PLL) (with a degree of polymerization $DP_n = 50$, corresponding to a molar masse of 8200 g/mol and a polydispersity index 1.04) was supplied by Alamanda Polymers (Huntsville, AL, USA). DGL (G2, G3, and G4) (batch numbers; DC 120902, DC 120103, DC 130604, respectively) were supplied by Colcom (Montpellier, France). Polydiallyldimethylammonium chloride (PDADMAC), $M_w = 400$–500 kDa, ammonium bicarbonate and sodium azide were purchased from Sigma Aldrich (St Quentin Fallavier, France). Tris(hydroxymethyl)aminomethane (Tris, $(CH_2OH)_3CNH_2$, 99.9%) was purchased from Merck (Darmstadt, Germany). Hydrochloric acid 37%, sodium hydroxide and sodium chloride were purchased from VWR (Leuven, Belgium). 2-2Bis(hydroxymethyl)-2,2',2''-nitrilotriethanol (Bis Tris, 99%) was purchased from Acros Organics (Geel, Belgium). Cellulose ester dialysis membrane of 100 Da (reference number: 131 018) was purchased from Spectrum Labs (Rancho Dominguez, CA, USA). Durapore© membrane filters were purchased from Merck Millipore (Darmstadt, Germany). Deionised water was further purified using a Milli-Q system (Millipore, Molsheim, France). All chemical (except DGL) were used without any further purification.

2.2. Intergeneration Purification of DGL by Semi-Preparative Size-Exclusion Chromatography

Intergeneration purification of DGL was realised by semi-preparative size exclusion chromatography (SEC) in order to remove any residual DGL molecules of the previous generations or any residual salts. SEC purifications were carried out on an Äkta purifier 100 GE healthcare system (Vélizy-Villacoublay, France) with a UV detector set at 210 nm. A Superdex 200 column (30 cm × 1 cm) with an exclusion domain between 1000 and 100,000 Da was used for DGL intergeneration separation. The particles of this column had a diameter between 22 and 44 μm. The mobile phase was ammonium bicarbonate 0.1 M at pH 11 at a flow rate of 1 mL/min. DGL samples were prepared at 30 g/L in mobile phase, and 500 μL were manually injected. The purified fractions were evaporated to eliminate the excess of the eluent and then freeze-dried. For DGL-PAMAMPS interactions, purified DGL were dissolved in a 10^{-4} M HCl solution, dialysed against the same HCl solution and then freeze-dried.

2.3. Measurement of DGL Refractive Index Increments

Before carrying out the refractive index measurements, a 2 g/L mother solution of each purified DGL generation dissolved in the eluent was dialysed against the eluent using a cellulose ester dialysis membrane of 100 Da. The eluent was the same for dialysis, refractive index increment measurements and molar mass distribution experiments. It was composed of 50 mM Bis Tris, 1 M sodium chloride, 0.3 g/L sodium azide. pH was adjusted to 6 using 1 M hydrochloric acid solution. The eluent was finally filtrated using Durapore membrane filters (0.1 μm cutoff, Millipore, Molsheim, France). The refractive index increments (dn/dC) of different DGL generations were determined at 35 °C using a Shimadzu RID-6A (Tokyo, Japan) refractive detector set at 690 nm. The instrument was calibrated using sodium chloride solutions of various known concentrations (2.0, 1.5, 1.0, 0.75, 0.5 and 0.25 g/L) to get a dn/dC value of 0.187. DGL solutions were prepared at 2.0, 1.5, 1.0, 0.75, 0.5 and 0.25 g/L by diluting the dialysed mother solution with the eluent. A volume of 2 mL of each DGL solution was injected into the refractometer. The refractive index increments (dn/dC) were calculated using the Astra software (v6.1.1.17, Wyatt Technology Corp., Santa Barbara, CA, USA).

2.4. Determination of the Molar Mass Distribution of DGL by Size-Exclusion Chromatography Coupled with Multi-Angle Laser Light Scattering (SEC–MALLS)

The weight-average molar masses (M_w) and polydispersity indexes (PDI) of purified DGL were determined using size-exclusion chromatography coupled with multi-angle laser light scattering (SEC–MALLS). Dialyzed samples at 2 g/L of DGL in the eluent (see previous section) were eluted using a Thermo Scientific Ultimate module 3000 separations at a flow rate of 0.8 mL/min equipped with column guard SHODEX OHpak SBG (50 × 6 mm) and two columns SHODEX SB-806M-HQ (300 × 8 mm) (Munich, Germany) connected in series. The eluent used for SEC-MALLS analyses was the same as described in the previous section. The eluted samples were detected using a mini DAWN-TREOS three-angles (45°, 90°, 135°) laser light scattering detector with a laser at 690 nm (Wyatt Technology Corp., Santa Barbara, CA, USA) and a RID-6A refractive index monitor Shimadzu (Tokyo, Japan) at a thermostated temperature of 35 °C. The data for molar mass determination were analysed using ASTRA software (v6.1.1.17, Wyatt Technology Corp., Santa Barbara, CA, USA).

2.5. Preparation of Polyelectrolytes Mixtures

PAMAMPS, PLL and DGL stock solutions were prepared in 12 mM Tris, 10 mM HCl and NaCl buffer (pH 7.4) at room temperature. The ionic strength of the buffer was adjusted by adding adequate amounts of NaCl. The concentrations of PAMAMPS stock solutions were 2 and 1.14 g/L for PAMAMPS 30% and 100%, respectively. The concentration of PLL and DGL stock solutions was 5 g/L. Diluted PLL and DGL solutions with concentrations from 0.1 to 4 g/L for polyelectrolytes mixtures and from 0.1 to 2 g/L for calibration curves, were prepared by dilution in the same Tris-HCl-NaCl buffer. PLL-PAMAMPS and DGL-PAMAMPS mixtures were prepared by adding 100 μL of the polyanionic stock solutions to 100 μL of the polycationic solutions (see Tables S1 and S2 for the concentrations of PAMAMPS, PLL, and DGL in the mixtures). The final mixtures with a volume of 200 μL were equilibrated by homogenizing with a vortex stirrer during 1 min. DGL-PAMAMPS mixtures were incubated for 12 h and then analysed by FACCE.

2.6. FACCE Procedure

Capillary electrophoresis experiments were carried out using an Agilent 3D system (Agilent, Waldbronn, Germany). Separations were realized using bare fused silica capillaries from Polymicro Technologies (Photonlines, Saint-Germain-en-Laye, France). Capillary dimensions were 50 μm internal diameter (i.d.) × 33.5 cm total length (8.5 cm to the detector). New capillaries were flushed for 30 min with a 1.0 M NaOH solution and with water for 20 min. The capillary inner surfaces were then coated with polycationic polymer by flushing the capillary for 20 min with a 0.2% *w/w* poly diallyldimethylammonium chloride (PDADMAC) solution prepared in a 2 M NaCl solution. Before each run the capillary was flushed with water for 2 min, PDADMAC 0.2% *w/w* in water for 3 min and finally Tris-HCl-NaCl buffer for 3 min. To reduce the migration times, samples were placed at the capillary end, which is the closest to the detection point (8.5 cm). The temperature of the capillary was kept at 25 °C and the detection wavelength was set at 200 nm. FACCE experiments were achieved by applying a continuous positive polarity voltage of +1 kV (from the injection end) and a *co*-pressure of 5 mbar (from the injection end) for PLL molecules and 4 mbar for DGL molecules, in order to allow the continuous electrokinetic entry of the free ligand molecules (free DGL or free PLL) contained in the equilibrated mixtures. These voltage and pressure conditions allowed the selective entry and the quantification of the free ligand preventing the entry of free PAMAMPS and PEC molecules.

3. Results and Discussions

3.1. Characterization of Purified DGL

Intergeneration purity of DGL turned out to be important in order to make accurate measurements of binding parameters between the DGL and the linear polyanions. For that reason, each DGL generation was purified by semi-preparative SEC as described in Section 2.2 and further analysed by SEC-MALLS analysis for the determination of molar mass distribution. In SEC analysis, the separation of DGL was exclusively governed by size exclusion. The refractive index increments were measured for each DGL-Cl generation after dialysis against the eluent to ensure that the chemical potential is constant in the eluent and in the injected DGL solution (see Section 2.3). Relatively narrow molar mass distributions were obtained (see Figure S1 in the Supplementary Materials for the molar mass distributions of the purified DGL). Table 1 reports the values of the refractive index increment (dn/dC), the weight-average molar mass (M_w) and the polydispersity index (*PDI*) for each DGL-Cl generation. It is known that, for any dendritic structure including DGL, the molar mass increases exponentially with increasing dendrimer generation number [33], which explains the high increase of the molar mass with the increase in generation number. The weight-average degree of polymerization DP_w

was obtained by dividing the weight-average molar masses (M_w) by the average molar mass of the monomer (M_0) taking into account the fraction of condensed counter-ions [39] according to:

$$M_0 = \left(M_{lys} \times \left(1 - \theta^+ \right) \right) + \left(M_{lys,Cl^-} \times \theta^+ \right) \tag{1}$$

where M_{lys} is the molar mass of a lysine residue (M_{lys} = 129 g/mol), M_{lys,Cl^-} is the molar mass of a lysine residue + Cl^- counter-ion (M_{lys,Cl^-} = 164.5 g/mol), θ^+ is the fraction of condensed counter-ions.

Table 1. Refractive index increment (dn/dC), weight-average molar masses (M_w) and polydispersity index (PDI) determined by SEC-MALLS for each purified DGL generation. M_0 is the average molar mass of a lysine monomer taking into account thefractionof cendensed counter-ions (see Equation (1)) and θ^+ is the fraction of condensed counter-ions [40]. Complete molar mass distributions are given in Figure S1.

Generation	θ^+	M_0 (g/mol)	DP_n	DP_w	Counter-Ion	dn/dC	M_w (g/mol)	PDI
G2	0.35	141	57	60		0.1756	8400	1.05
G3	0.65	152	136	137	Cl^-	0.1989	20723	1.01
G4	0.76	156	367	378		0.1749	58916	1.03

3.2. Determination of DGL-PAMAMPS Binding Parameters by FACCE

It is obvious that the interactions between DGL and PAMAMPS are expected to be highly dependent on the DGL generation and very different compared to poly(L-lysine)/PAMAMPS interactions, due to important changes in polycation topology. To examine the effect of dendrimer generation on the thermodynamic binding parameters, isotherms of adsorption were plotted for three successive generations of DGL (G2, G3, and G4) in interaction with linear polyanions of different charge densities (PAMAMPS 30% and PAMAMPS 100%). In this work, we have used a PAMAMPS 30% (instead of 15% in our former publication about PLL/PAMAMPS interactions [41]) because DGL/PAMAMPS 15% mixtures required very long equilibrium times (more than 24 h), while in the case of the PDGL/PAMAMPS 30% the equilibrium was reached in less than 12 h. For that purpose, FACCE methodology developed by Sisavath et al. [34] was used. In this method, a continuous voltage and a *co*-pressure were simultaneously applied in order to introduce selectively the free ligand (free DGL) in the capillary, avoiding the dynamic dissociation of the complex during electrophoretic migration. As a result, the free ligand continuously entering in the capillary is detected as a plateau, the height of which is proportional to the free ligand concentration at equilibrium in the mixture. The free ligand concentrations were determined using a calibration curve performed for each DGL generation in the same electrophoretic conditions. Examples of electropherograms obtained by FACCE for DGL G3 in the presence of PAMAMPS 30%, at 552 mM ionic strength are given in Figure 1A. Isotherms of adsorption were plotted by representing the average number of bound ligands (DGL) per substrate molecule (PAMAMPS) $\bar{n}$ (calculated according to Equation (2)) vs. the free ligand concentration [DGL] for different initial molar ratio [DGL]$_0$/[$PAMAMPS$]$_0$. The stoichiometry of interaction n expressed in term of bound DGL entities per PAMAMPS chain and the binding site constant k were determined by non-linear least square routine on Microsoft excel using the model of identical interacting sites [42] expressed by Equation (3).

$$\bar{n} = \frac{[DGL]_0 - [DGL]}{[PAMAMPS]_0} \tag{2}$$

$$\bar{n} = \frac{nk[DGL]}{1 + k[DGL]} \tag{3}$$

where [DGL]$_0$ and [DGL] are the initial and free ligand concentrations, respectively. [$PAMAMPS$]$_0$ is the concentration of PAMAMPS initially introduced in the mixture. It is worth noting that the intrinsic binding constant k refers to the association of one ligand (one DGL molecule) onto a binding site $-s$ corresponding to a small portion of PAMAMPS:

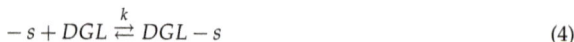

$$-s + DGL \overset{k}{\rightleftarrows} DGL - s \tag{4}$$

$$k = \frac{[DGL - s]}{[DGL][-s]} \tag{5}$$

where $[DGL]$ and $[DGL-s]$ are the free and complexed DGL concentrations at equilibrium respectively, $[-s]$ is the concentration of free sites.

An example of isotherm of adsorption for the interaction G3-PAMAMPS 30% at 552 mM ionic strength is given in Figure 1B (see Figures S2–S7 in the Supplementary Materials for the display of all the isotherms of adsorption for the other DGL generations and ionic strengths).

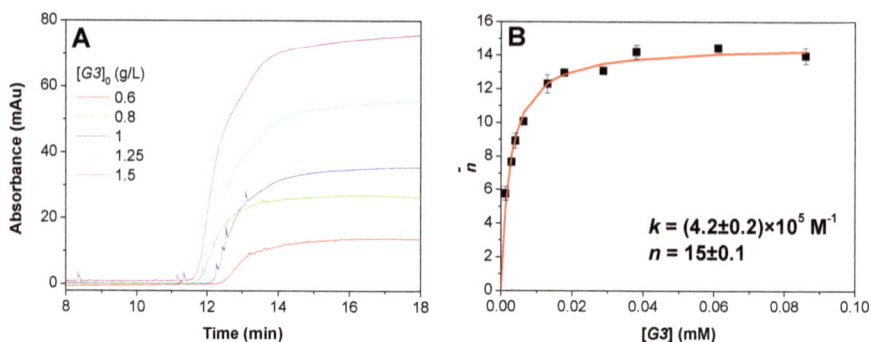

Figure 1. Example of determination of binding parameters (n and k) by FACCE for the interaction G3-PAMAMPS 30% at 552 mM ionic strength. Electropherograms obtained for different G3-PAMAMPS 30% equilibrated mixtures (**A**); and the corresponding isotherm of adsorption (**B**) representing the number of bound G3 entities per PAMAMPS 30% chain according to the free G3 concentration. Experimental conditions: PDADMAC coated capillary 33.5 cm (8.5 cm to the detector) × 50 µm i.d. Background electrolyte: 12 mM Tris, 10 mM HCl, 542 mM NaCl, pH 7.4. Applied voltage + 1 kV with a *co*-hydrodynamic pressure of +4 mbar. Detection at 200 nm. Samples were prepared in the background electrolyte by 50/50 *v/v* dilution of the following solutions: PAMAMPS 30% at 2 g/L with G3 at 5, 4, 3, 2.5, 2, 1.6, 1.2, 1, 0.8, 0.6 g/L.

In order to get the ionic strength dependence of the binding constants and stoichiometries, isotherms of adsorption were determined at different ionic strengths I for each DGL generation and for two PAMAMPS chemical charge density (30% and 100%). However, accurate measurements of the binding parameters by FACCE required the adjustment of the ionic strengths to avoid unmeasurable high binding constants at low ionic strength, or low affinities due to the dissociation of PEC at high ionic strength. For that, the ionic strength of recomplexation I_{recomp} (defined as the salt concentration at which a solid DGL-PAMAMPS complex previously destabilized at high ionic strength re-formed when water was added) was first determined by turbidimetry, as previously described [38]. Figure 2 shows the variation of I_{recomp} according to the molar masses of the purified DGL in double logarithmic scale. As can be seen in Figure 2, the logarithm of the salt concentration required to dissociate a DGL-PAMAMPS complex varied linearly with the logarithm of the DGL molar masses. The binding parameters n and k were measured in a range of ionic strength from 40% to 90% of I_{recomp} (i.e., between 220–747 mM for the interactions DGL-PAMAMPS 30%; and between 870–2011 mM for the interactions

DGL-PAMAMPS 100%) yielding measurable binding site constants k between 1.6×10^4 M^{-1} and 7.7×10^6 M^{-1}.

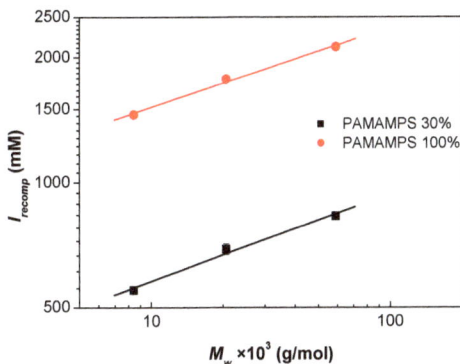

Figure 2. Variation of the ionic strength of recomplexation I_{recomp} as a function of the molar mass of the purified DGL.

3.3. Influence of the Ionic Strength and DGL Generation on the Stoichiometry of Interaction

The stoichiometry of interactions $n_{(DGL/PAMAMPS)}$ expressed in term of bound DGL molecules per PAMAMPS chain at saturation of the isotherms (i.e., in the presence of an excess of DGL) is presented in Figure 3. The ionic strength did not significantly influence the binding stoichiometry whatever the PAMAMPS chemical charge density and the DGL generation number. This result is similar to what was observed in a previous work for linear PLL [41–43]. The fluctuations in stoichiometry were attributed to experimental errors. It was found that the stoichiometry $n_{(DGL/PAMAMPS)}$ decreased when the generation number of DGL increased, in good agreement with the increase in molar mass, as observed for linear poly(L-lysine) [41].

The charge stoichiometry expressed in terms of lysine residues per AMPS monomers ($n_{(Lys/AMPS)}$) is given in Tables 2 and 3 for PAMAMPS 30% and 100%, respectively. As for PAMAMPS 30%, $n_{(Lys/AMPS)}$ is higher than one whatever the DGL generation number, and also for PLL 50. This is in good agreement with the general rule recently enounced [38] stating that, when the highest charged polyelectrolyte partner (PLL or DGL in that case) are introduced in excess, then the formed PEC has a stoichiometry in favour of the highest charged polyelectrolyte (here the polycation, i.e., $n_{(Lys/AMPS)} > 1$). DGL G2 and the linear PLL50 (*DP* 50) have almost the same molar masses (8400 and 8200 g/mol, respectively). It is thus interesting to consider the effect of the ramification on the PEC stoichiometry at comparable molar masses. Interestingly, the charge stoichiometry was, in average, slightly lower for DGL G2 (1.31) than for PLL50 (1.56), as if the charge parameter of DGL G2 was lower than for PLL50. On the other hand, for DGL G3 and G4, the $n_{(Lys/AMPS)}$ stoichiometry was higher than for PLL50, as if the DGL G3 and G4 were more charged compared to PLL50. Even if the charge parameter is hardly accessible for the DGL, or even not well-defined, one can compare the counter-ion condensation rates θ^+ that were previously determined by isotachophoresis [40], as an indication of the polyelectrolyte charge density (the higher the condensation rate, the higher the chemical charge density or charge parameter). Finally, it was observed that the charge stoichiometry correlated well with the counter-ion condensation rate, as shown in Table 2.

Table 2. Physico-chemical properties of oppositely charged polyelectrolytes (DGL G2 G3 or G4, PLL50 and PAMAMPS 30%) and the corresponding parameters of interaction obtained by FACCE. All these parameters were obtained by curve fitting of the isotherms of adsorption (see Figure S2–S4 in the Supplementary Materials for the isotherms of adsorption).

f (%)[a]	θ−[b]	N_{Na^+}[c]	Polycation	DP_n[d]	θ+[e]	I (M)	$n_{(DGL/PAMAMPS)}$	⟨n⟩[f]	$n_{(Lys/AMPS)}$	⟨n⟩[g]	$N_{counter-ions}$[h]	$-\frac{\partial\langle n\rangle\log k}{\partial\log k}$	Average% of Released Counter-Ions
30	0	0	PLL50	50	0.5	0.327	35	35	1.60	1.56	887	258 ± 113	29 ± 3
						0.37	32		1.42		792		
						0.458	37		1.56		916		
						0.49	37		1.66		919		
			G2	57	0.35	0.22	29	28	1.35	1.31	584	215 ± 21	38 ± 4
						0.25	29		1.36		587		
						0.33	27		1.24		537		
						0.44	28		1.3		560		
			G3	136	0.65	0.41	17	17	1.93	1.96	102	108 ± 20	7 ± 2
						0.48	20		2.45		1735		
						0.55	15		1.64		1290		
						0.62	16		1.82		1420		
			G4	367	0.76	0.5	6	6	1.92	1.97	1653	43 ± 2	3 ± 0.2
						0.58	7		2.12		1827		
						0.64	6		1.94		1673		
						0.75	6		1.89		1632		

[a] The DP_w of PAMAMPS 30% is 3689; [b] The fraction of condensed charged monomers on PAMAMPS 30% chain; see reference [44]; [c] The number of Na$^+$ counter-ions condensed onto a PAMAMPS 30% chain calculated as in reference [42]; [d] The degree of polymerisation of polycations; [e] The fraction of condensed charged monomers on PLL or DGL molecules; see reference [42]; [f] The average interactions stoichiometry expressed in term of PLL or DGL molecules bound per PAMAMPS chain; [g] The average interactions stoichiometries expressed in term of lysine residues bound per AMPS monomers; [h] The total entropic reserve of initially condensed counter-ions calculated as in reference [42].

Table 3. Physico-chemical properties of oppositely charged polyelectrolytes (DGL G2 G3 or G4, PLL50 and PAMAMPS 100%) and the corresponding parameters of the interactions obtained by FACCE. All these parameters were obtained by curve fitting of the isotherms of adsorption (See Figure S5 to S7 in the Supplementary Materials for the isotherms of adsorption).

f (%) [a]	θ− [b]	N_{Na^+} [c]	Polycation	DP_n [d]	θ+ [e]	I (M)	$n_{(DGL/PAMAMPS)}$	$\langle n \rangle$ [f]	$n_{(Lys/AMPS)}$	$\langle n \rangle$ [g]	$N_{counter-ions}$ [h]	$-\frac{\partial \langle n \rangle \log k}{\partial \log I}$	Average% of Released Counter-Ions
50	0.67	2794	PLL50	50	0.5	1.1	80	82	1.01	0.99	4801	876 ± 338	18 ± 2
						1.2	81		0.97		4820		
						1.3	88		1.05		4988		
						1.4	80		0.96		4799		
			G2	57	0.35	0.87	97	89	1.29	1.11	4722	372 ± 150	8 ± 3
						1.02	94		1.12		4674		
						1.16	78		0.94		4342		
100			G3	136	0.65	1.42	41	39	1.23	1.18	6403	226 ± 36	4 ± 1
						1.51	46		1.37		6827		
						1.6	34		1.07		5842		
						1.68	36		1.06		5940		
			G4	367	0.76	1.69	23	20	1.96	1.75	9153	204 ± 37	2 ± 0.5
						1.80	19		1.63		8071		
						1.91	19		1.65		8132		
						2.01	20		1.74		8440		

[a] The DP_w of PAMAMPS 100% is 4166; [b] The fraction of condensed charged monomers on PAMAMPS 100% chain; see reference [44]; [c] The number of Na+ counter-ions condensed onto a PAMAMPS 100% chain calculated as in reference [42]; [d] The degree of polymerisation of polycations; [e] The fraction of condensed charged monomers on PLL or DGL molecules; see reference [40]; [f] The average interactions stoichiometry expressed in term of PLL or DGL molecules bound per PAMAMPS chain; [g] The average interactions stoichiometries expressed in term of lysine residues bound per AMPS monomers; [h] The total entropic reserve of initially condensed counter-ions calculated as in reference [42].

As for the PLL50/PAMAMPS 100% interaction, the polyanion is the highest charged polyelectrolyte and it is introduced in default at saturation of the isotherm. In that case, the rule states that the charge stoichiometry tends to 1 [38], as observed experimentally for that system ($n_{(Lys/AMPS)}$ = 0.99). In the case of the DGL/PAMAMPS 100%, the $n_{(Lys/AMPS)}$ stoichiometry seems to be higher than one, and seems to increase with the increase in generation number, as if the charge density (or charge parameter) of the DGL increased with the DGL generation and becomes higher than the PAMAMPS 100% linear charge density. Again, these results correlate well with the increase of the counter-ion condensation with the DGL generation, i.e., an increase of the charge parameter with the DGL generation.

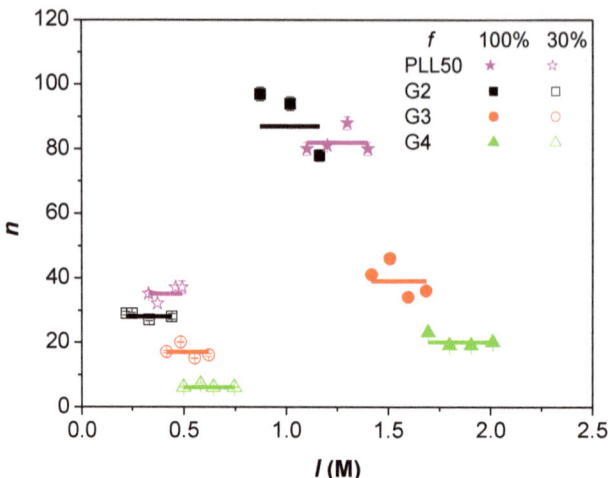

Figure 3. Variation of the interaction stoichiometry $n_{(Lys/AMPS)}$, expressed as the number of DGL molecules per PAMAMPS chain, as a function of the ionic strength for the interactions between DGL (or PLL50) with PAMAMPS 30% and PAMAMPS 100%.

3.4. Influence of the Ionic Strength and DGL Generation on the Binding Constants

According to the model of identical interacting sites previously described for the study of PLL/PAMAMPS interactions [41,42], the binding site constant k represents the equilibrium constant relative to the interaction between one DGL molecule (one ligand) with one interacting site $-s$ of a PAMAMPS chain (substrate). This interacting site $-s$ is defined as a fragment of PAMAMPS chain carrying electrostatic anchoring AMPS monomers. The variation of binding site constants k with the ionic strength is represented in Figure 4, where log k decreased linearly with log I. Moreover, for a given ionic strength, the higher the generation is, the higher the binding constant is, for both DGL-PAMAMPS 30% and DGL-PAMAMPS 100% systems. This increase of binding site constant can be qualitatively explained by the increase of the number of possible electrostatic anchoring points on each ligand, which grows exponentially with increasing DGL generation number.

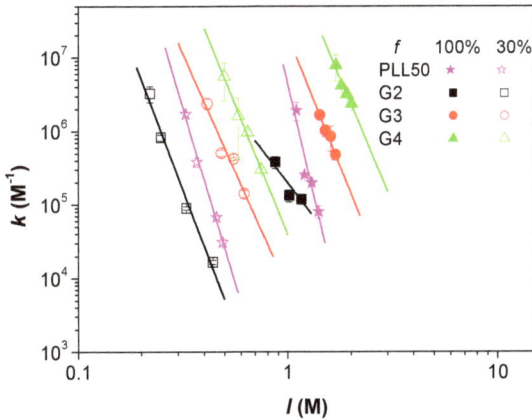

Figure 4. Variation of the binding site constant k as a function of the ionic strength I for the interactions between DGL (or PLL50) with PAMAMPS 30% and PAMAMPS 100%.

The global binding constant β_n, representing the equilibrium associated to the full binding of the n sites of the substrate [42] is related to the binding site constant by Equation (6)

$$\beta_n = k^n \tag{6}$$

As observed for PLL/PAMAMPS systems [41–43] and for other systems in the literature [45–53], the logarithm of β_n was found to decrease linearly with the ionic strength of the medium (see Figure 5) with an experimental slope representing the number of the counter-ions effectively released during the formation of one PEC composed of one PAMAMPS chain and n ligands. This double logarithmic dependence was attributed to the entropically dominant character of oppositely charged polyelectrolytes interactions. In Figure 5, log β_n was calculated using log β_n = <n> × log k, where <n> is the average value of stoichiometry on the four ionic strengths investigated for each DGL (or PLL50). It can be seen from Figure 5, that the interaction of PAMAMPS with higher dendrimer generations is less sensitive to the ionic strength than lower generations and linear PLL, this observation is in perfect agreement with the computer simulations in the literature [54].

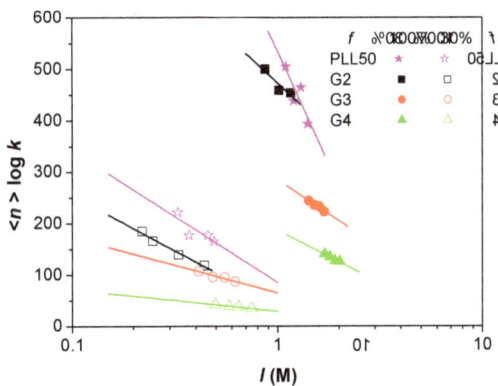

Figure 5. Variation of log β_n as a function of log I for the interactions between DGL (or PLL50) with PAMAMPS 30% and PAMAMPS 100%.

The number of counter-ions that are effectively released from the association of n ligands onto one PAMAMPS chain can be determined by calculating $-\frac{\partial <n> \log k}{\partial \log I}$ [41–43]. These numerical values are reported in Tables 2 and 3 for PAMAMPS 30% and 100% respectively, and are compared to the one obtained for linear PLL50-PAMAMPS interactions (see Figure 6). It can be observed that the higher the DGL generation number is, the lower the number of counter-ions effectively released is. For all DGL generations, the number of released counter-ions is lower than that observed for PLL50. It was previously observed for linear PLL of different molar masses that the number of released counter-ions decreases with increasing PLL molar mass, due to the formation of longer PAMAMPS loops in the PEC structure. In the case of DGL, the decrease of the number of released counter-ions may be due to the non-accessibility of the PAMAMPS chain into the DGL structure which may become even more dramatic for higher DGL generation. Figure 7 displays a schematic image of the interaction on one binding site in the case of PAMAMPS 30%, using the charge stoichiometry determined in this work. The *DP* of one binding site on the PAMAMPS 30% chain increased from 128 to 160 when going from linear PLL (*DP* = 50) to DGL 2nd generation, G2 (DP_n = 48), and increased with increasing DGL generation, 265 for G3 (DP_n = 123) and 750 for G4 (DP_n = 365). It clearly illustrates the steric constraints for the PAMAMPS chain to access the core of the DGL, especially for higher generations. Thus, despite the increase of the total number of condensed counter-ions before the association ($N_{counter-ions}$) given in Tables 2 and 3, (from ~4800 for PLL50 up to ~8500 for DGL G4 with PAMAMPS 100%; and from ~900 for PLL50 up to ~1700 for DGL G4 with PAMAMPS 30%), this huge entropic reservoir is not effectively released after the association. To better illustrate this effect, the percentage of released counter-ions compared to the total number of counter-ions condensed on both polyelectrolyte partners, decreases when the DGL generation number increases (from 38% to 3% for DGL-PAMAMPS 30% interactions, and from 8% to 2% for DGL-PAMAMPS 100% interactions). It is worth noting that, in this work, and as a first approximation, $N_{counter-ions}$ was estimated via the Manning condensation theory. Therefore, we did not consider here the possible variation of the number of condensed counter-ions according to the ionic strength. This topic is still discussed as some authors reported a decrease of the fraction of condensed counter-ions with an increase of the ionic strength using numerical simulations [55], while others predicted an increase of the condensed counter-ion fraction with increasing ionic strength [56].

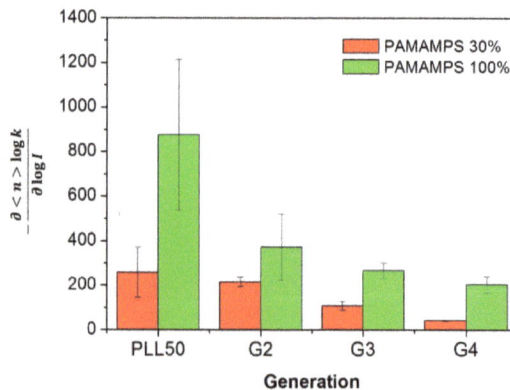

Figure 6. Variation of the number of counter-ions effectively released ($-\frac{\partial <n> \log k}{\partial \log I}$) for the linear PLL50 and for DGL G2, G3 and G4 in interaction with PAMAMPS 30% or PAMAMPS 100%.

Polymers **2018**, *10*, 45

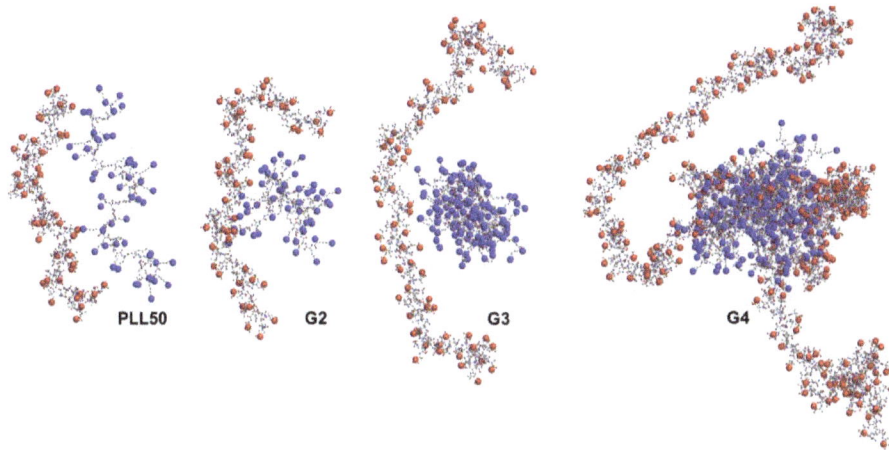

Figure 7. Schematic representation of the interaction between one ligand (PLL or DGL molecule) and one binding site –*s* of a PAMAMPS 30% chain. The *DP* of one binding site on the PAMAMPS substrate is: 128, 160, 265 and 750 for PLL50, G2, G3 and G4, respectively, as calculated by dividing the *DP* of PAMAMPS 30% by the stoichiometry $n_{(PLL \text{ or } DGL/PAMAMP)}$ of the interaction. Red dots represent the negative charge of the sulfonate groups and blue dots represent the positive charge of the ammonium groups.

4. Conclusions

FACCE method was applied to determine the binding parameters between linear and dendrigraft poly(L-Lysine) interacting with linear PAMAMPS. The influence of the ligand topology and the ionic strength on the interactions was studied. At a given ionic strength, it was found that higher DGL generations led to stronger binding site constants. DGL-PAMAMPS (*n*:1) stoichiometries decreased with the increase in DGL generation number and were almost independent of the ionic strength. Furthermore, a double logarithmic linear dependence between the global biding constants and the ionic strength was observed. The slope of each line is a direct experimental estimation of the number of released counter-ions from the association of *n* DGL molecules onto PAMAMPS chains. Surprisingly, the number of released counter-ions decreased with increasing generation number. This result was far from being intuitive since the entropic reservoir, corresponding to the number of initially condensed counter-ions, increases with increasing DGL generation. The fact that only a small proportion of the initially condensed counter-ions are released can be explained by a low accessibility of the PAMAMPS chain to the core of the DGL.

Supplementary Materials: The following are available online at www.mdpi.com/2073-4360/10/1/45/s1.

Acknowledgments: We thank the ANR for funding for the MESOPIC Project (2015–2019), No. ANR-15-CE07-0005. We thank Willy Vayaboury (Alamanda Polymers, Inc.) for the kind supply of PLL polymers. H.C. thanks the support from the Institut Universitaire de France (junior member, 2011–2016). F.M.L. thanks the Ministry of High Education and Scientific Research of Algeria for the research fellowship.

Author Contributions: Feriel Meriem Lounis performed the experiments, analyzed the data and wrote the paper. Philippe Gonzalez performed DGL purification by semi-preparative SEC and DGL analysis by SEC-MALLS. Joseph Chamieh, Laurent Leclercq and Hervé Cottet have interpreted the data and revised the paper. Joseph Chamieh and Jean-Christophe Rossi have performed the representation of the DGL/PAMAMPS interaction site given in Figure 7.

Conflicts of Interest: The authors declare no conflict of interest.

References

1. Abbasi, E.; Aval, S.F.; Akbarzadeh, A.; Milani, M.; Nasrabadi, H.T.; Joo, S.W.; Hanifehpour, Y.; Nejati-Koshki, K.; Pashaei-Asl, R. Dendrimers: Synthesis, applications, and properties. *Nanoscale Res. Lett.* **2014**, *9*, 247. [CrossRef] [PubMed]
2. Malkoch, M.; Malmström, E.; Nyström, A.M. Dendrimers: Properties and Applications. In *Polymer Science: A Comprehensive Reference*; Matyjaszewski, K., Möller, M., Eds.; Elsevier: Amsterdam, The Netherlands, 2012.
3. Kesharwani, P.; Jain, K.; Jain, N.K. Dendrimer as nanocarrier for drug delivery. *Prog. Polym. Sci.* **2014**, *39*, 268–307. [CrossRef]
4. Nanjwade, B.K.; Bechra, H.M.; Derkar, G.K.; Manvi, F.V.; Nanjwade, V.K. Dendrimers: Emerging polymers for drug-delivery systems. *Eur. J. Pharm. Sci.* **2009**, *38*, 185–196. [CrossRef] [PubMed]
5. Svenson, S. Dendrimers as versatile platform in drug delivery applications. *Eur. J. Pharm. Biopharm.* **2009**, *71*, 445–462. [CrossRef] [PubMed]
6. Wolinsky, J.; Grinstaff, M. Therapeutic and diagnostic applications of dendrimers for cancer treatment. *Adv. Drug Deliv. Rev.* **2008**, *60*, 1037–1055. [CrossRef] [PubMed]
7. Zhu, S.; Hong, M.; Zhang, L.; Tang, G.; Jiang, Y.; Pei, Y. PEGylated PAMAM Dendrimer-Doxorubicin Conjugates: In Vitro Evaluation and In Vivo Tumor Accumulation. *Pharm. Res.* **2010**, *27*, 161–174. [CrossRef] [PubMed]
8. Patil, M.L.; Zhang, M.; Taratula, O.; Garbuzenko, O.B.; He, H.; Minko, T. Internally Cationic Polyamidoamine PAMAM-OH Dendrimers for siRNA Delivery: Effect of the Degree of Quaternization and Cancer Targeting. *Biomacromolecules* **2009**, *10*, 258–266. [CrossRef] [PubMed]
9. Longmire, M.; Choyke, P.L.; Kobayashi, H. Dendrimer-based contrast agents for molecular imaging. *Curr. Top. Med. Chem.* **2008**, *8*, 1180–1186. [CrossRef] [PubMed]
10. Wijagkanalan, W.; Kawakami, S.; Hashida, M. Designing Dendrimers for Drug Delivery and Imaging: Pharmacokinetic Considerations. *Pharm. Res.* **2010**, *28*, 1500–1519. [CrossRef] [PubMed]
11. Yiyun, C.; Tongwen, X.; Rongqiang, F. Polyamidoamine dendrimers used as solubility enhancers of ketoprofen. *Eur. J. Med. Chem.* **2005**, *40*, 1390–1393. [CrossRef] [PubMed]
12. Yiyun, C.; Tongwen, X. Solubility of nicotinic acid in polyamidoamine dendrimer solutions. *Eur. J. Med. Chem.* **2005**, *40*, 1384–1389. [CrossRef] [PubMed]
13. Milhem, O.M.; Myles, C.; McKeown, N.B.; Attwood, D.; D'Emanuele, A. Polyamidoamine Starburst dendrimers as solubility enhancers. *Int. J. Pharm.* **2000**, *197*, 239–241. [CrossRef]
14. Zhang, Y.; Xu, M.-Y.; Jiang, T.-K.; Huang, W.-Z.; Wu, J.-Y. Low generational polyamidoamine dendrimers to enhance the solubility of folic acid: A "dendritic effect" investigation. *Chin. Chem. Lett.* **2014**, *25*, 815–818. [CrossRef]
15. Sato, K.; Anzai, J.-I. Dendrimers in Layer-by-Layer Assemblies: Synthesis and Applications. *Molecules* **2013**, *18*, 8440–8460. [CrossRef] [PubMed]
16. Wang, D.; Astruc, D. Dendritic catalysis—Basic concepts and recent trends. *Coord. Chem. Rev.* **2013**, *257*, 2317–2334. [CrossRef]
17. Astruc, D. Palladium catalysis using dendrimers: Molecular catalysts versus nanoparticles. *Tetrahedron Asymmetry* **2010**, *21*, 1041–1054. [CrossRef]
18. Astruc, D.; Chardac, F. Dendritic Catalysts and Dendrimers in Catalysis. *Chem. Rev.* **2001**, *101*, 2991–3024. [CrossRef] [PubMed]
19. Giri, J.; Diallo, M.S.; Simpson, A.J.; Liu, Y.; Goddard, W.A.; Kumar, R.; Woods, G.C. Interactions of Poly(amidoamine) Dendrimers with Human Serum Albumin: Binding Constants and Mechanisms. *ACS Nano* **2011**, *5*, 3456–3468. [CrossRef] [PubMed]
20. Froehlich, E.; Mandeville, J.S.; Jennings, C.J.; Sedaghat-Herati, R.; Tajmir-Riahi, H.A. Dendrimers Bind Human Serum Albumin. *J. Phys. Chem. B* **2009**, *113*, 6986–6993. [CrossRef] [PubMed]
21. Zou, T.; Oukacine, F.; Le Saux, T.; Cottet, H. Neutral Coatings for the Study of Polycation/Multicharged Anion Interactions by Capillary Electrophoresis: Application to Dendrigraft Poly-L-lysines with Negatively Multicharged Molecules. *Anal. Chem.* **2010**, *82*, 7362–7368. [CrossRef] [PubMed]
22. Örberg, M.L.; Schillen, K.; Nylander, T. Dynamic light scattering and fluorescence study of the interaction between double-stranded DNA and poly(amido amine) dendrimers. *Biomacromolecules* **2007**, *8*, 1557–1563. [CrossRef] [PubMed]

23. Ainalem, M.-L.; Campbell, R.A.; Khalid, S.; Gillams, R.J.; Rennie, A.R.; Nylander, T. On the Ability of PAMAM Dendrimers and Dendrimer/DNA Aggregates To Penetrate POPC Model Biomembranes. *J. Phys. Chem. B* **2010**, *114*, 7229–7244. [CrossRef] [PubMed]

24. Kłos, J.S.; Sommer, J.U. Monte Carlo simulations of charged dendrimer-linear polyelectrolyte complexes and explicit counterions. *J. Chem. Phys.* **2011**, *134*, 204902. [CrossRef] [PubMed]

25. Larin, S.V.; Darinskii, A.A.; Lyulin, A.V.; Lyulin, S.V. Linker Formation in an Overcharged Complex of Two Dendrimers and Linear Polyelectrolyte. *J. Phys. Chem. B* **2010**, *114*, 2910–2919. [CrossRef] [PubMed]

26. Tian, W.-D.; Ma, Y.-Q. Complexation of a Linear Polyelectrolyte with a Charged Dendrimer: Polyelectrolyte Stiffness Effects. *Macromolecules* **2010**, *43*, 1575–1582. [CrossRef]

27. Miura, N.; Dubin, P.L.; Moorefield, C.N.; Newkome, G.R. Complex Formation by Electrostatic Interaction between Carboxyl-Terminated Dendrimers and Oppositely Charged Polyelectrolytes. *Langmuir* **1999**, *15*, 4245–4250. [CrossRef]

28. Kabanov, V.A.; Zezin, A.B.; Rogacheva, V.B.; Gulyaeva, Z.G.; Zansochova, M.F.; Joosten, J.G.H.; Brackman, J. Interaction of Astramol Poly(propyleneimine) Dendrimers with Linear Polyanions. *Macromolecules* **1999**, *32*, 1904–1909. [CrossRef]

29. Leisner, D.; Imae, T. Polyelectrolyte Behavior of an Interpolyelectrolyte Complex Formed in Aqueous Solution of a Charged Dendrimer and Sodium Poly(L-glutamate). *J. Phys. Chem. B* **2003**, *107*, 13158–13167. [CrossRef]

30. Vasumathi, V.; Maiti, P.K. Complexation of siRNA with Dendrimer: A Molecular Modeling Approach. *Macromolecules* **2010**, *43*, 8264–8274. [CrossRef]

31. Froehlich, E.; Mandeville, J.S.; Weinert, C.M.; Kreplak, L.; Tajmir-Riahi, H.A. Bundling and Aggregation of DNA by Cationic Dendrimers. *Biomacromolecules* **2011**, *12*, 511–517. [CrossRef] [PubMed]

32. Yevlampieva, N.; Dobrodumov, A.; Nazarova, O.; Okatova, O.; Cottet, H. Hydrodynamic Behavior of Dendrigraft Polylysines in Water and Dimethylformamide. *Polymers* **2012**, *4*, 20–31. [CrossRef]

33. Sisavath, N.; Le Saux, T.; Leclercq, L.; Cottet, H. Effect of Dendrimer Generation on the Interactions between Human Serum Albumin and Dendrigraft Polylysines. *Langmuir* **2014**, *30*, 4450–4457. [CrossRef] [PubMed]

34. Sisavath, N.; Leclercq, L.; Le Saux, T.; Oukacine, F.; Cottet, H. Study of interactions between oppositely charged dendrigraft poly-L-lysine and human serum albumin by continuous frontal analysis capillary electrophoresis and fluorescence spectroscopy. *J. Chromatogr. A* **2013**, *1289*, 127–132. [CrossRef] [PubMed]

35. Salehi, A.; Larson, R.G. A Molecular Thermodynamic Model of Complexation in Mixtures of Oppositely Charged Polyelectrolytes with Explicit Account of Charge Association/Dissociation. *Macromolecules* **2016**, *49*, 9706–9719. [CrossRef]

36. Sing, C.E.; Zwanikken, J.W.; Olvera de la Cruz, M. Ion Correlation-Induced Phase Separation in Polyelectrolyte Blends. *ACS Macro Lett.* **2013**, *2*, 1042–1046. [CrossRef]

37. McCormick, C.L.; Chen, G.S. Water-soluble copolymers. IV. Random copolymers of acrylamide with sulfonated comonomers. *J. Polym. Sci. A Polym. Chem.* **1982**, *20*, 817–838. [CrossRef]

38. Lounis, F.M.; Chamieh, J.; Gonzalez, P.; Cottet, H.; Leclercq, L. Prediction of Polyelectrolyte Complex Stoichiometry for Highly Hydrophilic Polyelectrolytes. *Macromolecules* **2016**, *49*, 3881–3888. [CrossRef]

39. Gonzalez, P.; Leclercq, L.; Cottet, H. What is the Contribution of Counter-Ions to the Absolute Molar Mass of Polyelectrolytes Determined by SEC-MALLS? *Macromol. Chem. Phys.* **2016**, *217*, 2654–2659. [CrossRef]

40. Ibrahim, A.; Koval, D.; Kašička, V.; Faye, C.; Cottet, H. Effective Charge Determination of Dendrigraft Poly-L-Lysine by Capillary Isotachophoresis. *Macromolecules* **2013**, *46*, 533–540. [CrossRef]

41. Lounis, F.; Chamieh, J.; Leclercq, L.; Gonzalez, P.; Cottet, H. The Effect of Molar Mass and Charge Density on the Formation of Complexes between Oppositely Charged Polyelectrolytes. *Polymers* **2017**, *9*, 50. [CrossRef]

42. Lounis, F.M.; Chamieh, J.; Leclercq, L.; Gonzalez, P.; Cottet, H. Modelling and predicting the interactions between oppositely and variously charged polyelectrolytes by frontal analysis continuous capillary electrophoresis. *Soft Matter* **2016**, *12*, 9728–9737. [CrossRef] [PubMed]

43. Lounis, F.M.; Chamieh, J.; Leclercq, L.; Gonzalez, P.; Geneste, A.; Prelot, B.; Cottet, H. Interactions between Oppositely Charged Polyelectrolytes by Isothermal Titration Calorimetry: Effect of Ionic Strength and Charge Density. *J. Phys. Chem. B* **2017**, *121*, 2684–2694. [CrossRef] [PubMed]

44. Anik, N.; Airiau, M.; Labeau, M.P.; Vuong, C.T.; Reboul, J.; Lacroix-Desmazes, P.; Gerardin, C.; Cottet, H. Determination of Polymer Effective Charge by Indirect UV Detection in Capillary Electrophoresis: Toward the Characterization of Macromolecular Architectures. *Macromolecules* **2009**, *42*, 2767–2774. [CrossRef]

45. Lohman, T.M.; deHaseth, P.L.; Record, M.T., Jr. Pentalysine-deoxyribonucleic acid interactions: A model for the general effects of ion concentrations on the interactions of proteins with nucleic acids. *Biochemistry* **1980**, *19*, 3522–3530. [CrossRef] [PubMed]

46. Lohman, T.M.; Mascotti, D.P. Thermodynamics of ligand-nucleic acid interactions. *Methods Enzymol.* **1992**, *212*, 400–424. [CrossRef] [PubMed]

47. Mascotti, D.P.; Lohman, T.M. Thermodynamics of Oligoarginines Binding to RNA and DNA. *Biochemistry* **1997**, *36*, 7272–7279. [CrossRef] [PubMed]

48. Zhang, W.; Bond, J.P.; Anderson, C.F.; Lohman, T.M.; Record, M.T. Large electrostatic differences in the binding thermodynamics of a cationic peptide to oligomeric and polymeric DNA. *Proc. Natl. Acad. Sci. USA* **1996**, *93*, 2511–2516. [CrossRef] [PubMed]

49. Mascotti, D.P.; Lohman, T.M. Thermodynamic extent of counterion release upon binding oligolysines to single-stranded nucleic acids. *Proc. Natl. Acad. Sci. USA* **1990**, *87*, 3142–3146. [CrossRef] [PubMed]

50. Hattori, T.; Hallberg, R.; Dubin, P.L. Roles of Electrostatic Interaction and Polymer Structure in the Binding of β-Lactoglobulin to Anionic Polyelectrolytes: Measurement of Binding Constants by Frontal Analysis Continuous Capillary Electrophoresis. *Langmuir* **2000**, *16*, 9738–9743. [CrossRef]

51. Hattori, T.; Bat-Aldar, S.; Kato, R.; Bohidar, H.B.; Dubin, P.L. Characterization of polyanion-protein complexes by frontal analysis continuous capillary electrophoresis and small angle neutron scattering: Effect of polyanion flexibility. *Anal. Biochem.* **2005**, *342*, 229–236. [CrossRef] [PubMed]

52. Seyrek, E.; Dubin, P.L.; Tribet, C.; Gamble, E.A. Ionic Strength Dependence of Protein-Polyelectrolyte Interactions. *Biomacromolecules* **2003**, *4*, 273–282. [CrossRef] [PubMed]

53. Hileman, R.E.; Jennings, R.N.; Linhardt, R.J. Thermodynamic Analysis of the Heparin Interaction with a Basic Cyclic Peptide Using Isothermal Titration Calorimetry. *Biochemistry* **1998**, *37*, 15231–15237. [CrossRef] [PubMed]

54. Pandav, G.; Ganesan, V. Computer Simulations of Dendrimer–Polyelectrolyte Complexes. *J. Phys. Chem. B* **2014**, *118*, 10297–10310. [CrossRef] [PubMed]

55. Carrillo, J.-M.; Dobrynin, A. Salt Effect on Osmotic Pressure of Polyelectrolyte Solutions: Simulation Study. *Polymers* **2014**, *6*, 1897–1913. [CrossRef]

56. Muthukumar, M. Theory of counter-ion condensation on flexible polyelectrolytes: Adsorption mechanism. *J. Chem. Phys.* **2004**, *120*, 9343–9350. [CrossRef] [PubMed]

polymers

MDPI

Article

Preparation and Characterization of Antibacterial Polypropylene Meshes with Covalently Incorporated β-Cyclodextrins and Captured Antimicrobial Agent for Hernia Repair

Noor Sanbhal [1,2], Ying Mao [1], Gang Sun [1,3], Yan Li [1,*], Mazhar Peerzada [2] and Lu Wang [1,*]

1 Key Laboratory of Textile Science and Technology of Ministry of Education, Room 4023, College of Textiles, Donghua University, 2999 North Renmin Road, Songjiang, Shanghai 201620, China; Noor.Sanbhal@faculty.muet.edu.pk (N.S.); Maoying-dhu@163.com (Y.M.); gysun@ucdavis.edu (G.S.)
2 Department of Textile Engineering, Mehran University of Engineering and Technology, Jamshoro 76062, Sindh, Pakistan; Mazhar.peerzada@faculty.muet.edu.pk
3 Division of Textiles and Clothing, University of California, Davis, CA 95616, USA
* Correspondence: yanli@dhu.edu.cn (Y.L.); wanglu@dhu.edu.cn (L.W.); Tel./Fax: +86-(21)-67792637 (L.W.)

Received: 28 November 2017; Accepted: 5 January 2018; Published: 11 January 2018

Abstract: Polypropylene (PP) light weight meshes are commonly used as hernioplasty implants. Nevertheless, the growth of bacteria within textile knitted mesh intersections can occur after surgical mesh implantation, causing infections. Thus, bacterial reproduction has to be stopped in the very early stage of mesh implantation. Herein, novel antimicrobial PP meshes grafted with β-CD and complexes with triclosan were prepared for mesh infection prevention. Initially, PP mesh surfaces were functionalized with suitable cold oxygen plasma. Then, hexamethylene diisocyanate (HDI) was successfully grafted on the plasma-activated PP surfaces. Afterwards, β-CD was connected with the already HDI reacted PP meshes and triclosan, serving as a model antimicrobial agent, was loaded into the cyclodextrin (CD) cavity for desired antibacterial functions. The hydrophobic interior and hydrophilic exterior of β-CD are well suited to form complexes with hydrophobic host guest molecules. Thus, the prepared PP mesh samples, CD-TCL-2 and CD-TCL-6 demonstrated excellent antibacterial properties against *Staphylococcus aureus* and *Escherichia coli* that were sustained up to 11 and 13 days, respectively. The surfaces of chemically modified PP meshes showed dramatically reduced water contact angles. Moreover, X-ray diffractometer (XRD), differential scanning calorimeter (DSC), and Thermogravimetric (TGA) evidenced that there was no significant effect of grafted hexamethylene diisocyanate (HDI) and CD on the structural and thermal properties of the PP meshes.

Keywords: antibacterial meshes; guest molecule; cyclodextrin grafting; polypropylene; oxygen plasma

---

1. Introduction

Synthetic meshes have been used with the objective of delivering long-term reinforcement to the damaged tissues of the abdominal hernia [1]. An abdominal hernia is a "protrusion or rupture" of an organ through the abdomen wall, causing defect in the abdomen [2]. Numerous studies demonstrated that, among synthetic meshes, non-absorbable polypropylene (PP) warp-knitted, lightweight, and large pore size meshes are the most widely used implants to reduce hernia recurrence [3–7], due to their chemically inert property, light weight, excellent tissue incorporation ability, hydrophobicity, and stability that can withstand a maximum abdominal pressure of 170 mm Hg. The main drawback of these implants is that all synthetic devices favor infection due to the uneven topography of knitted meshes [8]. Thus, these medical mesh devices are prone to bacteria growth and biofilm formation [9,10]. PP mesh infection is a serious surgical complication and is one of the main causes of hernia repair failure [11]. The majority of the bacteria concerned in surgical wound infections are *Staphylococcus aureus* (SA). Mesh infection is difficult to cure when the infected wound is colonized. Mesh infection not only causes removal of infected mesh but health care costs and risks of morbidity and mortality are also increased after mesh infection [12].

One way to minimize mesh infection is pre-operative antibiotic prophylaxis or pre-soaking of mesh in antibiotics during hernia surgery [13]. Nevertheless, PP is a non-absorbable material; even after decades of such antibiotic treatments, mesh infection has remained a major problem [14,15]. Another way to control surgical site infection (SSI) may be to deliver a heavy dose of antibiotics by injections or orally in clinic. However, the distribution of such heavy antibiotic doses to the whole body causes severe side effects. Therefore, sustained release of antibiotics for a suitable duration might be the key to avoiding post-operative mesh infection [16].

The use of cyclodextrin (CD) in the designed preparation is due to its extraordinary properties of hydrophilicity, biocompatibility, and efficient sustained captive and delivery nature. Cyclodextrin monomers like β-CD are renowned as the most accessible cyclic oligosaccharides with unique properties to form complexes with drug molecules without covalent links in their hydrophobic cavities. CDs can be commonly cross-linked with various reactants due to their chemical structure, which contains multiple hydroxyl groups. Thus, they have the advantage of reacting with hexamethylene diisocyanate (HDI) [17].

Triclosan is a proven antibacterial agent with properties similar to antibiotics, with low solubility in water [18]. However, there have been concerns about the use of triclosan in medical devices. Nevertheless, the application of triclosan as a model antimicrobial agent can still present the designed functions of the chemically modified polymers and triclosan has been used to be captured into the system [19,20].

In the past, polypropylene nonwoven implant devices have been finished with cyclodextrin at a higher temperature and loaded with antibacterial substances for mesh infection prevention [21], or surface activation of PP devices with atmospheric pressure plasma and subsequent coating of chitosan/ciprofloxacin has been performed for four days of antibacterial release [22]. The selection of material type (lightweight knitted meshes) is a key factor for hernia repair and in later cases chitosan could deliver antibacterial properties for a limited time.

Cold oxygen plasma is a suitable surface treatment process to modify the surfaces of polymers without changing their bulk properties [23]. Thus, low-pressure cold oxygen plasma treatment is one of the best approaches available so far to increase the adhesion and wettability of polymer surfaces [24], and also has good reproducibility [25]. Grafting reactions onto PP fibers' surfaces could be introduced by surface activation with oxygen plasma to change the chemistry of PP fibers' surfaces, which could also increase the wettability and adhesion for grafting [26]. Overall, the PP surface is prone to oxygen plasma treatments and results in C–H bond cleavage, which may lead to the introduction of carboxyl and hydroxyl groups [27,28]. These results suggest that low-pressure cold oxygen plasma is an effective pretreatment process to modify the surfaces of PP fibers. To the best of our knowledge, no study has been published to graft β-cyclodextrin onto PP knitted mesh devices for mesh infection prevention. However, chemical covalent connection of cyclodextrin onto the inert polypropylene mesh fibers is a key step in the modification reactions.

Therefore, the aim of this study was to prepare cyclodextrin-grafted PP meshes, which could make complexes with triclosan and demonstrate sustained antibacterial release for mesh infection prevention. For this reason, low-pressure cold oxygen plasma was used to functionalize the surfaces of PP fibers to create a hydroxyl or carboxyl group on PP surfaces before two simple grafting steps were performed. In the first step, hexamethylene diisocyanate (HDI) reacted with plasma-activated PP meshes and in the second step CD covalently bonded with PP-HDI meshes. Moreover, grafted CD samples were loaded with triclosan for antibacterial released function, as illustrated in Figure 1a. Characterization of modified PP meshes such as surface morphology, element analysis, and structural, thermal,

Polymers **2018**, *10*, 58

and hydrophilicity were investigated using scanning electron microscopy (SEM), Energy Despersive X-ray spectroscopy (EDX), Fourier Transform Infrared Spectroscopy (FTIR), X-ray diffractometer (XRD), differential scanning calorimeter (DSC), and Thermogravimetric (TGA). Additionally, antibacterial properties were assessed using a suitable sustained efficacy test. Thus, results showed that CD was successfully grafted onto the surfaces of PP meshes and captured triclosan. Moreover, modified PP mesh devices provided prolonged and stable antimicrobial properties without affecting the original properties of PP mesh devices, which could be used in the prevention of mesh infection.

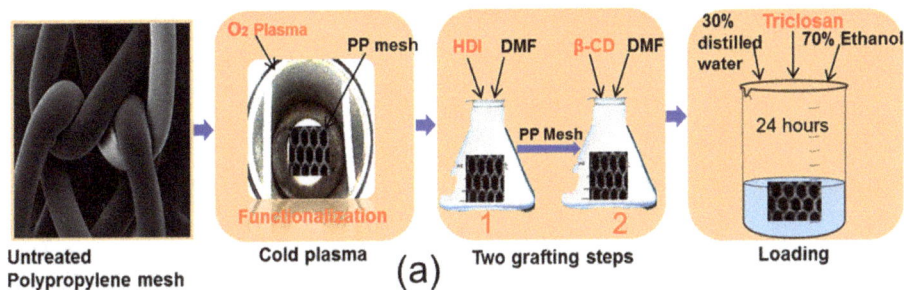

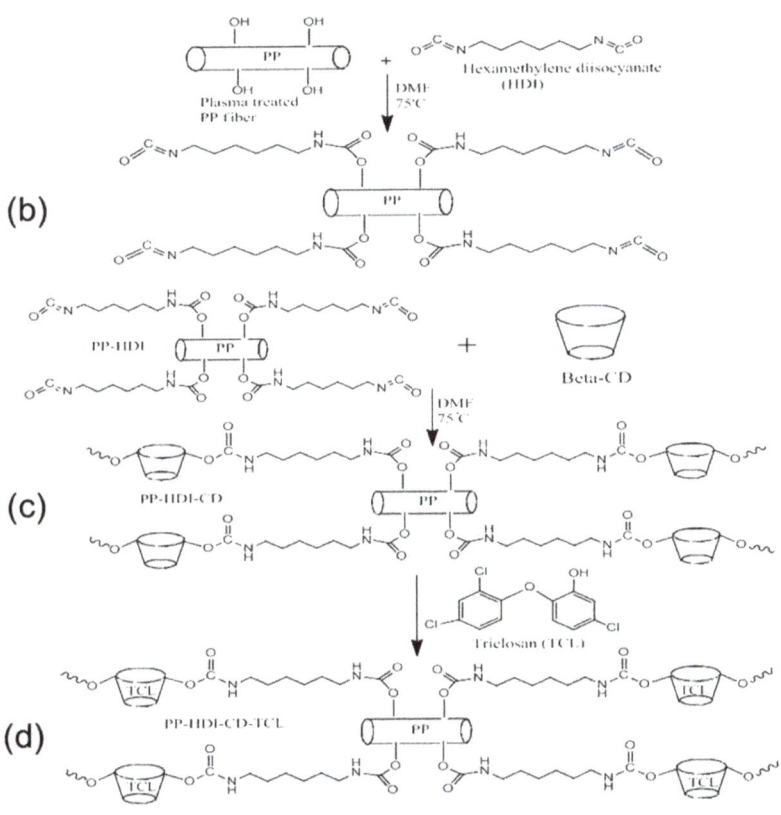

Figure 1. (a) Illustrations of experimental design for β-CD grafted antibacterial surgical PP meshes; Schemes (**b**) step-1, PP-HDI grafting; (**c**) step-2, PP-HDI-CD grafting; and (**d**) triclosan loaded meshes.

2. Materials and Methods

2.1. Materials

Polypropylene (PP) warp-knitted mesh, made of fine diameter (0.1 mm) monofilament with a large pore size (3.5 mm × 2.5 mm) and light weight (27 g/m^2), was received from Nantong Newtec Textile and Chemical Fiber Co. Ltd., Nantong, China. For water contact angle measurement, a PP nonwoven melt-blown fabric of 23 g/m^2 wasn selected. Hexamethylene diisocyanate (HDI) (≥98%) CAS: 822-06-0 and β-Cyclodextrin (CD) (≥97%) CAS: 7585-39-9 was purchased from Sigma Aldrich, Shanghai, China. Triclosan CAS: 3380-34-5 was purchased from Aladdin Company Shanghai, China, while *N-N* dimethylformamide (DMF) Case No. 68-12-12 was received from Shanghai Ling feng Company. Ethanol (≥99.7%) was purchased from Yangyuan Chemical Engineering, Changshu, Jiangsu, China.

2.2. Methods

2.2.1. Cold Plasma Surface Functionalization

A cold plasma machine (HD-300) at low pressure was used for the surface functionalization of PP mesh samples. The machine is consisted of Power (500 W), radio frequency (13.56 MHz) of plasma and a vacuum reaction chamber (300 cm × 300 cm) for the treatment of samples. All PP mesh samples (10 cm × 10 cm) were surface activated at constant powered (40 W) and pressure (10 pa) of oxygen gas. Moreover, optimized treatment time (300 s) was selected for all samples.

2.2.2. Two Grafting Steps and PP-HDI-CD Incorporation

Plasma treated PP meshes was grafted with β-cyclodextrins by simple two grafting step method. First HDI (2–6%) was reacted with the plasma treated samples. Afterwards, the surface modified PP-HDI samples were further reacted with CD (2–8%). For both processes the liquor ratio was (mesh:solution) 1:100. The time and temperature were 30–90 min and 65–85 °C, respectively. Afterward, the PP surface modified meshes (HDI-CD) were washed several times with warm distilled (50 °C) water to eliminate residues of unconnected chemical before drying. HDI grafted PP samples were named PP-HDI-2 (2%), PP-HDI-4 (4%), PP-HDI-6 (6%) and CD-HDI modified polypropylene mesh samples were respectively named as HDI-CD-2 (2%), HDI-CD-4 (4%), HDI-CD-6 (6%), and HDI-CD-8 (8%) according to the CDs concentrations in the solutions.

2.2.3. Loading of Triclosan

Cyclodextrin (HDI-CD)-modified samples (0.5 g) were immersed in the 50 mL mixture (70% ethanol and 30% water) containing 0.6% (0.3 g) triclosan as a guest molecule. All samples were soaked for 24 h before drying. The modified polypropylene samples after triclosan loading were named as CD-TCL-2, CD-TCL-4, and CD-TCL-6, and CD-TCl -8, respectively.

2.3. Characterization Techniques

2.3.1. SEM and EDX

A coating machine (E-1045, Hitachi, Tokyo, Japan) was first used to coat untreated and modified samples with platinum (pt.) ion. Subsequently, all coated samples were tested by a scanning electron microscope (JEOL JSM 6330, Tokyo, Japan) and Energy Despersive X-ray spectroscopy (EDX) (Oxford Instrument ISIS 300, Oxfordshire, UK) connected with SEM.

2.3.2. FTIR

Chemical compositions of untreated and modified samples were measured by a Fourier Transform Infrared Spectroscopy (FTIR) Attenuated Total Reflection mode (ATR) (Nicolet 6700, Downers Grove,

IL, USA). All samples were characterized with resolution of 4.0 cm^{-1} and wave number ranging from 500 to 4000 cm^{-1}.

2.3.3. Contact Angle

Due to the fine filament diameter (0.1 mm) and larger pore size of the PP meshes, a nonwoven PP was preferred as a model fabric to reflect the differences between the contact angles of the PP control and the modified meshes. Two methods, captive bubble (static contact angle) and sessile drop (dynamic contact angle), were used to measure the water contact angle, using WCA 20 software to estimate the hydrophilic/hydrophobic behavior of samples. A Teli CCD camera (Data physics instrument, San Jose, CA, USA) was used to capture images of water contact angle and at least five drops of 5 μL for each sample were added to get an average contact angle.

2.3.4. XRD

Powders of PP control and modified samples were prepared for X-ray diffractometer (XRD) examination. An XRD machine (Rigaku D/MAX 2550/PC, Tokyo, Japan) was used for scanning rate of 0.02/min at 40 kV, 200 mA and a range of 0°–60° (2θ) to characterize all samples.

2.3.5. Thermal Analysis

PP control and modified mesh samples were scanned on a differential scanning calorimeter (DSC) (Perkin Elmer 4000, Downers Grove, IL, USA) at a rate of 20 °C /min over the range of 20–250 °C; the cooling rate was set to 20 °C/min.

Thermogravimetric analysis was conducted to evaluate the percentage residual weight loss of PP control and modified samples. A Perkin Elmer thermogravimetric analyzer (TGA) 4000 (Downers Grove, IL, USA) device was used to scan all samples over a temperature range of 250–550 °C at 30 °C/min. Thus, nitrogen gas was used at a flow rate of 20 mL/min with a pressure of 2 bars during characterization.

2.3.6. Antibacterial Activity Assessment

Agar Diffusion Test Method (Qualitative Analysis)

Antibacterial activity was conducted using the agar diffusion test method described in a previously published research paper [16]. First, 400 μL of bacterial suspension (1 × 10^8 CFU/mL) (*Staphylococcus aureus* ATCC 25923 and *Escherichia coli* ATCC25922) was evenly spread on to the agar plate with the help of pre-sterilized disposable swabs. Then, PP control and modified samples of size 1 cm × 1 cm were placed on an agar plate and incubated for 24 h at 37 °C. The zone of inhibition of incubated samples was measured with a digital vernier caliper in four directions and reported as the mean value. The zone of inhibition diameter was measured by the following formula:

$$L = (D - T)/2$$

where L = zone of inhibition, T = diameter of original sample, and D = inhibition zone diameter after 24 h incubation.

Sustained Efficacy Test (Serial Plate Transfer Test)

The objective of this test was to evaluate the antibacterial activity of CD-grafted samples with reference to the antibacterial release time in days. After each 24 h, the agar plate was changed and mesh samples (1 cm × 1 cm) were transferred to a new agar plate containing similar colony forming units (1 × 10^8 CFU/mL) of bacteria (*Staphylococcus aureus* ATCC 25923 and *Escherichia coli* ATCC25922) after contact with the previous one. The test was continued until modified meshes sustained their antibacterial activity.

Polymers **2018**, *10*, 58

2.3.7. Statistical Analysis

The data in the figures are described as mean and standard deviations. However, the error bars in the figures indicate standard deviations. ANOVA one-way analysis (single-factor) was used to determine the differences between each sample and the data in the figures are labeled, such as $p < 0.001$ (***), $p < 0.01$ (**), and $p < 0.05$ (*). Although a p value of (*) < 0.05 was considered as the confidence interval, variances were first confirmed to be significant.

3. Results and Discussion

3.1. Cyclodextrin Grafting and Loading of Triclosan

The grafting of β-cyclodextrin (CD) onto the PP fibers was achieved by a two-step reaction, as shown in Figure 1. Before grafting, low-pressure oxygen plasma was used to create a hydroxyl group on the surfaces of the PP meshes. Afterward, two grafting steps were performed. In first step hexamethylene diisocyanate (HDI) was reacted (Figure 1b) with the plasma-treated PP fibers and then in second step PP-HDI samples were further reacted with hydroxyl groups of β-CD (Figure 1c). However, cyclodextrins are well suited to form complexes with triclosan. Thus, triclosan was loaded (Figure 1d) into HDI-CD grafted samples.

The process parameters, reaction time, and temperature of the cyclodextrin grafting reaction (Figure 2) were optimized. The effect of temperature on CD grafting (Figure 2a) indicates that weight % increased with an increase in temperature from 65 to 75 °C. However, an additional temperature increase from 75 to 85 °C did not have any effect on the weights of the treated samples. Grafting reaction time was also optimized, as illustrated in Figure 2b, and the grafted weight of the products increased with an increase in reaction time from 30 to 60 min. Thus, a temperature of 75 °C and time of 60 min were considered the optimum conditions for HDI and cyclodextrin grafting. Moreover, Figure 2c displays the total weight increase for HDI, CD, and triclosan. HDI (6%) showed better results for the grafting reaction on to the PP meshes and demonstrated a total weight increase of 1.63%. All CD-grafted samples were treated with the same concentration of HDI. However, it was difficult to find the real increased weight of the HDI samples due to the fact that PP-HDI samples were advanced to the second step of CD grafting without washing to save the unreacted diisocyanate group on the HDI-treated PP surface. The average weight increases of CD grafting for HDI-CD-2, HDI-CD-4, HDI-CD-6, and HDI-CD-8 were 5.130%, 6.410%, 7.27%, and 7.17%, respectively. Nevertheless, after triclosan loading the average corresponding weight increased to 2.025%, 2.268%, 2.414%, and 2.40%, respectively. HDI-CD-6 was considered the best sample, which gained more weight of CD and triclosan. Thus, CD-HDI-6 and CD-TCL-6 samples were included in further testing. Moreover, PP meshes without plasma treatment were also used for HDI grafting, but unfortunately we did not succeed with HDI grafting. Therefore, plasma treatment was key to increasing the surface roughness of PP meshes for better grafting of HDI.

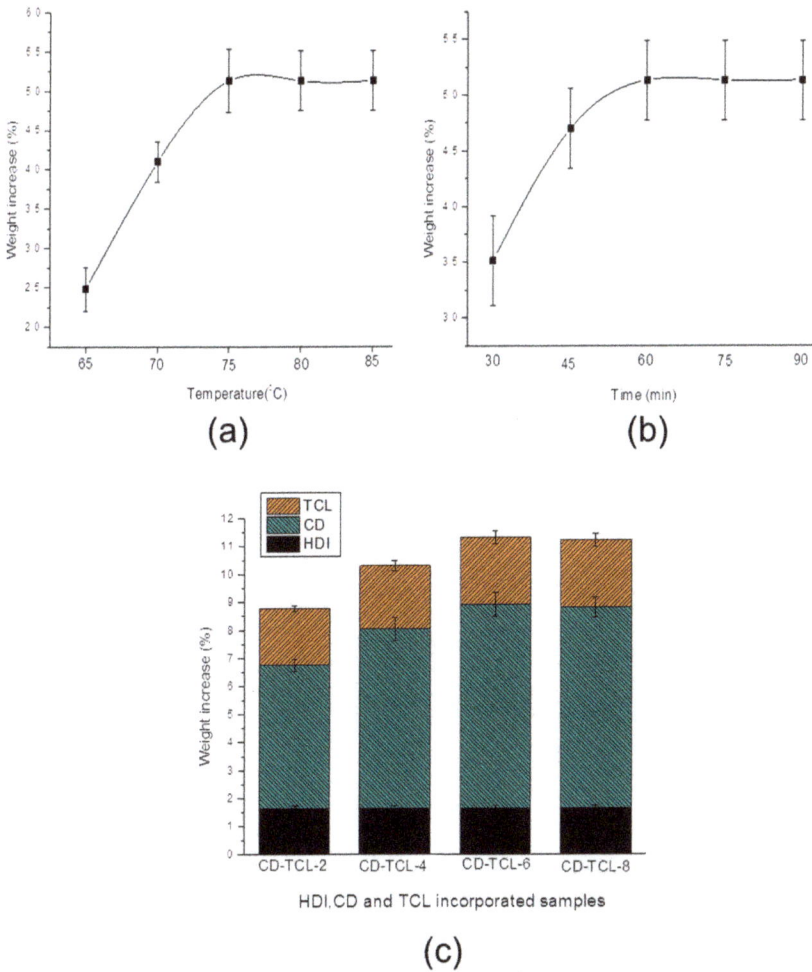

Figure 2. Optimization of two-step grafted PP meshes; (**a**) Temperature for grafting; (**b**) reaction time; (**c**) weight increase % of HDI-CD-TCL samples. Data of each sample expressed as averages ($n = 3$).

3.2. Surface Morphology of PP Meshes

Polypropylene (PP) knitted mesh surfaces before plasma treatment (Figure 3a1,a2) were very smooth, but after plasma treatment (Figure 3b1,b2) the PP surface became slightly rough. The samples after the plasma treatment and reaction with HDI showed a thin layer of coating (Figure 3c1,c2) on the surfaces of PP fibers. Moreover, the surfaces of the HDI-CD grafted samples (Figure 3d1,d2) are covered with small hills and valleys. It can be observed (Figure 3e1,e2) that after triclosan loading the HDI-CD grafted samples showed similar but a little more swollen coating on the PP fibers, indicating that the loading of triclosan did not affect the grafted structure on the PP fibers.

Fauland reported that low-pressure oxygen plasma can be a suitable pretreatment process to introduce polar groups onto the surfaces of PP fibers and increase their wettability and adhesion for improved grafting of hexamethyl-disiloxane (HMDSO) [26]. Similarly, using low-pressure cold oxygen plasma, HDI-CD structures are successfully grafted onto the chemically inert PP surfaces. The SEM results confirmed that a layer of HDI coated was further connected with β-CD.

Figure 3. SEM micrographs; (**a1,a2**) PP control; (**b1,b2**) oxygen plasma treated; (**c1,c2**) HDI grafted PP-HDI-6; (**d1,d2**) HDI-CD-6; and (**e1,e2**) CD-TCL-6 (top row 200×, bottom row 1000×).

3.3. Characterization of HDI, CD, and Triclosan on Modified PP Mesh

The PP control and grafted samples were characterized using EDX (Figure 4) for elemental analysis.

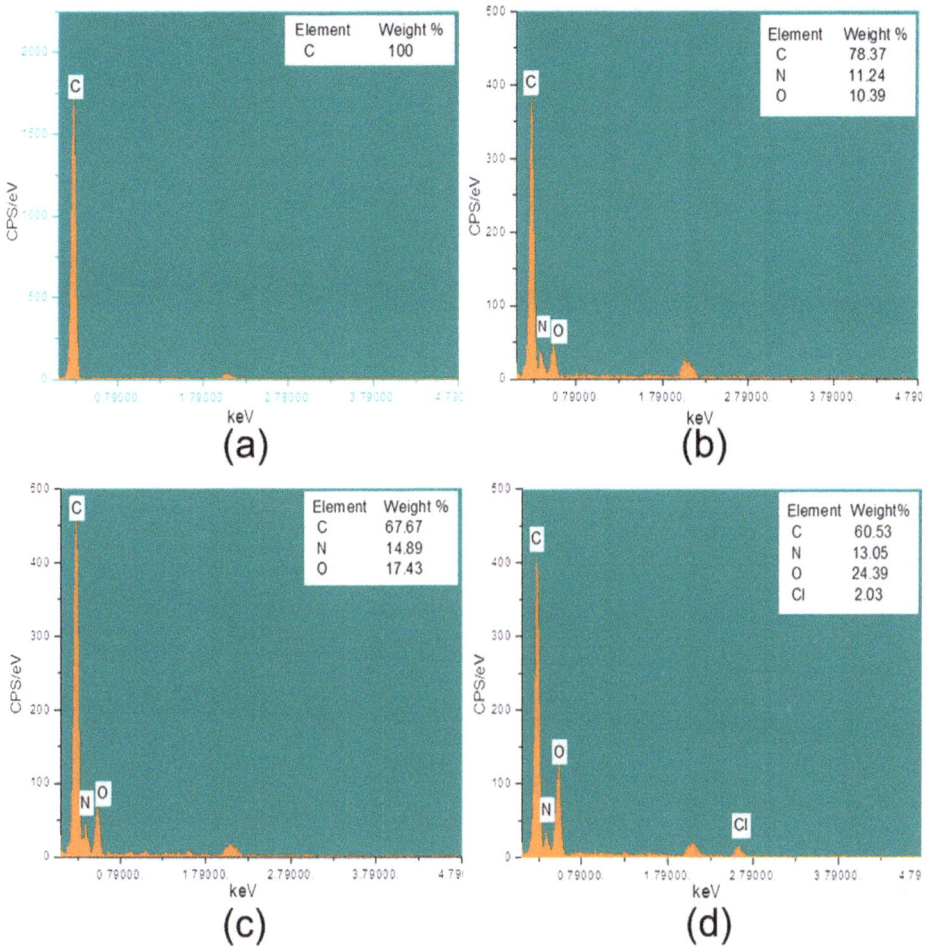

Figure 4. EDX spectra and atomic percent of PP control and modified meshes (**a**) PP control; (**b**) HDI grafted, PP-HDI-6; (**c**) HDI-CD-6; (**d**) CD-TCL-6.

It can be observed that the PP control (Figure 4a) sample shows the presence of a (C) carbon atom that is within 0.5 keV. Nevertheless, PP-HDI-6 (Figure 4b) shows two additional peaks (O) and (N) in addition to the carbon atom and all three peaks are under 0.5 keV, confirming the presence of hexamethylene diisocyanate (HDI) on PP surfaces. Moreover, HDI-CD-6 displayed the existence of the same carbon (C), oxygen (O), and nitrogen (N) (Figure 4c) but at different peak heights, which confirmed the presence of cyclodextrin. It can be observed that CD-TCL-6 showed (Figure 4d) an additional peak of chlorine (Cl) atom within 3.0 keV in addition to carbon, nitrogen, and oxygen atoms, confirming the successful loading of triclosan onto the CD-grafted samples.

Structures of the PP control, PP-HDI, and HDI-CD-grafted samples at different reaction stages were analyzed using FTIR-ATR. The PP control exhibits absorbance bands at 2950 cm^{-1} (va CH$_3$), 2916 cm^{-1} (va CH$_2$), 1452 cm^{-1} (δCH$_3$), and 1376 cm^{-1} (δCH$_3$) [29,30]. Nevertheless, all PP-HDI grafted samples (Figure 5a) showed additional absorbance peaks at 3321 cm^{-1} (OH), 1620 cm^{-1} (ν C=O), and 1576 cm^{-1} (δNH) due to hydroxyl, amide I, and amide II formations, respectively. CD-grafted (Figure 5b) polypropylene samples maintained all previous peaks of HDI-grafted samples

Polymers **2018**, *10*, 58

and additionally showed absorbance peaks at 1701 cm^{-1} (carbonyl group, ν C=O) and C–H (νCH) stretch at 1256 cm^{-1}. However, C–O groups of CD also appear at 1030 cm^{-1}, consistent with a previously published paper [31]. The absorbance band of amide I 1620 cm^{-1} (ν C=O) and amide II 1576 cm^{-1} (δNH) can be also observed in all CD-grafted samples, which confirmed the formation of urethane, consistent with the literature on CD [17,32]. Moreover, we did not find any absorbance peak differences between HDI-CD (CD-grafted) and CD-TCL (triclosan-loaded) samples, which is due to the fact that triclosan was not reacted but captured by the CD cavity.

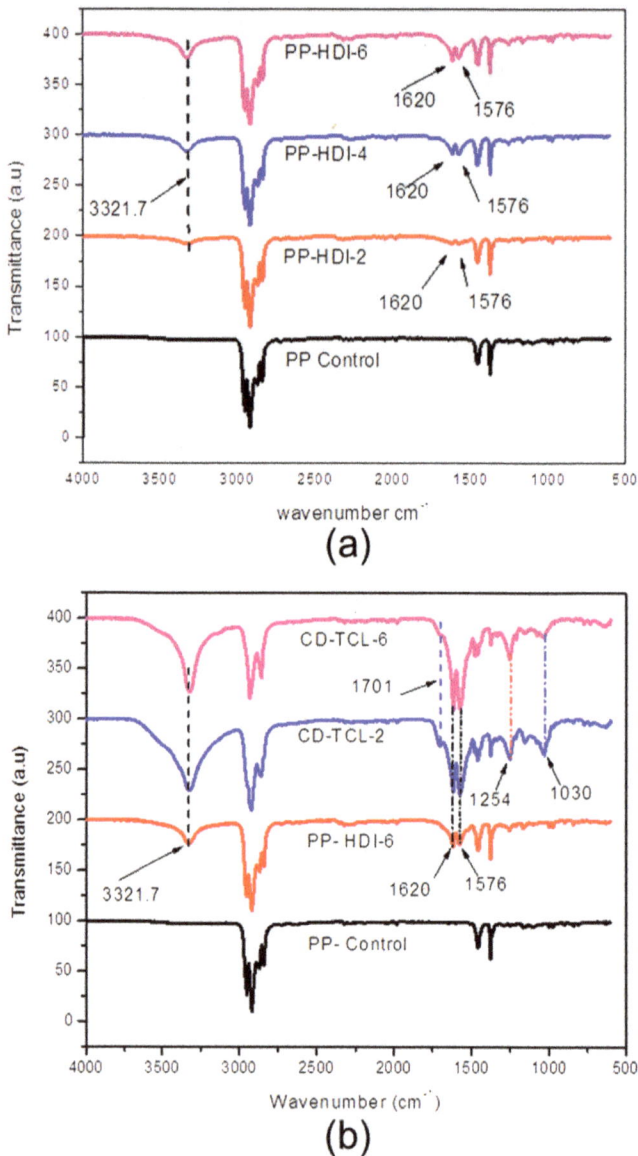

Figure 5. FTIR (ATR) spectra; (**a**) PP-HDI grafted samples (PP-HDI-2), (PP-HDI-4) and PP-HDI-6; and (**b**) PP control, PP-HDI-6-grafted, and triclosan-loaded (CD-TCL-2 and CD-TCL-6) samples.

3.4. Hydrophilicity of PP Meshes (Water Contact Angle)

Water contact angles of PP control and surface-modified samples (Figure 6) were measured by both captive bubble and sessile drop methods. After CD grafting, all modified samples became much more hydrophilic. Thus, the captive bubble method was mainly used to measure the water contact angles of the modified surfaces. Nevertheless, the PP control still retains a hydrophobic surface; therefore, sessile drop was used to measure its water contact angle. As shown in Figure 6a, the contact angles of all treated samples reduced dramatically. The average water contact angle of the PP control was 138.66° and decreased to 64.5° (113.3%), 59.82 (131.8%), and 57.7° (140.7%) for CD-TCL-2, CD-TCL-6, and HDI-CD-6, respectively.

The average water contact angle difference between CD-TCL-2 (64.57°) and CD-TCL-6 (59.82°) was 7.94% ($p < 0.001$), which reveals the effect of more CD grafted onto CD-TCL-6. Likewise, the difference between water contact angles of HDI-CD-6 (59.82°) and triclosan-loaded CD-TCL-6 (57.6°) was 3.84%. This indicates the hydrophobicity of host molecule triclosan loaded in comparison to plain CD. These results are in consensus with a paper published in 2017 [33]. The average water contact angle drops for the PP control and the grafted samples are shown in Figure 6b; it can be observed that CD-TCL-2 shows a slightly increased contact angle compared to CD-TCL-6 and HDI-CD-6.

β-cyclodextrin makes complexes with a wide range of molecules due to its cavity and multi-hydroxyl groups [34]. As expected, all cyclodextrin-grafted PP samples showed remarkably decreased water contact angles. Nevertheless, the hydrophobicity of grafted CD increases with the increase in complexes between CD and triclosan.

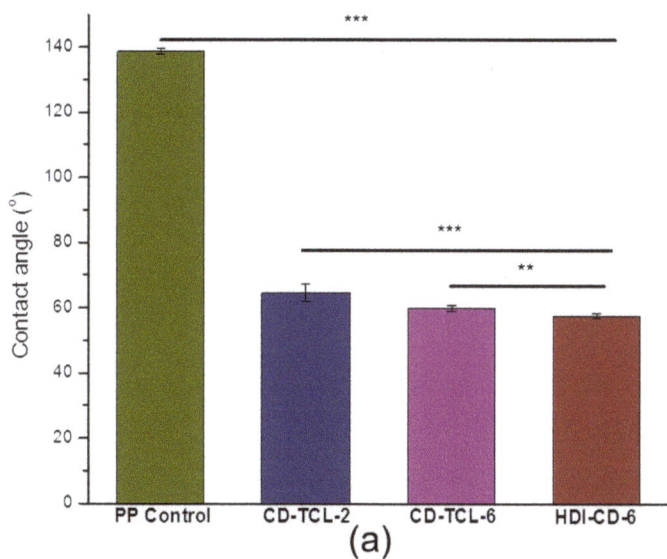

Figure 6. *Cont.*

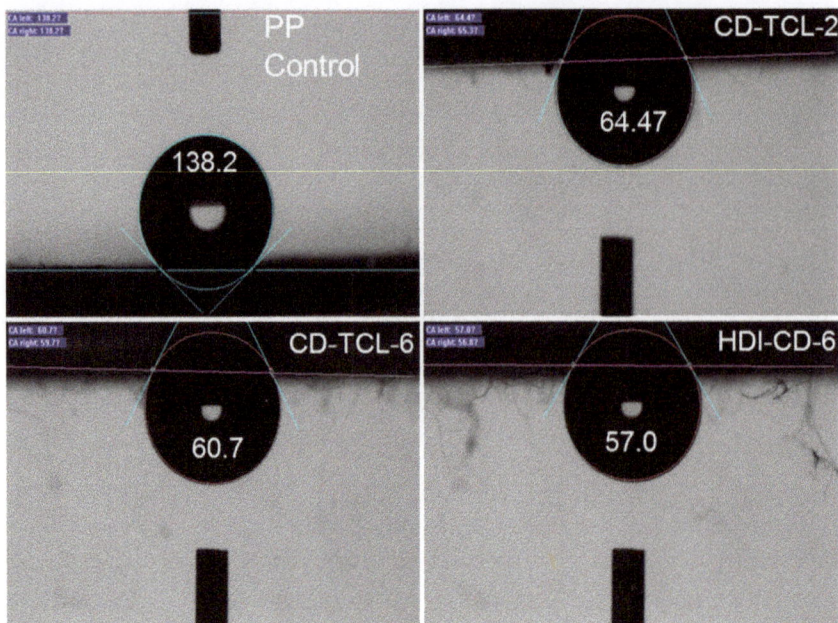

(b)

Figure 6. (**a**) Contact angles measured by sessile drop method and captive bubble (**b**) average contact angle drops, PP control (sessile drop), triclosan-loaded CD-TCL-2, triclosan-loaded CD-TCL-6 and CD-grafted HDI-CD-6. Data for each sample are expressed as an average of five measurements ($n = 5$). Statistical differences are indicated with (***) for $p < 0.001$ and (**) for $p < 0.01$.

3.5. Structural and Thermal Properties

The PP control and surface-modified samples were analyzed by XRD, as shown in Figure 7a. It can be observed that the PP control shows five peak lattices (14.21°, 17.14°, 18.93°, 21.40°, and 25°) up to 25° [35], and all modified samples maintained the same peaks and crystal structure. Furthermore, the crystallinity of CD-TCL-2 (61.30%) and CD-TCL-4 (63.04%) samples were slightly increased as compared to that of the PP control (61.05%). However, the change was minimal; therefore, cyclodextrin grafting has no significant influence on the crystal structure of the modified PP mesh devices.

As shown in Figure 7b, the PP control has a melting temperature of 148.08 °C, whereas CD-TCL-2 (148.52 °C) and CD-TCL-6 (148.90 °C) showed slightly increased melting temperatures due to HDI-CD grafting, consistent with a previously published research paper [21]. Similarly, the endothermic temperature for the PP control (94.2 °C) was slightly higher than CD-TCL-2 (90.9 °C) and CD-TCL-6 (92 °C). Figure 7c reveals the residual weight loss and thermal stability of the PP control and the grafted samples. It can be observed that the PP control was more thermally stable at a temperature of 400 °C and lost only 2% of its total weight. CD-TCL-2 and CD-TCL-6 were less thermally stable and lost their weights of 5% and 8% at the same temperature, consistent with a previously published paper [21]. This may be due to the less stable structure of CD on the grafted samples. Overall, there is no significant difference in the decomposition temperature of PP control and grafted samples.

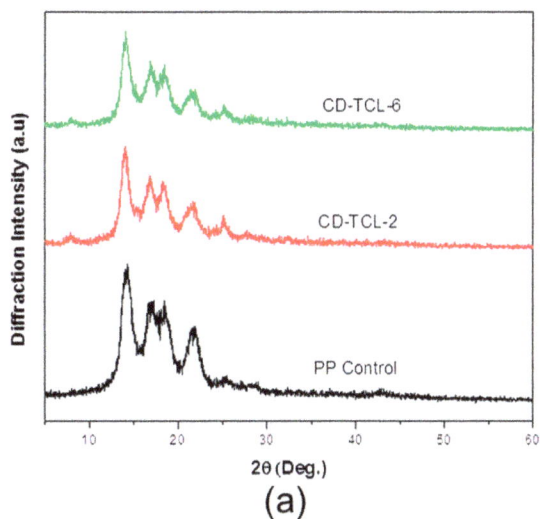

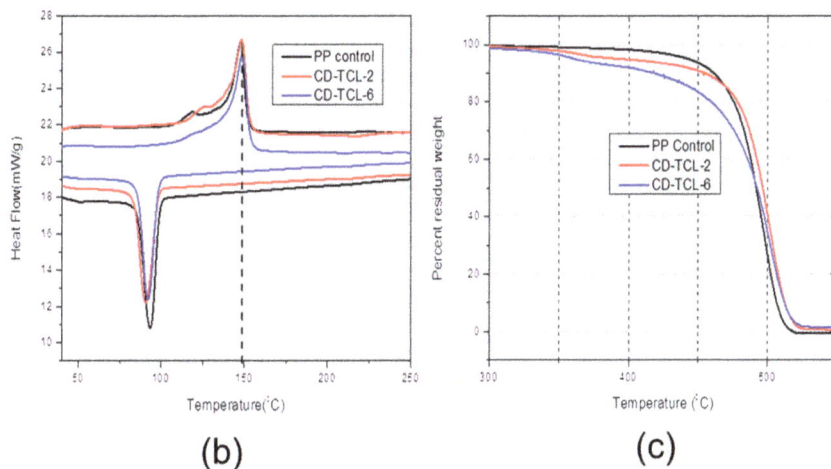

Figure 7. Structural and thermal properties: (**a**) XRD pattern of treated and untreated PP meshes; (**b**) DSC of PP control and modified samples; (**c**) TGA: percentage residual weight changes of PP control and modified samples.

3.6. Antibacterial Activity

The antibacterial properties of the modified and PP control meshes were measured using qualitative analysis by inhibition zone diameter. Figure 8 displays the inhibition zone diameters of the untreated PP control and the modified (CD-TCL-2 and CD-TCL-6) meshes. The PP control and HDI-CD (without triclosan) samples did not show antibacterial properties. However, the PP-modified and triclosan-loaded samples demonstrated good inhibition zone diameters. However, CD-TCL-2 displayed an average inhibition zone of 5.06 and 5.133 mm for *S. aureus* and *E. coli*, respectively. CD-TCL-6 exhibited maximum inhibition zones of 6.2 and 6.40 mm to *S. aureus* and *E. coli*, respectively.

Moreover, CD-TCL-2 and CD-TCL-6 (Figure 9) were further evaluated using a sustained efficacy tests to assess their durable antibacterial properties with respect to number of days. A sustained efficacy test is a very important test to analyze the release of antibacterial agent after each 24 h.

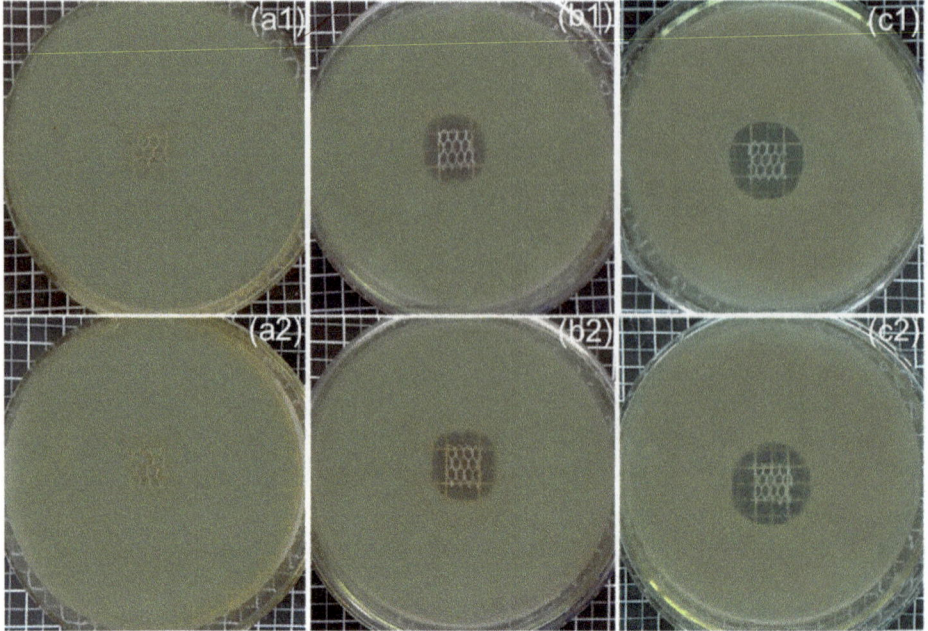

Figure 8. Inhibition zone diameter of (**a1,a2**) PP control, (**b1,b2**) CD-TCL-2, (**c1,c2**) CD-TCL-6, (top row) *S. aureus* and (bottom row) *E. coli*.

The average inhibition zone diameters of CD-TCL-2 and CD-TCL-6 for both bacteria (*E. coli* and *S. aureus*) decreased each day. CD-TCL-2 ended its antibacterial function after 11 days for both *S. aureus* and *E. coli*, and CD-TCL-6 ended its antibacterial function after 13 days. As shown in Figure 9a, CD-TCL-2 demonstrated different inhibition zones of diameters for both *S. aureus* and *E. coli*. It can be observed that *E. coli* samples demonstrated greater zones of inhibition diameters than *S. aureus* during all 11 days.

However, the zones of inhibition were decreased quickly after six days for both types of bacteria. The minimum inhibition zone for *S. aureus* and *E. coli* was 0.7 and 0.8 mm, respectively. In the case of CD-TCL-6 (Figure 9b), inhibition zone diameters of *S. aureus and E. coli* were also different to each other, and *E. coli* had greater inhibition zone diameters throughout all 13 days and the zone of inhibition decreased regularly with respect to days. The minimum inhibition zone for *S. aureus and E. coli* was 0.8 and 1 mm, respectively.

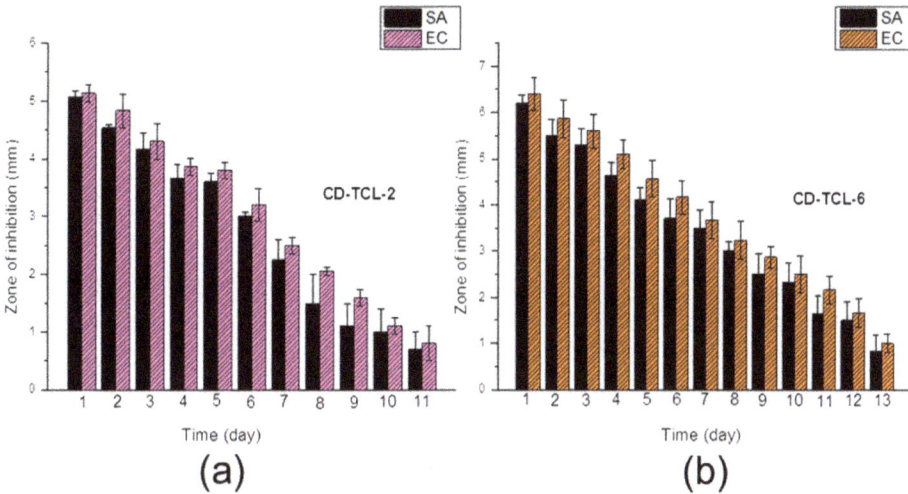

Figure 9. Sustained efficacy test (**a**) CD-TCL-2, (**b**) CD-TCL-6. Data for antimicrobial test samples are expressed as an average of three samples (*n* = 3).

Overall, due to the fact that CD-TCL-6 samples carried more CD and triclosan as compared to the CD-TCL-2, CD-TCL-6 always demonstrated more powerful antimicrobial properties than CD-TCL-2. Generally, *E. coli* had a greater inhibition zone diameter for CD-TCL-2 and CD-TCL-6.

Previously, polypropylene implants have been coated with silver nano-clusters to prevent mesh infection [36]. However, the sustained release of antibacterial agent is a suitable method to ensure the release of antibiotic for a prolonged duration. Furthermore, Majumdar et al. also performed antibacterial properties for a polyester implant by (AATCC 100-2004) colony counting units a quantitative test method [37], which may not be sufficient to prevent mesh infection. Unfortunately, mesh infection is a major problem in hernia surgery and may require continuous antibacterial activity for at least 10 days. Thus, β-CD inclusion complexes with antibiotics could be a suitable technique to capture the antimicrobial agent and release it for a prolonged duration to prevent mesh infection. This work could be further carried out to perform in vitro and in vivo experiments.

4. Conclusions

This study proven that β-CD was covalently bonded onto PP mesh devices without affecting the original properties of the PP fibers. The products captured triclosan and provided excellent sustained released antibacterial properties. The low-pressure, cold oxygen plasma was able to activate the surfaces of PP meshes effectively, and created hydroxyl and carboxyl groups on PP surfaces for hexamethylene diisocyanate (DHI) grafting. HDI was successfully grafted on the oxygen plasma activated PP fibers' surfaces. Moreover, β-CD was connected to the HDI-grafted PP meshes, and the modified PP meshes were successfully loaded with triclosan as a model antimicrobial agent. All samples were characterized using SEM, FTIR, EDX, XRD, TGA, DSC, and water contact angle.

Supplementary Materials: The following are available online at www.mdpi.com/2073-4360/10/1/58/s1. Data for graphs and figures for the following are attached in a supplementary file in the form of CSV data. Graphs for Figure 2a–c, Figures 6a and 9b.

Acknowledgments: This work was supported by the National Key Research and Development Program of China (Grant No. 2016YFB 0303300-03), the 111 project "Biomedical Textile Material Science and Technology" (grant No. B07024), and the Fundamental Research Funds for the Central Universities (Grant No. 17D110111).

Polymers **2018**, *10*, 58

Author Contributions: Noor Sanbhal and Yan Li conceived and designed the experiments. Noor Sanbhal and Mao Ying conducted all experiments. Lu Wang guided the experiments and discussed the results with Noor Sanbhal. Gang Sun and Mazhar Peerzada made some valuable comments. Noor Sanbhal and Lu Wang wrote the article and Gang Sun edited the manuscript.

Conflicts of Interest: The authors declare no conflict of interest.

References

1. Chughtai, B.; Thomas, D.; Mao, J.; Eilber, K.; Anger, J.; Clemens, J.Q.; Sedrakyan, A. Hernia repair with polypropylene mesh is not associated with an increased risk of autoimmune disease in adult men. *Hernia* **2017**, *21*, 637–642. [CrossRef] [PubMed]

2. Miao, L.; Wang, F.; Wang, L.; Zou, T.; Brochu, G.; Guidoin, R. Physical characteristics of medical textile prostheses designed for hernia repair: A comprehensive analysis of select commercial devices. *Materials (Basel)* **2015**, *8*, 8148–8168. [CrossRef] [PubMed]

3. Brown, C.N.; Finch, J.G. Which mesh for hernia repair? *Ann. R. Coll. Surg. Engl.* **2010**, *92*, 272–278. [CrossRef] [PubMed]

4. Bilsel, Y.; Abci, I. The search for ideal hernia repair; mesh materials and types. *Int. J. Surg.* **2012**, *10*, 317–321. [CrossRef] [PubMed]

5. Hazebroek, E.J.; Ng, A.; Yong, D.H.; Berry, H.; Leibman, S.; Smith, G.S. Evaluation of lightweight titanium-coated polypropylene mesh (timesh) for laparoscopic repair of large hiatal hernias. *Surg. Endosc.* **2008**, *22*, 2428–2432. [CrossRef] [PubMed]

6. Klinge, U.; Klosterhalfen, B.; Birkenhauer, V.; Junge, K.; Conze, J.; Schumpelick, V. Impact of polymer pore size on the interface scar formation in a rat model. *J. Surg. Res.* **2002**, *103*, 208–214. [CrossRef] [PubMed]

7. Gil, D.; Rex, J.; Cobb, W.; Reukov, V.; Vertegel, A. Anti-inflammatory coatings of hernia repair meshes: A pilot study. *J. Biomed. Mater. Res. B* **2018**, *106*, 589–597. [CrossRef] [PubMed]

8. Sanbhal, N.; Miao, L.; Xu, R.; Khatri, A.; Wang, L. Physical structure and mechanical properties of knitted hernia mesh materials: A review. *J. Ind. Text.* **2017**. [CrossRef]

9. Kulaga, E.; Ploux, L.; Balan, L.; Schrodj, G.; Roucoules, V. Mechanically responsive antibacterial plasma polymer coatings for textile biomaterials. *Plasma Process. Polym.* **2014**, *11*, 63–79. [CrossRef]

10. Knetsch, M.L.W.; Koole, L.H. New strategies in the development of antimicrobial coatings: The example of increasing usage of silver and silver nanoparticles. *Polymers* **2011**, *3*, 340–366. [CrossRef]

11. Nisticò, R.; Magnacca, G.; Martorana, S. Surface science in hernioplasty: The role of plasma treatments. *Appl. Surf. Sci.* **2017**, *419*, 860–868. [CrossRef]

12. Perez-Kohler, B.; Fernandez-Gutierrez, M.; Pascual, G.; Garcia-Moreno, F.; San Roman, J.; Bellon, J.M. In vitro assessment of an antibacterial quaternary ammonium-based polymer loaded with chlorhexidine for the coating of polypropylene prosthetic meshes. *Hernia* **2016**, *20*, 869–878. [CrossRef] [PubMed]

13. Labay, C.; Canal, J.M.; Modic, M.; Cvelbar, U.; Quiles, M.; Armengol, M.; Arbos, M.A.; Gil, F.J.; Canal, C. Antibiotic-loaded polypropylene surgical meshes with suitable biological behaviour by plasma functionalization and polymerization. *Biomaterials* **2015**, *71*, 132–144. [CrossRef] [PubMed]

14. Erdas, E.; Medas, F.; Pisano, G.; Nicolosi, A.; Calo, P.G. Antibiotic prophylaxis for open mesh repair of groin hernia: Systematic review and meta-analysis. *Hernia* **2016**, *20*, 765–776. [CrossRef] [PubMed]

15. Hedrick, T.L.; Smith, P.W.; Gazoni, L.M.; Sawyer, R.G. The appropriate use of antibiotics in surgery: A review of surgical infections. *Curr. Probl. Surg.* **2007**, *44*, 635–675. [CrossRef] [PubMed]

16. Zhang, Z.; Tang, J.; Wang, H.; Xia, Q.; Xu, S.; Han, C.C. Controlled antibiotics release system through simple blended electrospun fibers for sustained antibacterial effects. *ACS Appl. Mater. Interfaces* **2015**, *7*, 26400–26404. [CrossRef] [PubMed]

17. Thatiparti, T.R.; Shoffstall, A.J.; von Recum, H.A. Cyclodextrin-based device coatings for affinity-based release of antibiotics. *Biomaterials* **2010**, *31*, 2335–2347. [CrossRef] [PubMed]

18. Perez-Alvarez, L.; Matas, J.; Gomez-Galvan, F.; Ruiz-Rubio, L.; Leon, L.M.; Vilas-Vilela, J.L. Branched and ionic beta-cyclodextrins multilayer assembling onto polyacrylonitrile membranes for removal and controlled release of triclosan. *Carbohydr. Polym.* **2017**, *156*, 143–151. [CrossRef] [PubMed]

19. Gomez-Galvan, F.; Perez-Alvarez, L.; Matas, J.; Alvarez-Bautista, A.; Poejo, J.; Duarte, C.M.; Ruiz-Rubio, L.; Vila-Vilela, J.L.; Leon, L.M. Preparation and characterization of soluble branched ionic beta-cyclodextrins and their inclusion complexes with triclosan. *Carbohydr. Polym.* **2016**, *142*, 149–157. [CrossRef] [PubMed]

20. Fidaleo, M.; Zuorro, A.; Lavecchia, R. Enhanced antibacterial and anti-quorum sensing activities of triclosan by complexation with modified beta-cyclodextrins. *World J. Microbiol. Biotechnol.* **2013**, *29*, 1731–1736. [CrossRef] [PubMed]

21. Laurent, T.; Kacem, I.; Blanchemain, N.; Cazaux, F.; Neut, C.; Hildebrand, H.F.; Martel, B. Cyclodextrin and maltodextrin finishing of a polypropylene abdominal wall implant for the prolonged delivery of ciprofloxacin. *Acta Biomater.* **2011**, *7*, 3141–3149. [CrossRef] [PubMed]

22. Avetta, P.; Nisticò, R.; Faga, M.G.; D'Angelo, D.; Boot, E.A.; Lamberti, R.; Martorana, S.; Calza, P.; Fabbri, D.; Magnacca, G. Hernia-repair prosthetic devices functionalised with chitosan and ciprofloxacin coating: Controlled release and antibacterial activity. *J. Mater. Chem. B* **2014**, *2*, 5287–5294. [CrossRef]

23. Zhang, Z.; Zhang, T.; Li, J.; Ji, Z.; Zhou, H.; Zhou, X.; Gu, N. Preparation of poly(L-lactic acid)-modified polypropylene mesh and its antiadhesion in experimental abdominal wall defect repair. *J. Biomed. Mater. Res. B* **2014**, *102*, 12–21. [CrossRef] [PubMed]

24. Sorrentino, L.; Carrino, L.; Napolitano, G. Oxygen cold plasma treatment on polypropylene: Influence of process parameters on surface wettability. *Surf. Eng.* **2013**, *23*, 247–252. [CrossRef]

25. Shishoo, R. *Plasma Technologies for Textiles*; Woodhead Publishing Limited: Cornwall, UK, 2007.

26. Fauland, G.; Constantin, F.; Gaffar, H.; Bechtold, T. Production scale plasma modification of polypropylene baselayer for improved water management properties. *J. Appl. Polym. Sci.* **2015**, *132*, 41294. [CrossRef]

27. Jelil, R.A. A review of low-temperature plasma treatment of textile materials. *J. Mater. Sci.* **2015**, *50*, 5913–5943. [CrossRef]

28. Lai, J.; Sunderland, B.; Xue, J.; Yan, S.; Zhao, W.; Folkard, M.; Michael, B.D.; Wang, Y. Study on hydrophilicity of polymer surfaces improved by plasma treatment. *Appl. Surf. Sci.* **2006**, *252*, 3375–3379. [CrossRef]

29. Nava-Ortiz, C.A.; Alvarez-Lorenzo, C.; Bucio, E.; Concheiro, A.; Burillo, G. Cyclodextrin-functionalized polyethylene and polypropylene as biocompatible materials for diclofenac delivery. *Int. J. Pharm.* **2009**, *382*, 183–191. [CrossRef] [PubMed]

30. Sarau, G.; Bochmann, A.; Lewandowska, R.; Christianse, S. From micro- to macro-raman spectroscopy: Solar silicon for a case study. In *Advanced Aspects of Spectroscopy*; InTech: Lodz, Poland, 2012.

31. Gawish, S.M.; Matthews, S.R.; Wafa, D.M.; Breidt, F.; Bourham, M.A. Atmospheric plasma-aided biocidal finishes for nonwoven polypropylene fabrics. I. Synthesis and characterization. *J. Appl. Polym. Sci.* **2007**, *103*, 1900–1910. [CrossRef]

32. Choi, J.M.; Park, K.; Lee, B.; Jeong, D.; Dindulkar, S.D.; Choi, Y.; Cho, E.; Park, S.; Yu, J.H.; Jung, S. Solubility and bioavailability enhancement of ciprofloxacin by induced oval-shaped mono-6-deoxy-6-aminoethylamino-beta-cyclodextrin. *Carbohydr. Polym.* **2017**, *163*, 118–128. [CrossRef] [PubMed]

33. Pérez-Álvarez, L.; Ruiz-Rubio, L.; Lizundia, E.; Hernáez, E.; León, L.M.; Vilas-Vilela, J.L. Active release coating of multilayer assembled branched and ionic β-cyclodextrins onto poly(ethylene terephthalate). *Carbohydr. Polym.* **2017**, *174*, 65–71. [CrossRef] [PubMed]

34. Pinho, E.; Henriques, M.; Soares, G. Cyclodextrin/cellulose hydrogel with gallic acid to prevent wound infection. *Cellulose* **2014**, *21*, 4519–4530. [CrossRef]

35. Lin, J.-H.; Pan, Y.-J.; Liu, C.-F.; Huang, C.-L.; Hsieh, C.-T.; Chen, C.-K.; Lin, Z.-I.; Lou, C.-W. Preparation and compatibility evaluation of polypropylene/high density polyethylene polyblends. *Materials* **2015**, *8*, 8850–8859. [CrossRef] [PubMed]

36. Muzio, G.; Miola, M.; Perero, S.; Oraldi, M.; Maggiora, M.; Ferraris, S.; Vernè, E.; Festa, V.; Festa, F.; Canuto, R.A.; et al. Polypropylene prostheses coated with silver nanoclusters/silica coating obtained by sputtering: Biocompatibility and antibacterial properties. *Surf. Coat. Technol.* **2017**, *319*, 326–334. [CrossRef]

37. Majumdar, A.; Butola, B.S.; Thakur, S. Development and performance optimization of knitted antibacterial materials using polyester-silver nanocomposite fibres. *Mater. Sci. Eng. C Mater. Biol. Appl.* **2015**, *54*, 26–31. [CrossRef] [PubMed]

polymers

MDPI

Article

α-Cyclodextrins Polyrotaxane Loading Silver Sulfadiazine

Sa Liu [1,2], Chunting Zhong [1,2], Weiwei Wang [1,2], Yongguang Jia [1,2], Lin Wang [1,2] and Li Ren [1,2,*]

1   School of Materials Science and Engineering, South China University of Technology, Guangzhou 510641,
     China; sliu@scut.edu.cn (S.L.); msctzhong@mail.scut.edu.cn (C.Z.); wei880719@163.com (W.W.);
     ygjia@scut.edu.cn (Y.J.); wanglin3@scut.edu.cn (L.W.)
2   National Engineering Research Center for Tissue Restoration and Reconstruction, Guangzhou 510006, China
*   Correspondence: psliren@scut.edu.cn; Tel.: +86-20-2223-6528

Received: 30 November 2017; Accepted: 5 February 2018; Published: 14 February 2018

Abstract: As a drug carrier, polyrotaxane (PR) has been used for targeted delivery and sustained release of drugs, whereas silver sulfadiazine (SD-Ag) is an emerging antibiotic agent. PR was synthesized by the use of α-cyclodextrin (CD) and poly(ethylene glycol) (PEG), and a specific antibacterial material (PR-(SD-Ag)) was then prepared by loading SD-Ag onto PR with different mass ratios. The loading capacity and the encapsulation efficiency were 90% at a mass ratio of 1:1 of PR and SD-Ag. SD-Ag was released stably and slowly within 6 d in vitro, and its cumulative release reached more than 85%. The mechanism of PR loading SD-Ag might be that SD-Ag attached to the edge of α-CD through hydrogen bonding. PR-(SD-Ag) showed a higher light stability than SD-Ag and held excellent antibacterial properties against *Escherichia coli* (*E. coli*) and *Staphylococcus aureus* (*S. aureus*).

Keywords: polyrotaxane; silver sulfadiazine; mechanism; antibacterial

1. Introduction

To improve the health of the population, a variety of antimicrobial agents have been used to suppress and kill harmful microorganisms in people's living environments [1–3]. After the discovery and application of antibiotics, antibacterial and bactericidal properties seem increasingly prominent. However, the emergence of super-resistant strains of bacteria threatens human health, resulting in the extensive usage of antibiotics [4]. The research on broad-spectrum antibacterial materials showing nontoxicity and bio-heat resistance has become extremely urgent [5–7].

Compared to other kinds of antibacterial agents, silver-based antimicrobial materials are effective and safe. They have been widely used and have become a focus of commercialization in antimicrobial research [8,9]. For example, silver sulfadiazine (SD-Ag) has received wide-spread acceptance as a topical agent to control bacterial infections [10]. The key of bacteriostatic agents is silver ions [11], which significantly affect the treatment of burn wounds and the promotion of wound healing and infection control. However, crystalline silver sulfadiazine easily deteriorates under the influence of light or heat. It can dissolve in acid or ammonia aqueous media, but not in ethanol, chloroform, and some other organic solvents. These drawbacks limit the application of these silver-based antimicrobial agents.

Polyrotaxane (PR) has a necklace-like molecular structure, where many cyclic molecules are threaded by a linear molecule ending in two bulky groups to avoid the dissociation of the cyclic molecules [12]. Recently, PR has attracted a great deal of interest in the production of various raw materials [13–15] due to its specific characteristics of free sliding and/or rotating threaded cyclic molecules. Moreover, PR can be easily metabolized by the body to reduce cytotoxicity [16–18]. As a drug carrier, PR has been used for targeted drug-delivery and the sustained release of drugs, such as peptides [19–21] and genes [22–24]. Circumstantially, this strategy can improve the biological properties of drugs, such as low haemolysis, lasting pharmacodynamics, biodegradable properties, and low cytotoxicity. Thus, these PRs could be excellent carriers for SD-Ag-based antibacterial materials.

In this paper, the antibacterial agent SD-Ag was loaded onto PR and antibacterial complexes PR-(SD-Ag), and different mass ratios of PR/SD-Ag were prepared. The drug loading content, the in vitro drug release, light stability, and loading mechanism are discussed. Antibacterial properties of PR-(SD-Ag) against *Escherichia coli* (*E. coli*) and *Staphylococcus aureus* (*S. aureus*) were also tested.

2. Materials and Methods

2.1. Materials

α-Cyclodextrins (α-CDs) were purchased from Shandong BinzhouZhiyuan Bio-Technology Co., Ltd. (Binzhou, China) (α-CD content > 99%); poly(ethylene glycol) (PEG 35,000 and PEG 20,000) was purchased from Sigma (Shanghai, China). The free-base forms of 1-adamantanamine and silver sulfadiazine (SD-Ag) were purchased from the Aladdin Reagent Company (Shanghai, China). All reagents were of experimental grade and were used without further purification. All solvents were dehydrated.

2.2. Synthesis of PR

PEG (10 g) was oxidized using 2,2,6,6-Tetramethylpiperidine 1-Oxyl (TEMPO) (100 mg), NaBr (100 mg), and aqueous NaClO (10 mL, available chlorine > 5.0%) in water (100 mL). After stirring at room temperature (pH 10–11) for 15 min, the oxidation reaction was quenched by adding ethanol (10 mL). Then, HCl solution was used to lower the pH to less than 2, and the solution was extracted three times using CH_2Cl_2. The organic phase was dried in vacuo. The crude residue was dissolved in hot ethanol (250 mL) and precipitated at $-20\,°C$ overnight. PEG-COOH was obtained after the second recrystallization with ethanol. After titration with 0.01 mol·L^{-1} NaOH and a percent conversion of >99%, it was indicated that PEG (20,000 and 35,000) were converted to the corresponding PEG-COOH.

PEG-COOH (1.5 g) and α-CD (6 g) were dissolved in water (100 mL). After stirring at 70 °C for 20 h and 4 °C for 48 h in succession, a white precipitate appeared. After freeze-drying, the pseudo-polyrotaxane was mixed with adamantanamine (0.8 g), (benzotriazol1-yloxy) tris-(dimethylamino) phosphonium hexafluorophosphate (0.24 g), and ethyldiisopropylamine (EDIPA) (0.1 mL), and was then dissolved in dehydrated dimethylformamide (DMF) (100 mL). The slurry-like mixture was stirred at room temperature overnight, then washed with DMF/methanol (1:1) and methanol in succession. The PR solid was obtained after the white residue was dissolved in dimethyl sulfoxide (DMSO) (40 mL), precipitated, and washed with water (400 mL), and, finally, freeze-dried. PR1 t and PR2 t (based on molecular weight 20,000 and 35,000) were prepared. ^{1}H NMR (400 MHz, DMSO-d_6) δ (ppm) 5.63 (s, 1H, O_2–H), 5.48 (s, 1H, O_3–H), 4.90 (s, 1H, C_1–H), 4.44 (s, 1H, O_6–H), 3.69 (d, *J* = 32.4 Hz, 3H, $C_{3,5,6}$–H), 3.51 (s, 5H, C_{PEG}–H), 2.01 (s, 1H, N–H), 1.94 (s, 1H, N–H), 1.62 (s, 1H, N–H).

2.3. Preparation of PR-(SD-Ag)

According to the mass ratios 1:1, 1.5:1, and 2:1 of PR and SD-Ag, the different mixtures were each dissolved in DMSO and protected from light. After stirring for 6 h at room temperature and dialyzing in deionized water for 4 d at 25 °C, the liquid was freeze-dried, generating white solid PR-(SD-Ag). ^{1}H NMR (400 MHz, DMSO-d_6) δ (ppm) 8.38 (d, *J* = 4.9 Hz, 1H, P_1–H), 7.62 (d, *J* = 8.6 Hz, 1H, B_1–H), 6.77 (s, 1H, B_2–H), 6.51 (d, *J* = 8.6 Hz, 1H, P_2–H), 5.68 (s, 1H, O_2–H), 5.49 (s, 1H, O_3–H), 4.80 (s, 1H, C_1–H), 4.45 (s, 1H, O_6–H), 3.69 (d, *J* = 37.5 Hz, 3H, $C_{3,5,6}$–H), 3.51 (s, 3H, C_{PEG}–H), 3.28 (s, 2H, $C_{2,4}$–H).

2.4. Fourier Transform Infrared (FTIR) Spectroscopy

Absorbance spectra were recorded using the KBr pellet method with a VECTOR-22 FTIR spectrometer (Bruker, Ettlingen, Germany). The spectra were collected in the region of 4000 to 400 cm^{-1}.

2.5. X-ray Diffraction

X-ray diffraction was recorded on powdered samples using a D8 ADVANCE X-ray diffractometer (Rigaku, Tokyo, Japan). The conditions were set to the following: 40 kV (voltage), 40 mA (current), and 1.54 A° (wavelength). The samples were mounted on a circular sample holder and then sealed with Scotch tape. The proportional counter detector collected data at a rate of $2\theta = 10°\cdot min^{-1}$ over the following range: $2\theta = 5–90°$.

2.6. ^1H NMR Spectroscopy

^{1}H NMR spectra were recorded on a Bruker AVANCE 400 MHz spectrometer (Bruker, Ettlingen, Germany) at room temperature, using DMSO-d6 as solvent and tetramethylsilane (TMS) as internal standard.

2.7. Light Stability

Different PR-(SD-Ag) systems were dissolved in DMSO. Then, the solutions were placed under fluorescent light for an irradiation period, after which a change in color could be observed. The solution concentration was 0.02 g/mL.

2.8. Analysis of Drug Content

The sample was placed in a small beaker; then, a small amount of concentrated nitric acid was added for dissolution. After pouring into a 50 mL volumetric flask, phosphate-buffered saline (PBS) solution was added as a diluent. The absorbance values were obtained using a UV/visible spectrophotometer (Persee, Beijing, China) at 245 nm. All the experiments were done in triplicate.

2.9. In Vitro Release Studies

Quantitative PR-(SD-Ag) was placed in a dialysis bag immersed in phosphate-buffered saline solution (PBS, 50 mL, pH 7.4), and drug release was performed at 37 °C with stirring. Supernatants were removed at every desired time interval to determine the amount of SD-Ag released from PR, which was then analyzed using a UV/visible spectrophotometer at 245 nm. Then, the supernatant was added again to maintain a constant volume. The amount of SD-Ag was previously quantified using a built analytical curve, which was designed using the absorbance versus standard SD-Ag solution, varying from 2.5 to 20 $\mu g\cdot mL^{-1}$. The linear correlation coefficient (R^2) was 0.9988.

3. Results and Discussion

3.1. Preparation of the PR-(SD-Ag) Complex

FTIR spectra of SD-Ag, PR, and their complex PR-(SD-Ag), as well as the physical mixture in the region from 500 to 4000 cm^{-1}, are presented in Figure S1. In the spectrum of SD-Ag, peaks at 3394, 3344, and 3261 cm^{-1} were attributed to N–H stretching vibrations. Peaks at 1654, 1593, and 1583 cm^{-1} were attributed to the phenyl and pyrimidinyl rings. Asymmetric and symmetric vibrations of γ (SO$_2$) appear at 1354 and 1137 cm^{-1}, respectively. This is in good agreement with literature data. In the spectrum of pure PR, the broad band at 3356 cm^{-1} and the weak band at 1643 cm^{-1} were assigned to the stretching vibration of the hydroxyl groups in α-CD and the hydration of water, respectively, which is similar to that of pseudo-PR between PEG and α-CD (supporting information) [25]. The spectrum of PR-(SD-Ag) contains the characteristic absorption peaks of PR, but lost some of SD-Ag. A new peak appeared at 1586 cm^{-1}, slightly higher than 1583 cm^{-1} in the SD-Ag spectrum, which is related to the unusual interactions around the phenyl and pyrimidinyl rings. The IR spectrum of the physical mixture of PR and SD-Ag clearly showed almost all the characteristic absorption peaks. These results indicate that PR-(SD-Ag) is not a simple mixture of PR and SD-Ag.

PR, SD-Ag, and PR-(SD-Ag) were characterized using XRD (Figure S2). There was a broad diffraction peak at 2θ 11.10°, a distinct peak at 2θ 19.84°, and a relatively insignificant peak at 2θ 13.32° for PR. The peaks of SD-Ag were more complicated, and were mainly at 2θ 8.80°, 10.21°, and 18.49°. Compared to SD-Ag and PR, the peaks of PR-(SD-Ag) moved slightly, appearing at 2θ 8.86°, 10.38°, and 19.91°, respectively, and a new peak was observed at 2θ 11.46°. This may suggest that there is a specific interaction between SD-Ag and α-CD in PR.

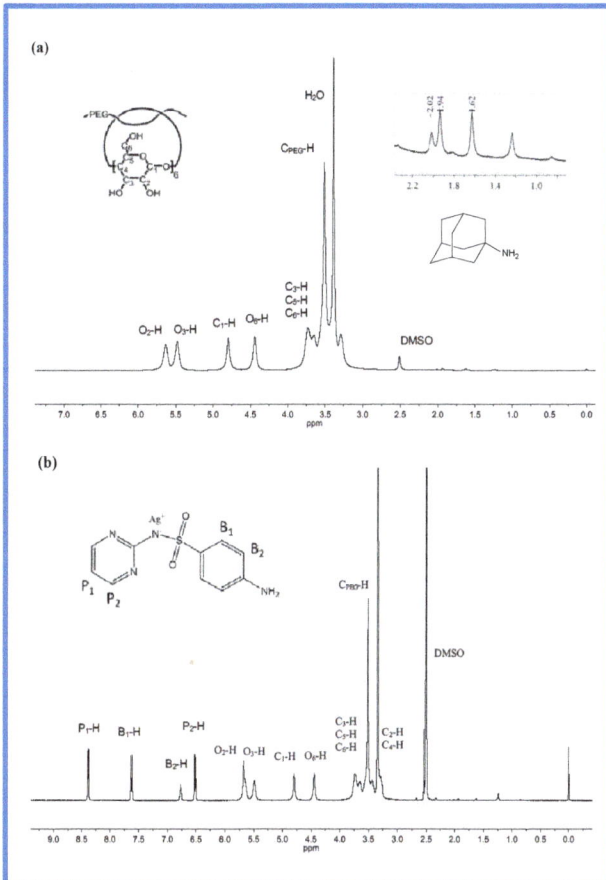

Figure 1. ^{1}H NMR spectra of (**a**) polyrotaxane (PR) and (**b**) polyrotaxane-silver sulfadiazine complex (PR-(SD-Ag)) in DMSO-d_6.

PRs were then systematically characterized using ^{1}H NMR analysis (Figure 1a). All peaks of the protons on PEG and α-CD were confirmed. The signal peaks of PEG methylene were around 3.51 ppm. The chemical shifts of protons on α-CD appeared at 5.63 ppm (O_2–H), 5.48 ppm (O_3–H), 4.44 ppm (O_6–H), and 4.80 (C_1–H), 3.80–3.20 ppm ($C_{2,3,4,5,6}$–H). Additionally, peaks at 2.02, 1.94, and 1.62 ppm showed the protons of the carbon skeleton in amantadine. This manifested the formation of PR after the end-capping reaction between amino groups in amantadine and terminal carboxyl groups in PEG. ^{1}H NMR analysis was used to confirm the structure of PR-(SD-Ag) (Figure 1b). The characteristic peaks at 5.68 ppm (O_2–H), 5.49 ppm (O_3–H), and 4.45 ppm (O_6–H) express the information of hydroxyl in α-CD. Peaks at 4.80 and 3.20–3.80 ppm showed the information of CH in α-CD, and the peak at 3.51 ppm was attributed to the methylene of PEG. The new peaks around 6.50–8.39 ppm were attributed to the benzene rings and pyrimidine rings in SD-Ag. However, the signals from the amino groups in SD-Ag disappeared and the chemical shifts of the hydroxyl group in α-CD of PR-(SD-Ag) shifted downward slightly compared with that of PR. These results also indicated that the complex formed upon mixing PR and SD-Ag.

3.2. Drug Loading and In Vitro Release

Strategies were used through designing different mass ratios of PR/SD-Ag to study the drug loading and encapsulation efficiency of $PR_{1 t}$-(SD-Ag) and $PR_{2 t}$-(SD-Ag) (Figure 2). They presented similar variational tendencies in these histograms. Maintaining a certain mass ratio of PR/SD-Ag, the increased content of α-CD in PR led to an increase in drug loading capacity and encapsulation rates. Maintaining a certain content of α-CD, the higher the mass ratio of PR/SD-Ag, the lower the drug loading capacity, while the encapsulation efficiency maintained a sustainable growth with an increased quantity ratio. As both the drug loading capacity and the encapsulation efficiency were higher than 78.2%, a 1:1 mass ratio of PR/SD-Ag is desired. The greatest drug loading capacity was 89.0% for $PR_{1 t}$-(SD-Ag) and 91.2% for $PR_{2 t}$-(SD-Ag). The encapsulation rates were 78.2%–90% for $PR_{1 t}$-(SD-Ag) and 80.4%–98.4% for $PR_{2 t}$-(SD-Ag). $PR_{2 t}$-(SD-Ag) showed higher drug loading capacity and encapsulation rates than $PR_{1 t}$-(SD-Ag). This could be attributed to the PEG chain length. The longer the PEG chain, the more the α-CD are bunched, which causes more silver sulfadiazine to be encapsulated. In summary, a 1:1 mass ratio of PR/SD-Ag was optimal, and more SD-Ag was encapsulated using a smaller amount of PR.

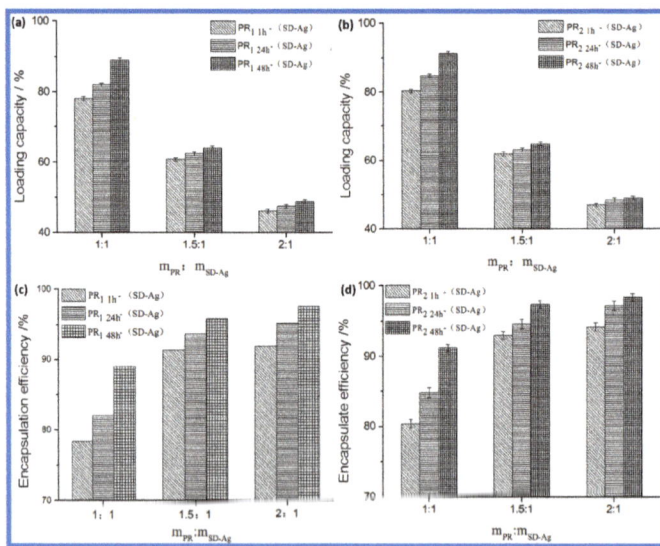

Figure 2. The drug loading capacity of (**a**) $PR_{1 t}$-(SD-Ag), (**b**) $PR_{2 t}$-(SD-Ag), and the encapsulation efficiency of (**c**) $PR_{1 t}$-(SD-Ag), (**d**) $PR_{2 t}$-(SD-Ag).

Figure 3 shows the process of drug release for $PR_{1 t}$-(SD-Ag) and $PR_{2 t}$-(SD-Ag) in different mass ratios of PR/SD-Ag in a PBS solution (pH at 7.4) at 37 °C. All samples exhibited a similar release tendency. At an earlier stage, the accumulating release rates increased slowly and linearly at a 1:1 mass ratio under vibration. However, the linear regularities became unclear at 1.5:1 and 2:1 mass ratios. After 72 h, the accumulation release rates slowed down gradually until reaching the maximum value, which was about 88% for $PR_{1 t}$-(SD-Ag) and 93% for $PR_{2 t}$-(SD-Ag). Circumstantially, the accumulating release rates of $PR_{1\ 1h}$-(SD-Ag), $PR_{1\ 24h}$-(SD-Ag), and $PR_{1\ 48h}$-(SD-Ag) increased successively (Figure 3a–c). Maintaining a certain time, the relationships for different mass ratios were $PR_{1\ t}$-(SD-Ag) 1:1 > $PR_{1\ t}$-(SD-Ag) 1.5:1 > $PR_{1\ t}$-(SD-Ag) 2:1. These phenomena were attributed to the change in viscosity. The higher the amount of PR, the higher the viscosity of the solution. In proportion, the free volume in the system decreased, which restricted the movement of α-CD. The relation of $PR_{2\ t}$-(SD-Ag) was consistent with that of $PR_{1\ t}$-(SD-Ag). In summary, PR-(SD-Ag) was expected to exhibit a slow and steady upper release and a high release ratio without an initial burst release.

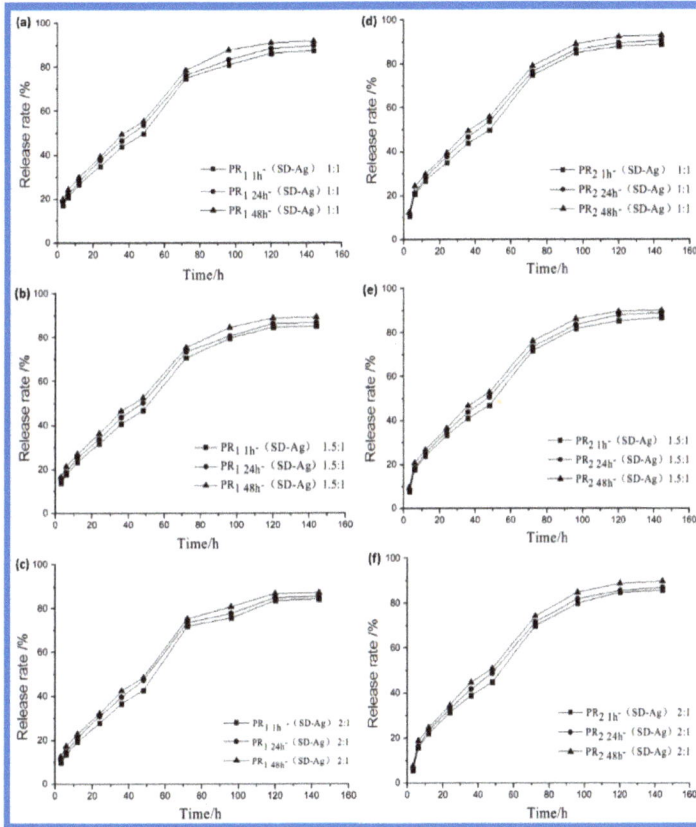

Figure 3. In vitro release profile of (**a–c**) $PR_{1\ t}$-(SD-Ag) and (**d–f**) $PR_{2\ t}$-(SD-Ag) with different mass ratios of polyrotaxane and silver sulfadiazine.

3.3. Antibacterial Studies and Light Stability

Figure 4 shows the antibacterial effects of PR-(SD-Ag) on *E. coli* and *S. aureus* (1×10^7 CFUmL^{-1}), respectively. Compared with the blank group's PR (Figure 4A(a),B(a)), all experimental groups expressed good antibacterial effects. The four-hour antibacterial rate of $PR_{1\ t}$-(SD-Ag) and $PR_{2\ t}$-(SD-Ag) for *E. coli* and *S. aureus* were tested, and the results showed that the antibacterial rates of all groups were higher than 98%. It was found that the antibacterial effects of PR-(SD-Ag) with different mass ratios of polyrotaxane and silver sulfadiazine were slightly different. When the mass ratio was 1:1, the antibacterial effect of PR-(SD-Ag) was the best. The maximum antibacterial rate reached 99.9% for *E. coli* and 99.6% for *S. aureus*, respectively. This is consistent with the results of in vitro drug release for PR-(SD-Ag).

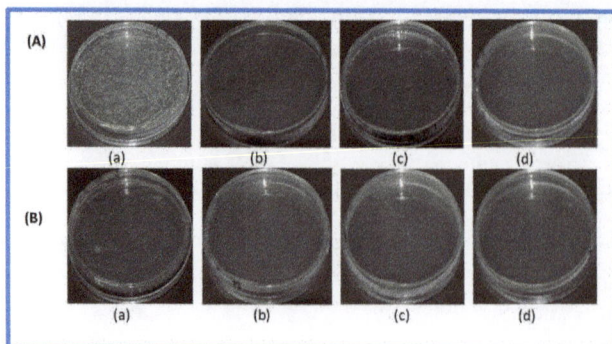

Figure 4. The antibacterial effect of PR-(SD-Ag) on (**A**) *E. coli* and (**B**) *S. aureus*: (**a**) blank groups (PR$_{2\,24\,h}$); (**b**) PR$_{2\,24\,h}$-(SD-Ag)2:1; (**c**) PR$_{2\,24\,h}$-(SD-Ag)1.5:1; (**d**) PR$_{2\,24\,h}$-(SD-Ag)1:1.

Light stability of SD-Ag and PR-(SD-Ag) was compared under fluorescent light for different amounts of time (Figure 5). Except for manually-added light conditions, experiments proceeded at room temperature, and were acquiescently protected from light. PR-(SD-Ag) and SD-Ag were each dissolved in a small amount of ammonia water. Then, they were respectively diluted with DMSO. The color of the SD-Ag solution became brown after being exposed to fluorescent light for 30 min (Figure 5a). After being exposed to fluorescent lamp irradiation for 10 d, the color of PR-(SD-Ag) did not change (Figure 5b). After 29 d, it became slightly transparent and red, and turned dark brown after 45 d. These results indicate that the light stability of SD-Ag in a complex improved greatly, perhaps due to the formation of an inclusion with PR.

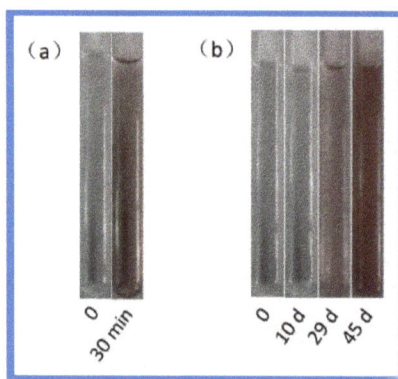

Figure 5. Comparison of light stability: (**a**) The color of SD-Ag solution became brown after being exposed to fluorescent light for 30 min; (**b**) the color of PR-(SD-Ag) did not change until day 29, when it became slightly transparent and red, and dark brown after day 45.

3.4. The Mechanism of PR Loading SD-Ag

According to FTIR, XRD, and ^{1}H NMR spectroscopic analyses, it was confirmed that PR-(SD-Ag) was not a simple mixture of PR and SD-Ag. Numerous studies have clarified the size-matching law in PR formation [25]. The sizes of α-CD cavities (4.70–5.30 Å) and SD-Ag (4.20–4.70 Å) match [26]. However, it is almost impossible to form complete host–guest inclusion between α-CD and SD-Ag due to PEG chains occupying the α-CD cavity. It is probable that complexes were formed by hydrogen bonding between SD-Ag and the α-CD on the PEG. The interactions of protons in pyrimidine and benzene rings with the α-CD cavity protons lead to a downfield shift of δ [26,27]. The above ¹H NMR information (a small downfield shift of O-H in α-CD, Figure 1) demonstrate the formation of hydrogen bonds between α-CD and SD-Ag. The molar ratios of α-CD and SD-Ag were estimated to be approximately 2:5.4, based on the NMR integration ratio. Accordingly, the model of each of the two α-CD molecules including four-to-six SD-Ag molecules showed a great advantage because of the hydrophobic cavity environment and abundant exocoel hydroxyls in α-CD. Herein, we describe the model of drug release (Figure 6A), as well as the main complex mechanisms (Figure 6B). First, after the aggregation of four SD-Ag molecules through N···H···N hydrogen bonds between –NH₂, four pyrimidine rings can be partially included by the α-CD cavity [26] under the protection of O···H···O hydrogen bonds between –OH and O=S=O (Figure 6B(a)). This might be the most stable combination. Second, the combination through N···H···N hydrogen bonds between –NH₂ and O···H···O hydrogen bonds between –OH and O=S=O might also form a stable complex (Figure 6B(b)). Thirdly, the combination of numerous O···H···N hydrogen bonds between –OH and –NH₂ showed relative stability (Figure 6B(c)). In these cases, more SD-Ag molecules might attach to the small edge of α-CD through O···H···N hydrogen bonds between –OH and –NH₂. However, non-covalent bonding between PEG and SD-Ag at all these stages—which might facilitate a boost in drug loading—could not be neglected. When releasing SD-Ag, the weak binding patterns might first be disrupted at 37 °C. Due to the balance between the breaking and linking of partial hydrogen bonds, there was no initial burst release (Figure 3). Later, the stable states were disrupted step-by-step, presenting a slow and steady upper release. Under the protection of PR, SD-Ag became more stable, even when exposed to light (Figure 5).

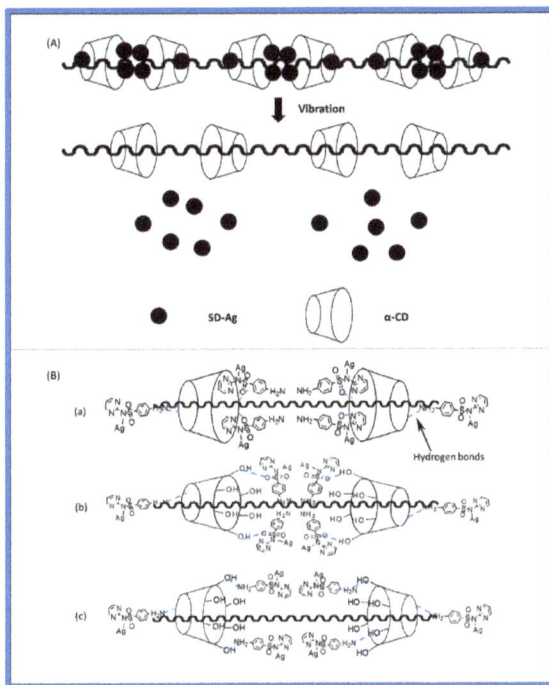

Figure 6. The mechanism of PR loading SD-Ag: (**A**) Drug release model; (**B**) combination through (a) N···H···N hydrogen bonds between –NH₂ in SD-Ag, O···H···O hydrogen bonds between –OH in α-cyclodextrin (CD) and O=S=O in SD-Ag and partial inclusion of pyrimidine rings by α-CD cavity; (b) N–H–N hydrogen bonds between –NH₂ in SD-Ag and O···H···O hydrogen bonds between–OH in α-CD and O=S=O in SD-Ag; (c) O···H···N hydrogen bonds between –OH in α-CD and –NH₂ in SD-Ag.

4. Conclusions

In this article, two kinds of PR of PEG toward α-CD were synthesized, and the antibacterial agent SD-Ag was loaded into these PRs in a controlled manner. The loading capacity and encapsulation efficiency of the resulting antibacterial materials, PR-(SD-Ag), were both near 90% when the mass ratio of PR/SD-Ag was 1:1. SD-Ag could be stably and slowly released within 6 d in vitro, and showed release rates of over 85%. PR-(SD-Ag) held excellent antibacterial properties against *E. coli* and *S. aureus*. Moreover, PR-(SD-Ag) had a better light stability compared with SD-Ag alone. The method of using PR as an antibacterial drug carrier will provide more choices in the preparation of antibacterial materials. These PR-(SD-Ag) can be used as a potential antibacterial drug in the future.

Supplementary Materials: The following are available online at http://www.mdpi.com/2073-4360/10/2/190/s1, Figure S1: The FTIR spectra of PR, SD-Ag, PR+SD-Ag and PR-(SD-Ag), Figure S2: The XRD patterns of PR-(SD-Ag), SD-Ag and PR, Table S1: Antibacterial rate of PR$_{1t}$-(SD-Ag) and PR$_{2t}$-(SD-Ag) on *E. coli* and *S. aureus*.

Acknowledgments: This work was supported by the National Natural Science Foundation of China (51673071, 51232002), Natural Science Foundation of Guangdong Province (2016A030313509), Guangzhou Important Scientific and Technological Special Project (201508020123), Guangdong Scientific and Technological Project (2016B090918040, 2014B090907004), which are gratefully acknowledged.

Author Contributions: Sa Liu and Li Ren conceived and designed the experiments; Weiwei Wang performed the experiments; Sa Liu, Chunting Zhong and Weiwei Wang analyzed the data; Sa Liu and Li Ren contributed materials and analysis tools; Chunting Zhong, Sa Liu, Yongguang Jia and Lin Wang wrote the paper.

Conflicts of Interest: The authors declare no conflict of interest.

References

1. Heydari, B.; Khalili, H.; Karimzadeh, I.; Emadi-Kochak, H. Clinical, paraclinical, and antimicrobial resistance features of community-acquired acute bacterial meningitis at a large infectious diseases ward in Tehran, Iran. *Iran. J. Pharm. Res.* **2016**, *15*, 347–354. [PubMed]

2. Niu, F.G.; Pan, W.C.; Su, Y.J.; Yang, Y.J. Physical and antimicrobial properties of thyme oil emulsions stabilized by ovalbumin and gum Arabic. *Food Chem.* **2016**, *212*, 138–145. [CrossRef] [PubMed]

3. Przybylski, R.; Firdaous, L.; Chataigne, G.; Dhulster, P.; Nedjar, N. Production of an antimicrobial peptide derived from slaughterhouse by-product and its potential application on meat as preservative. *Food Chem.* **2016**, *211*, 306–313. [CrossRef] [PubMed]

4. Vespa, P.M. Fever in critical neurologic illness. *JAMA-J. Am. Med. Assoc.* **2014**, *312*, 1456–1457. [CrossRef] [PubMed]

5. Stretton, S.; Gopinathan, U.; Willcox, M.D.P. Corneal ulceration in pediatric patients. *Paediatr. Drugs* **2002**, *4*, 95–110. [CrossRef] [PubMed]

6. Kong, M.; Chen, X.G.; Xing, K.; Park, H.J. Antimicrobial properties of chitosan and mode of action: A state of the art review. *Int. J. Food Microbiol.* **2010**, *144*, 51–63. [CrossRef] [PubMed]

7. Jiang, L.; Wang, F.; Han, F.; Prinyawiwatkul, W.; No, H.K.; Ge, B. Evaluation of diffusion and dilution methods to determine the antimicrobial activity of water-soluble chitosan derivatives. *J. Appl. Microbiol.* **2013**, *114*, 956–963. [CrossRef] [PubMed]

8. Yang, W.M.; Fang, W.Y. Antimicrobial plastics. *Fine Petrochem.* **1998**, *6*, 14–18.

9. Rai, M.; Yadav, A.; Gad, A. Silver nanoparticles as a new generation of antimicrobials. *Biotechnol. Adv.* **2009**, *27*, 76–83. [CrossRef] [PubMed]

10. Fan, L.H.F.; Zhao, J.J.; Huang, J.; Xu, Y.M. Polyelectrolyte sponges with antimicrobial functions based on chitosan and sodium alginate. *J. Wuhan Univ. Technol.* **2006**, *28*, 25–28.

11. Rosenkranz, H.S.; Carr, H.S. Silver sulfadiazine: Effect on the growth and metabolism of bacteria. *Antimicrob. Agents Chemother.* **1972**, *2*, 367–372. [CrossRef] [PubMed]

12. Araki, J.; Kataoka, T.; Ito, K. Preparation of a "sliding graft copolymer", an organic solvent-soluble polyrotaxane containing mobile side chains, and its application for a crosslinked elastomeric supramolecular film. *Soft Matter* **2008**, *4*, 245–249. [CrossRef]

13. Araki, J.; Ito, K. Recent advances in the preparation of cyclodextrin-based polyrotaxanes and their applications to soft materials. *Soft Matter* **2007**, *3*, 1456–1473. [CrossRef]

14. Li, J.; Loh, X.J. Cyclodextrin-based supramolecular architectures: Syntheses, structures, and applications for drug and gene delivery. *Adv. Drug Deliv. Rev.* **2008**, *60*, 1000–1017. [CrossRef] [PubMed]

15. Harada, A.; Hashidzume, A.; Yamaguchi, H.; Takashima, Y. Polymeric rotaxanes. *Chem. Rev.* **2009**, *109*, 5974–6023. [CrossRef] [PubMed]

16. Lin, L.; Dong, M.; Liu, C.; Wei, C.; Wang, Y.; Sun, H.; Ye, H. A supramolecular strategy for self-mobile adsorption sites in affinity membrane. *Macromol. Rapid Commun.* **2014**, *35*, 1587–1591. [CrossRef] [PubMed]

17. Tardy, B.L.; Dam, H.H.; Kamphuis, M.M.; Richardson, J.J.; Caruso, F. Self-assembled stimuli-responsive polyrotaxane core-shell particles. *Biomacromolecules* **2014**, *15*, 53–59. [CrossRef] [PubMed]

18. Tan, S.; Nam, E.; Cui, J.; Xu, C.; Fu, Q.; Ren, J.M.; Wong, E.H.H.; Ladewig, K.; Caruso, F.; Blencowe, A.; et al. Fabrication of ultra-thin polyrotaxane-based films via solid-state continuous assembly of polymers. *Chem. Commun.* **2015**, *51*, 2025–2028. [CrossRef] [PubMed]

19. Moon, C.; Kwon, Y.M.; Lee, W.K.; Park, Y.J.; Yang, V.C. In vitro assessment of a novel polyrotaxane-based drug delivery system integrated with a cell-penetrating peptide. *J. Control. Release* **2007**, *124*, 43–50. [CrossRef] [PubMed]

20. Zhang, J.X.; Ma, P.X. Cyclodextrin-based supramolecular systems for drug delivery: Recent progress and future perspective. *Adv. Drug Deliv. Rev.* **2013**, *65*, 1215–1233. [CrossRef] [PubMed]

21. Inoue, Y.; Ye, L.; Ishihara, K.; Yui, N. Preparation and surface properties of polyrotaxane-containing tri-block copolymers as a design for dynamic biomaterials surfaces. *Colloids Surf. B* **2012**, *89*, 223–227. [CrossRef] [PubMed]

22. Ooya, T.; Choi, H.S.; Yamashita, A.; Yui, N.; Sugaya, Y.; Kano, A.; Maruyama, A.; Akita, H.; Ito, R.; Kogure, K.; et al. Biocleavable polyrotaxane-plasmid DNA polyplex for enhanced gene delivery. *J. Am. Chem. Soc.* **2006**, *128*, 3852–3853. [CrossRef] [PubMed]

23. Yamashita, A.; Yui, N.; Ooya, T.; Kano, A.; Maruyama, A.; Akita, H.; Kogure, K.; Harashima, H. Synthesis of a biocleavable polyrotaxane-plasmid DNA (pDNA) polyplex and its use for the rapid nonviral delivery of pDNA to cell nuclei. *Nat. Protoc.* **2006**, *1*, 2861–2869. [CrossRef] [PubMed]

24. Zhang, L.; Su, T.; He, B.; Gu, Z. Self-assembly polyrotaxanes nanoparticles as carriers for anticancer drug methotrexate delivery. *Nano-Micro Lett.* **2014**, *6*, 108–115. [CrossRef]

25. Harada, A.; Li, J.; Kamachi, M. Preparation and properties of inclusion complexes of Poly(ethyleneglycol) with α-cyclodextrin. *Macromolecules* **1993**, *26*, 5698–5703. [CrossRef]

26. Rajendiran, N.; Venkatesh, G.; Saravanan, J. Supramolecular aggregates formed by sulfadiazine and sulfisomidine inclusion complexes with α- and β-cyclodextrins. *Spectrochim. Acta A* **2014**, *129*, 157–162. [CrossRef] [PubMed]

27. Rajendiran, N.; Thulasidhasan, J. Interaction of sulfanilamide and sulfamethoxazole with bovine serum albumin and adenine: Spectroscopic and molecular docking investigations. *Spectrochim. Acta A* **2015**, *144*, 183–191. [CrossRef] [PubMed]

polymers

MDPI

Article

α-Cyclodextrin and α-Cyclodextrin Polymers as Oxygen Nanocarriers to Limit Hypoxia/Reoxygenation Injury: Implications from an In Vitro Model

Saveria Femminò [1,†], Claudia Penna [1,†], Federica Bessone [2], Fabrizio Caldera [3], Nilesh Dhakar [3], Daniele Cau [1], Pasquale Pagliaro [1], Roberta Cavalli [2] and Francesco Trotta [3,*]

[1] Department of Clinical and Biological Sciences, University of Turin, 10043 Orbassano, Italy; saveria.femmino@unito.it (S.F.); claudia.penna@unito.it (C.P.); daniele.cau@edu.unito.it (D.C.); pasquale.pagliaro@unito.it (P.P.)

[2] Department of Drug Science and Technology, University of Turin, 10125 Turin, Italy; f.bessone@unito.it (F.B.); roberta.cavalli@unito.it (R.C.)

[3] Department of Chemistry, University of Turin, 10125 Turin, Italy; fabrizio.caldera@unito.it (F.C.); dhakar.neelesh89@gmail.com (N.D.)

* Correspondence: francesco.trotta@unito.it; Tel.: +39-011-670-7550

† These authors contributed equally to this work.

Received: 15 December 2017; Accepted: 16 February 2018; Published: 22 February 2018

Abstract: The incidence of heart failure (HF) is increasing worldwide and myocardial infarction (MI), which follows ischemia and reperfusion (I/R), is often at the basis of HF development. Nanocarriers are interesting particles for their potential application in cardiovascular disease. Impaired drug delivery in ischemic disease is challenging. Cyclodextrin nanosponges (NS) can be considered innovative tools for improving oxygen delivery in a controlled manner. This study has developed new α-cyclodextrin-based formulations as oxygen nanocarriers such as native α-cyclodextrin (α-CD), branched α-cyclodextrin polymer (α-CD POLY), and α-cyclodextrin nanosponges (α-CD NS). The three different α-CD-based formulations were tested at 0.2, 2, and 20 µg/mL to ascertain their capability to reduce cell mortality during hypoxia and reoxygenation (H/R) in vitro protocols. H9c2, a cardiomyoblast cell line, was exposed to normoxia (20% oxygen) or hypoxia (5% CO_2 and 95% N_2). The different formulations, applied before hypoxia, induced a significant reduction in cell mortality (in a range of 15% to 30%) when compared to samples devoid of oxygen. Moreover, their application at the beginning of reoxygenation induced a considerable reduction in cell death (12% to 20%). α-CD NS showed a marked efficacy in controlled oxygenation, which suggests an interesting potential for future medical application of polymer systems for MI treatment.

Keywords: α-cyclodextrin; α-cyclodextrin nanosponges; α-cyclodextrin polymers; oxygen delivery; myocardial infarction; ischemia; reperfusion

1. Introduction

Advanced and smart polymer-based nanocarriers were designed for tissue-specific targeted delivery, triggered release, and co-delivery of synergistic drug combinations to develop safer and more efficient therapeutics. The heart is a metabolically demanding organ and mismatches between supply and demand of oxygen may occur, which could lead to hypoxia and cardiac cell distress. Hypoxia is a pathological condition for many diseases and it particularly occurs in acute coronary occlusion, which leads to cardiac cell death. This is the first phase of acute myocardial infarction (AMI). Hypertrophic surviving cells and heart failure are consequences of AMI. These serious outcomes of coronary occlusion can be limited by early and beneficial reperfusion. Full reperfusion is usually performed in primary angioplasty, but it may be deleterious (reperfusion injury) [1]. Actually, reperfusion/reoxygenation could be a hazardous practice. During ischemia, oxygen shortage may reduce oxidative phosphorylation and lead to a drop in intracellular pH as well as mitochondrial membrane depolarization. However, reperfusion might induce the massive formation of radical oxygen species (ROS) and a cascade of deleterious events leading to cardiac death and the inflammatory process known as redox stress. Moreover, ROS may be involved in triggering cardioprotective pathways through redox signalling [2–5]. Therefore, the production of devices that allow controlled delivery of oxygen can be of paramount importance for avoiding stress and favoring redox signalling.

Indeed, in order to reduce injury associated with ischemia/reperfusion (I/R), several studies have proposed a series of cardio-protective maneuvers including the application of brief periods of ischemia before or after prolonged ischemia. Ischemic preconditioning or post-conditioning might reduce significant I/R damage [2–4,6,7]. While preconditioning procedures may be applied to certain programmed clinical conditions that can jeopardise the heart, treatment after ischemia (post-conditioning mode) is clinically feasible even in unpredictable conditions. Preconditioning and post-conditioning protection can be induced by several stimuli besides ischemia such hypoxia, "oxygen tension control", and several endogenous and exogenous compounds [2–4,8] all of which may act via redox signaling. Since post-conditioning depends on redox signaling, I/R injury may be conditioned by the reoxygenation modality in early reperfusion [9,10]. Hence, oxygen tension and the oxygen delivery patterns may play a crucial role in determining I/R injury.

Slow re-flow procedures have been proposed as experimental protective strategies. Intermittent re-flow (post-conditioning mode) has gained much more attention because, compared to slow re-flow, it does not present the problem of stagnant blood and consequent white blood cell infiltration [2,11]. However, intermittent reperfusion has raised several concerns in the clinical scenario and full reperfusion is still the treatment of choice [12].

Several studies have examined the ability to target nanoparticles for drug delivery to regions affected by organ ischemia. However, to the best of our knowledge, no study has tested a polymer nanoparticle approach for delivering oxygen in a controlled manner to limit I/R injury. Since nanoparticle approaches for oxygen delivery are theoretically feasible, this study aimed at testing the protective potential of cyclodextrin-based nanoparticles that can deliver oxygen in a highly controlled manner. Cyclodextrin-based nanosponges are innovative nanosized polymer systems consisting of cross-linked cyclodextrins nanostructured within a three-dimensional network.

Cyclodextrins are a class of cyclic glucopyranose oligomers obtained from starch by enzymatic action. Cyclodextrins have a characteristic toroidal shape that forms a truncated cone-shaped lipophilic cavity. The main common native cyclodextrins are named α, β, and γ which comprise six, seven, and eight glucopyranose units, respectively. Cyclodextrins may include compounds whose size and polarity are compatible with that of their cavity. Moreover, due to the lipophilicity of the internal cavities, cyclodextrins are able to extract cholesterol from cell membranes [13]. As a consequence, the cytotoxicity of cyclodextrins was often tested in vitro through their hemolytic activity. The ability to induce hemolysis depends on the types of the cyclodextrin ring.

Cyclodextrin-based nanosponges exhibited a marked capability of incorporating many types of molecules including small molecules, macromolecules, and gases [14].

Concerning the safety aspect, β-CD-based NS proved to be biocompatible, biodegradable, and had negligible toxicity in preclinical studies in various animal models [15–17]. We took into account α-cyclodextrin and its ability to synthesize two polymer derivatives. α-CD contains six glucopyranose units and consequently a smaller internal cavity. Its safety profile was deeply investigated as candidate for pharmaceutical applications [13,18]. Recently, its cytotoxicity was evaluated on murine microvascular endothelial cells and compared with modified cyclodextrins [19,20].

Cyclodextrin nanosponges (NS) are suitable for oxygen transport, which was previously shown [21]. The major advantage of nanosponges is that they allow to "plan" the oxygen delivery pattern. The main objective of this study is to verify whether oxygenated or non-oxygenated (nitrogen filled) NS given in pre- or post-conditioning mode at different concentrations can reduce cell mortality or whether it can exacerbate cell death in a hypoxia/reoxygenation (H/R) in vitro model in H9c2 cardiomyocytes.

2. Materials and Methods

α-Cyclodextrin was kindly provided by Wacker Chemie AG (Munich, Germany) and desiccated in the oven at 80 °C for at least 24 h before use. All other chemicals employed for the synthesis of CD polymers were purchased from Sigma-Aldrich (Steinheim, Germany) and used as received.

2.1. Synthesis of Soluble α-CD Polymer

A soluble hyper-branched α-CD polymer (α-CD POLY) was prepared following the procedure previously reported by Trotta et al. [22] with minor changes. In detail, 0.977 g (1.00 mmol) of α-CD were solubilized in 6.00 mL of dimethyl sulfoxide (DMSO) while stirring at an ambient temperature in a 20 mL vial. Subsequently, 1.00 mL (7.17 mmol) of triethylamine and 2.629 g (12.05 mmol) of pyromellitic dianhydride (PMDA) were added. After 24 h, the polymerization reaction was quenched by pouring the viscous solution into an excess of ethyl acetate. After precipitation, the synthesised polymer was recovered by vacuum filtration, washed with an excess of ethyl acetate, and left to dry. Finally, the polymer was solubilized in deionized water (approximately 100 mL) and freeze-dried. The polymer was stored in a desiccator after lyophilization.

2.2. Synthesis of Insoluble α-CD Nanosponge

An insoluble hyper-cross-linked α-CD polymer named α-CD nanosponge (α-CD NS) was synthesized by dissolving 1.954 g (2.01 mmol) of α-CD in 8.00 mL of DMSO while stirring at ambient temperatures in a 20 mL vial. Then, 2.00 mL (14.35 mmol) of triethylamine and 1.752 g (8.03 mmol) of PMDA were added [23]. In a few minutes, the cross-linking reaction led to the formation of a rigid gel. During the next several days, the gel was ground in a mortar and washed with an excess of deionized water under vacuum filtration. Finally, the NS was rinsed twice with acetone and left to dry. Further purification was performed by Soxhlet extraction by extracting the α-CD NS in acetone for approximately 24 h to remove the unreacted reagents.

2.3. Preparation of α-CD-based Formulations

Three types of α-CD-based formulations were prepared as oxygen delivery systems. For this purpose, α-CD, α-CD POLY, and α-CD NS were weighed and suspended in a NaCl solution (0.9%, w/v) to obtain an α-CD solution at a concentration of 1.3 mg/mL. Both the α-CD POLY solution and the α-CD NS suspension had a concentration of 40 mg/mL. Since α-CD NS is insoluble, it was first homogenized with a high-shear mixer (ltra-Turrax®, Konigswinter, Germany) at 24,000 rpm for 20 min to obtain a nano-suspension that presented a uniform particle size. The three α-CD-based formulations were saturated with oxygen by using an oxygen purge with a gas concentration up to 35 mg/mL.

2.4. Characterization of α-CD-Based Formulations

All α-CD formulations were characterized by determining pH and tonicity. The surface morphology of α-CD-based polymers in powder was determined by scanning electron microscopy (SEM) using a Leica Stereoscan410 (Wetzlar, Germany). The samples were observed after Au metallization. The voltage used was in the 5 kV to 10 kV range.

α-CD NS formulation was in vitro characterized to evaluate the size and surface charge. The average diameter and polydispersity index of the formulation was measured by using photocorrelation spectroscopy (PCS) with a 90 Plus instrument (Brookhaven, NY, USA) at a fixed angle of 90° and a temperature of 25 °C after dilution with filtered water. Each value represents the average of five measurements of three different sample batches. The zeta potential was determined by using electrophoretic mobility with a 90Plus instrument (Brookhaven, NY, USA). The zeta potential determinations were performed at blood plasma ionic strength (0.15 M). For zeta potential determination, diluted samples of α-CD-based formulations were placed in an electrophoretic cell to which a rounded 15 V/cm electric field was applied. Each value reported is the average of 10 measurements of 3 different formulations.

2.5. Hemolytic Assay

To determine hemolytic activity, 100 microliters of α-CD-based formulations were incubated at 37 °C for 90 min with blood (1:4 v/v) obtained after suitable dilution with freshly prepared phosphate buffer saline (pH 7.4). After incubation, sample-containing blood was centrifuged at 1000 rpm for five minutes to separate plasma. The amount of hemoglobin released due to hemolysis was determined spectrophotometrically (absorbance readout at 543 nm using a DU® 730, Beckman Coulter, (Brea, CA, USA). Hemolytic activity was calculated using 0.9% NaCl solution (negative control) as reference. However, complete hemolysis (positive control) was induced by the addition of ammonium sulphate (20% w/v).

2.6. In Vitro Oxygen Release Determination

In vitro oxygen release from the three different types of α-CD-based formulations was investigated using the dialysis bag technique. The donor phase consisted of 3 mL of different α-CD formulations placed in a dialysis bag (cellulose membrane with molecular weight 12,000 Da to 14,000 Da) that was hermetically sealed and placed in 45 mL of the receiving phase. The receiving phase consisted of 0.9% NaCl solution with oxygen concentration previously reduced using an N_2 purge up to 0 mg/L in order to mimic hypoxic conditions. Then, oxygen release kinetics from α-CD formulations were monitored for 48 h using an oximeter (HQ40d model, Hach, Loveland, CO, USA) at 25 °C.

2.7. Cell Culture

Commercially available H9c2 samples were obtained from the American Type Culture Collection (ATCC; Manassas, VA, USA). H9c2 samples were grown in Dulbecco's modified Eagle's medium nutrient mixture F-12 HAM (DMEM) and supplemented with 10% fetal bovine serum (FBS) and 1% (v/v) streptomycin/penicillin (Wisent Inc., Quebec, QC, Canada) at 37 °C and 5% CO_2 [24]. When the cells reached 80% confluence, they were removed from the flask, counted in the Burker chamber, and plated in a 96-well plate with 5000 cells/well density. After 48 h, the cells were used for dose–response analysis and hypoxia–reoxygenation (H/R) protocols (see Figure 1).

Upon completion of the protocols, the cells were washed with DMEM for three hours and an 3-(4,5-dimethylthiazol-2-yl)-2,5-diphenyltetrazolium bromide (MTT) assay was performed (see below).

2.8. Stability Determination of α-CD-based Formulations in Cell Culture Medium and Ischemic Buffer

The stability of α-CD-based formulations was determined in cell culture medium (DMEM HAM F12, 2% FBS), ischemic buffer, and NaCl 0.9% used as a negative control. α-CD-based formulations were incubated with three different solutions at a concentration of 100 µg/ml at 37 °C. Z-potential values were measured as described before over time (t = 0 h, 3 h, and 6 h).

2.9. Protocols

2.9.1. Normoxic Experimental Conditions (Dose–Response Studies)

To verify either oxygenated or non-oxygenated (Nitrogen) cell toxicity, α-CD-based formulations were tested at different concentrations (0.2 µg/mL, 2 µg/mL, and 20 µg/mL) (see Figure 1A).

Therefore, cell vitality of the Untreated Control Group (cells with DMEM HAM F12, 2% bovine serum only, CTRL) was compared against cell vitality of cells exposed to the following α-CD-based formulations, which are comprised of two types of α-CD nanosponges including either soluble (α-CD POLY) or insoluble (α-CD NS) and native α-CD in normoxic conditions (21% O_2 and 5% CO_2).

- Oxygenated α-CD POLY Groups (O-α-CD POLY) and Nitrogen α-CD POLY Groups (N-α-CD POLY)
- Oxygenated α-CD NS Groups (O-α-CD NS) and Nitrogen α-CD NS Groups (N-α-CD NS)
- Oxygenated α-CD Groups (O-α-CD) and Nitrogen α-CD Groups (N-α-CD)

2.9.2. Hypoxia/Reoxygenation Experimental Conditions

To study the response to the hypoxia–reoxygenation (H/R) in vitro protocol, the cells were exposed to hypoxia (5% CO_2 and 95% N_2) for two hours and subsequently reoxygenated (21% O_2 and 5% CO_2) for one hour. In particular, during hypoxia, H9c2 cells were exposed to simulated ischemia by replacing the medium with an "ischemic buffer". This buffer contained 137 mM NaCl, 12 mM KCl, 0.49 mM $MgCl_2$, 0.9 mM $CaCl_2$ $2H_2O$, 4 mM HEPES, and 20 mM sodium lactate (pH 6.2) [25]. At the end of the hypoxic period, the cells were subjected to reoxygenation with a new medium (DMEM HAM F-12 2%FBS) for 1 h. At the end of reoxygenation, the cell vitality was assessed using the MTT test (see Figure 1B).

2.9.3. Pre-Treatment with α-CD-Based Formulations

To mimic preconditioning, the cells were exposed to α-CD-based formulations that were either oxygenated or which contained nitrogen as described above and then subjected to the hypoxia–reoxygenation protocol, which is mentioned above (Figure 1C).

2.9.4. Post-Treatment with α-CD-Based Formulations

To mimic post-conditioning, the oxygenated or nitrogen NSs were added to the cells at the beginning of reoxygenation (see Figure 1D).

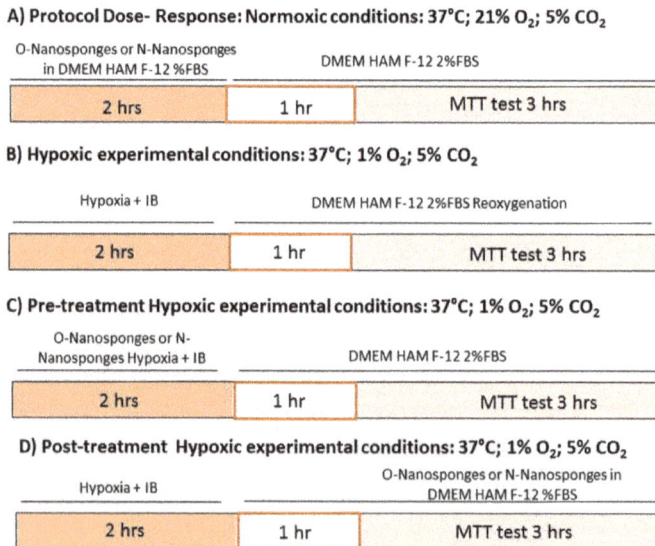

A) Protocol Dose- Response: Normoxic conditions: 37°C; 21% O_2; 5% CO_2

O-Nanosponges or N-Nanosponges in DMEM HAM F-12 %FBS		DMEM HAM F-12 2%FBS	
2 hrs	1 hr	MTT test 3 hrs	

B) Hypoxic experimental conditions: 37°C; 1% O_2; 5% CO_2

Hypoxia + IB		DMEM HAM F-12 2%FBS Reoxygenation	
2 hrs	1 hr	MTT test 3 hrs	

C) Pre-treatment Hypoxic experimental conditions: 37°C; 1% O_2; 5% CO_2

O-Nanosponges or N-Nanosponges Hypoxia + IB		DMEM HAM F-12 2%FBS	
2 hrs	1 hr	MTT test 3 hrs	

D) Post-treatment Hypoxic experimental conditions: 37°C; 1% O_2; 5% CO_2

Hypoxia + IB		O-Nanosponges or N-Nanosponges in DMEM HAM F-12 %FBS	
2 hrs	1 hr	MTT test 3 hrs	

Figure 1. Experimental protocols and 3-(4,5-dimethylthiazol-2-yl)-2,5-diphenyltetrazolium bromide (MTT) test. IB: ischemic buffer. (**A–D**) are different experimental conditions.

2.10. MTT Assay

When all experiments were completed, the cell vitality was assessed using the 3-(4,5-Dimethylthiazol-2-yl)-2,5-diphenyltetrazolium bromide (MTT) kit. MTT (10 μL/well, Sigma, St. Louis, MO, USA) was added to each well. Cells were incubated for another three hours at 37 °C. Then 100 μL of dimethyl sulfoxide (DMSO, Sigma, St Louis, MO, USA) samples were added to each well and the plates were shaken for five minutes. Each experiment was performed three times. The plates were read by a spectrophotometer at 570 nm to obtain optical density values [26].

2.11. Statistical Analysis

All values were expressed as a mean ± SEM and were analyzed using the Analysis of Variance (ANOVA) test followed by Bonferroni's post-test and the *t*-test. A value in the range of $p < 0.05$ was considered statistically significant.

3. Results

3.1. Physicochemical Characterisation of α-CD-Based Formulations

The three α-CD-based formulations were prepared in NaCl (0.9% w/v), which are suitable for biological studies. The pH value of the three different α-CD-based formulations was 5.50 and tonicity was about 300 mOSM.

The α-CD POLY showed an average molecular weight of 26 KDa and a content of α-CD macrocycles corresponding to 27.10%. The aqueous solubility of α-CD and α-CD POLY is 145 mg/mL and more than 250 mg/mL while α-CD NS entities are insoluble.

The results of the physicochemical characterization of the α-CD NS formulation are stated in Table 1.

Table 1. Physio-chemical characterization of an α-CD formulation. Each point represents the mean ± SD of the three different formulations ($n = 3$).

Formulation	Average diameter ± SD (nm)	PDI *	Z-potential ± SD (mV)
α-CD **	-	-	−29.12 ± 4.74
α-CD POLY **	-	-	−12.87 ± 4.60
α-CD NS	850.55 ± 57.90	0.17	−36.37 ± 3.58

* PDI = polydispersity index; ** Soluble in aqueous media.

α-CD NS was about 850 nm in size and had a negative surface charge with a high value of zeta potential (about −36 mV) that was suitable for avoiding the nano-sponge aggregation phenomenon. PDI values lower than 0.2 indicate a nearly homogeneous particle size distribution. Figure 2 reports SEM images of α-CD-based polymers in comparison with α-CD.

Figure 2. Scanning electron microscopy (SEM) images of (**a**) α-cyclodextrin (**b**) α-cyclodextrin polymer, and (**c**) α-cyclodextrin nanosponges (Magnification 1000×).

The pictures underlined the different morphology of α-CD, α-CD POLY, and α-CD NS in powder form.

3.2. Evaluation of α-CD-Based Formulation Biocompatibility

Red blood cell (RBC) hemolysis was performed to assess the formulation's biocompatibility. Negligible hemolysis was observed (see Figure 3) for α-CD-based formulations at the concentration used in the biological assays, which confirms their biocompatibility. These results are consistent with literature data, which report that α-CD exhibits hemolysis starting from 6 mM [27].

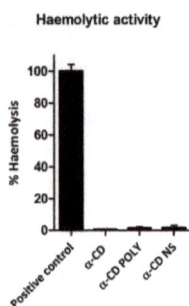

Figure 3. α-CD-based formulations presented negligible hemolytic activity. Results were expressed as percentage of total hemolysis (positive control), which was obtained when red blood cells were incubated with ammonium sulphate (20% *w/v*). Each bar represents the mean ± Standard Deviation (SD) of the three experiments.

3.3. In Vitro Oxygen Release Kinetics

The in vitro oxygen release profile obtained from the three different α-CD-based formulations is noted in Figure 4. All oxygen-loaded α-CD-based formulations were able to store and slowly release oxygen with prolonged and constant release kinetics. The release profile has two phases for all the systems. α-CD showed a faster release than the other two formulations. This behavior underscores the role played by the presence of the polymer.

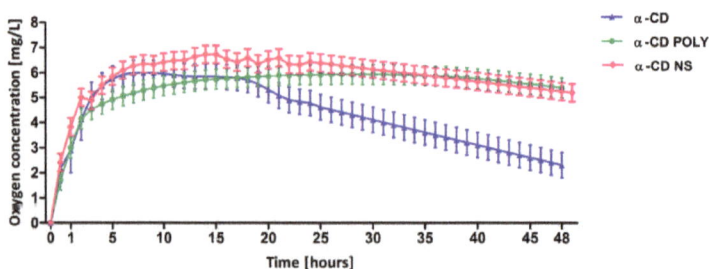

Figure 4. In vitro oxygen release from α-CD-based formulations over time measured by an oximeter. Each point represents the mean ± SD of the three experiments.

3.4. Stability Determination of α-CD-Based Formulations in Cell Culture Medium and Ischemic Buffer

The zeta potential behavior of the three formulations was monitored up to six hours in cell culture medium and ischemic buffer in order to confirm the system stability during the time range used in the following biological assays (see Figure 5).

No significant changes in zeta potential values were observed, which suggests that no aggregation phenomena occurred. Preliminary experiments and data reported in the literature [14] demonstrated the chemical stability of α-CD POLY and α-CD-NS over time. In addition, no changes of oxygen concentration were seen with regard to storing NS aqueous suspensions for 30 days in sealed vials [21].

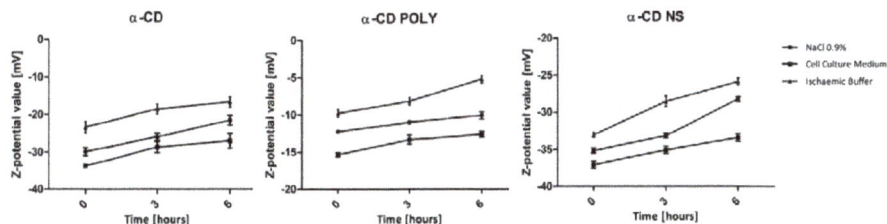

Figure 5. Zeta potential values of α-CD-based formulations in NaCl 0.9% (negative control) cell culture medium in an ischemic buffer over time. Each point represents the mean ± SD of three experiments.

3.5. Dose–Response in Normoxic Conditions

Results obtained with different concentrations (0.2, 2, and 20 μg/mL) of oxygenated (upper panels) and non-oxygenated (nitrogen in lower panels) α-CD-based formulations in normoxic conditions are available in Figure 6. The exposure of cells to different types of nitrogen-loaded (flushed) α-CD-based formulations for two hours did not affect cell vitality in all tested concentrations. On the other hand, treatment with oxygenated samples had a slight proliferative effect depending on concentration and type of α-CD-based formulations. In particular, O-α-CD POLY induced a significant increase in cell vitality when compared to untreated cells (CTRL) at two different concentrations (0.2 and 20 μg/mL, $p < 0.001$ and $p < 0.05$ vs. CTRL, respectively) (see Figure 6A). O-α-CD NS induced a significant increase in vitality in all tested concentrations (0.2 μg/mL and 2 μg/mL $p < 0.05$, 20 μg/mL $p < 0.001$ vs. CTRL; Figure 6B). For O-α-CD, we observed a significant increase in vitality when the cells were treated with 20 μg/mL only ($p < 0.01$ vs. CTRL, see Figure 6C upper panel).

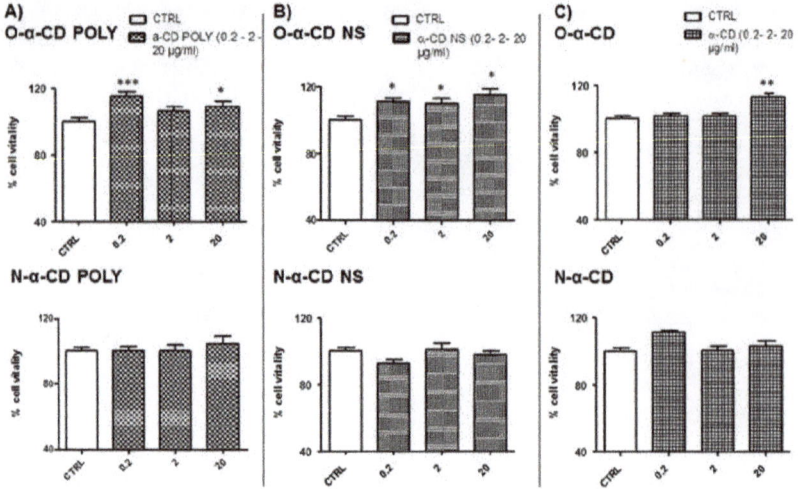

Figure 6. Dose–response in normoxic conditions. (**A**) Treatment with O-α-CD POLY (0.2, 2, 20 μg/mL) and N-α-CD POLY (0.2, 2, 20 μg/mL); (**B**) Treatment with O-α-CD NS (0.2, 2, 20 μg/mL) and N-α-CD NS (0.2, 2, 20 μg/mL); (**C**) Treatment with O-α-CD (0.2, 2, 20 μg/mL) and N-α-CD (0.2, 2, 20 μg/mL), compared to the untreated control group (CTRL). Data were normalized to the mean value in control conditions and expressed as a percentage.

3.6. Untreated Cells: Normoxic and H/R Conditions

When comparing the first two bar graphs of all panels in Figures 7–9, the untreated cells (not exposed to NS) at the end of the H/R protocol (CTRL H/R) displayed a 30% reduction in vitality ($p < 0.05$ CTRL vs. CTRL H/R in all cases).

3.7. NS-Treated Cells and H/R Conditions

3.7.1. α-CD POLY

All concentrations of Oxygenated-α-CD POLY limited H/R damage when used during pre-treatment (see Figure 7A) or post-treatment (see Figure 7B). In all cases, vitality was significantly higher than in CTRL H/R conditions (Figure 7A,B). It must be mentioned that, in all cases (pre-and post-treatment), cell vitality resembled that of the normoxic CTRL.

Nitrogen-α-CD POLY were not protective when used during pre-treatment (see Figure 7C) or post-treatment (see Figure 7D). In all cases, vitality was not significantly higher than in CTLR H/R conditions (see Figure 7C,D). Moreover, in post-treated conditions, cells displayed reduced vitality when compared to normoxic conditions for all concentrations (CTRL; Figure 7D) whereas in pre-treatment conditions, cells only at 2 μg/mL H/R showed significantly reduced cell vitality when compared to the CTLR ($p < 0.05$, see Figure 7C).

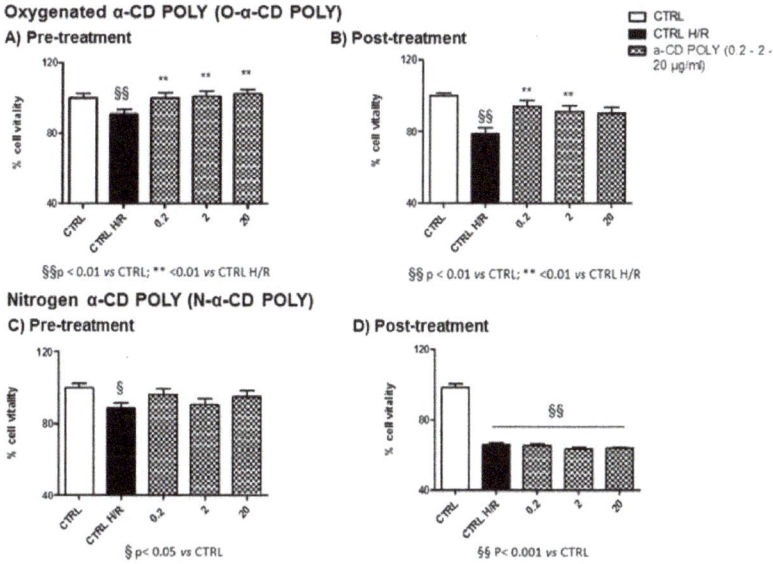

Figure 7. Cell vitality of α-CD POLY treated H9c2 cells and H/R conditions. (**A**) Pre-treatment with O-α-CD POLY (0.2, 2, and 20 μg/mL); (**B**) Post-treatment with O-α-CD POLY (0.2, 2, and 20 μg/mL); (**C**) Pre-treatment with N-α-CD POLY (0.2, 2, and 20 μg/mL); (**D**) Post-treatment with O-α-CD POLY (0.2, 2, and 20 μg/mL) compared to the untreated control group (CTRL) and the hypoxia–reoxygenation group (CTRL H/R). Data were normalized to mean value in control conditions and expressed as a percentage.

3.7.2. α-CD NS

All concentrations of oxygenated-α-CD NS limited H/R damage when used during pre-treatment (see Figure 8A). However, when used during post-treatment (see Figure 8B), the vitality of cells exposed to a higher concentration of NS did not differ from the CTRL H/R. At all pre-treated concentrations, cells displayed a cell vitality similar to the normoxic CTRL (see Figure 6A). However, when used during post-treatment, the cells displayed vitality resembling the normoxic CTRL only for the 2 μg/mL concentration (Figure 8B).

Nitrogen-α-CD NS were not protective when used during pre-treatment (see Figure 8C) or post-treatment (see Figure 8D). In all cases, vitality was not significantly higher than in CTLR H/R conditions (see Figure 8C,D). In post-treated conditions, cells displayed reduced vitality compared to normoxic conditions for all concentrations (CTLR, see Figure 8D) whereas in pre-treatment conditions, no concentration significantly reduced cell vitality compared to the CTLR (see Figure 8C).

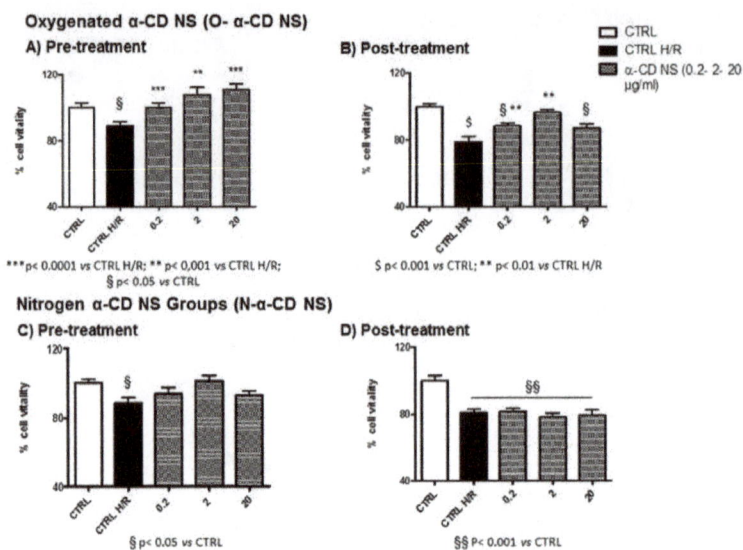

Figure 8. Cell vitality of α-CD NS treated H9c2 cells and H/R conditions. (**A**) Pre-treatment with O-α-CD NS (0.2, 2, and 20 μg/mL); (**B**) Post-treatment with O-α-CD NS (0.2, 2, and 20 μg/mL); (**C**) Pre-treatment with N-α-CD NS (0.2, 2, and 20 μg/mL); (**D**) Post-treatment with O-α-CD NS (0.2, 2, and 20 μg/mL) compared to the untreated control group (CTRL) and the hypoxia–reoxygenation group (CTRL H/R). Data are normalized to the mean value in control conditions and expressed as a percentage.

3.7.3. α-CD

All concentrations of oxygenate-α-CD limited H/R damage when used during pre-treatment (see Figure 9A). However, when used during post-treatment (see Figure 9B), the vitality of cells exposed to an NS concentration of 2 μg/mL did not differ from the CTRL H/R. Moreover, in all pre-treated concentrations, cells displayed vitality resembling the normoxic CTRL (see Figure 9A). However, when used during post-treatment, no group displayed vitality resembling that of the normoxic CTRL (see Figure 9B).

Nitrogen-α-CD particles were not protective when used during pre-treatment (see Figure 8C) or post-treatment (see Figure 9D). On the other hand, in pre-treated conditions, vitality was significantly higher than in CTLR H/R conditions for higher concentrations (see Figure 9C). In post-treated conditions, cells displayed reduced vitality when compared to normoxic conditions for all concentrations (CTLR, see Figure 9D). Cell vitality was significantly reduced when compared to the CTLR for all concentrations (see Figure 9C).

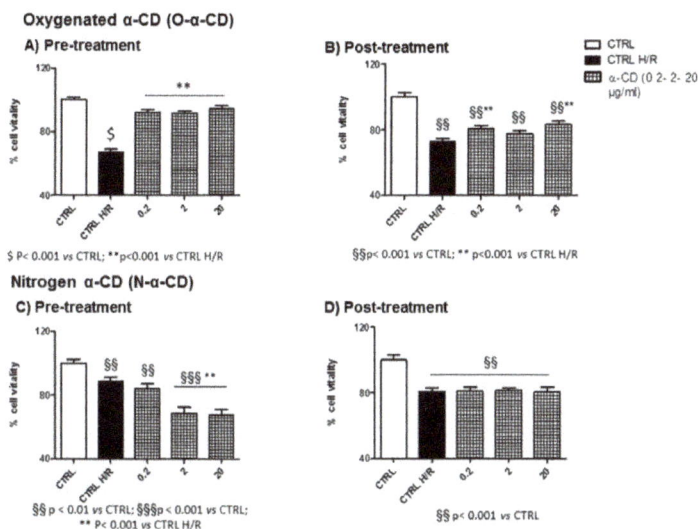

Figure 9. Cell vitality of α-CD treated H9c2 cells and H/R conditions. (**A**) Pre-treatment with O-α-CD (0.2, 2, and 20 µg/mL); (**B**) Post-treatment with O-α-CD (0.2, 2, and 20 µg/mL); (**C**) Pre-treatment with N-α-CD (0.2, 2, and 20 µg/mL); (**D**) Post-treatment with O-α-CD (0.2, 2, and 20 µg/mL), compared to the untreated control group (CTRL) and the hypoxia–reoxygenation group (CTRL H/R). Data were normalized to the mean value in control conditions and expressed as a percentage.

4. Discussion

Nanocarriers for oxygen delivery have been the focus of extensive research for the treatment of hypoxic tissues, infectious diseases, regenerative medicine, and wound healing [21,28–33].

In this study, two new α-CD-based polymers were synthesised as oxygen carriers. We had previously developed cyclodextrin-based nanosponges as oxygen carriers by cross-linking α, β, and γ-CD using carbonyldiimidazole [21]. α-CD nano-sponges showed the capability to store and release oxygen. Cyclodextrins can store gases in their cavity. Molecular encapsulation of gases in cyclodextrins was first studied by Cramer and Henglein [34]. Since gases have low molecular weight and a small size, α-CD was the most suitable compound for this application while, due to the higher dimension of the inner cavity, β-CD generally did not fit the requirements to host gases [35].

This paper investigates two new types of polymers obtained using α-CD and pyromellitic dianhydride (PMDA) as a cross-linker to obtain either a soluble polymer (hyperbranched) or an insoluble one (cross-linked) called a nanosponge (NS). The rationale for using α-CD is two-fold.

First, α-CD has a permitted daily exposure (PDE) of 0.2 mg/kg, which corresponds to 14 mg for a 70 kg human body. Based on the data, we can theorize that α-CD-based polymers could have an α-CD concentration that is suitable for future in vivo administration. Moreover, oxygen release kinetics might be slower since the α-CD cavity is smaller than those of other native cyclodextrins. α-CD cross-linked with PMDA were able to encapsulate, store, and release oxygen for a prolonged period of time. In addition, the polymer network may play a key role in oxygen storage and release. Treatment of myocardial infarction with nanotechnology has provided evidence of beneficial effects in both AMI and clinical studies.

This is a first-time study that evaluates different preparations of α-CD-based polymers incorporated with oxygen in a biological system and compared with native α-CD. These preparations provide clear protection against H/R-induced cell death when administered either before or after hypoxia. We also demonstrated that, at a different time of administration such as before or after the hypoxia protocol, these oxygen carriers induce protection. Our data suggest that in vitro oxygenated nanosponge treatment protects against reperfusion injury with a dose response. Considering a potential in vivo administration, the nanoparticle sizes are crucial parameters in the study. The dimensions significantly affect the pharmacokinetics and the bio-distribution of nano-delivery systems [36]. It is possible to reduce at nanoscale level a coarse material using milling processes. Top down methods such as high pressure homogenization can be exploited to reduce sizes of nanosponges and obtain an almost homogenous nanoparticle distribution, which was previously noted in [37,38]. Therefore, the size of α-CD-NS might be significantly decreased to obtain an aqueous nano-suspension suitable for future intravenous administration.

This study demonstrates a beneficial effect of controlled oxygen delivery with nano-sponges. Since these may be directly injected into the myocardial wall before starting full blood reperfusion, physicians may now possess a new tool for controlled oxygen delivery to limit ischemia/reperfusion injury.

Growing evidence supports the role of oxygenation levels and associated ROS generation in signal transduction pathways involved in heart conditioning [2–4,8]. Previous studies have demonstrated that a few minutes of both hypoxic and hyperoxic preconditioning might trigger beneficial ROS-dependent signaling, which may limit myocardial I/R injury. Similar beneficial ROS signaling has been described in the post-conditioning procedure with a few seconds of intermittent ischemia [2,9,10]. However, during the reperfusion phase, the effects of ROS are more complex and high levels of ROS may be detrimental [3]. Additionally, the dose–response curve we observed with oxygenated NS given during reoxygenation reveals a type of inverted U curve in which the middle dose (2 μg/mL) is the most effective and the higher dose (20 μg/mL) is not protective.

In conclusion, α-cyclodextrin and α-cyclodextrin polymers used as oxygen nanocarriers limit hypoxia–reoxygenation injury in a cardiac cell model both during pre-conditioning and post-conditioning. As such, these nanocarriers might be used to limit hypoxia–reoxygenation injury before a scheduled period of ischemia, which occurs during organ surgery, or during reperfusion treatment for infarction to provide controlled oxygenation. They possess interesting potential for future medical applications.

Acknowledgments: The authors would like to thank University of Turin funds (ex-60%) and Roquette Italia for their support.

Author Contributions: Saveria Femminò, Federica Bessone, Fabrizio Caldera, Nilesh Dhakar and Daniele Cau performed the experiments. Pasquale Pagliaro analyzed the data. Claudia Penna designed the experiments. Roberta Cavalli and Francesco Trotta conceived the work and wrote the paper.

Conflicts of Interest: The authors declare that there is no conflict of interest.

References

1. Hausenloy, D.J.; Yellon, D.M. Ischaemic conditioning and reperfusion injury. *Nat. Rev. Cardiol.* **2016**, *13*, 193–209. [CrossRef] [PubMed]
2. Pagliaro, P.; Penna, C. Redox signaling and cardioprotection: Translatability and mechanism. *Br. J. Pharmacol.* **2015**, *172*, 1974–1995. [CrossRef] [PubMed]
3. Tullio, F.; Angotti, C.; Perrelli, M.G.; Penna, C.; Pagliaro, P. Redox balance, and cardioprotection. *Basic Res. Cardiol.* **2013**, *108*, 392. [CrossRef] [PubMed]
4. Penna, C.; Granata, R.; Gabriele Tocchetti, C.; Pia Gallo, M.; Alloatti, G.; Pagliaro, P. Endogenous cardioprotective agents: Role in pre and postconditioning. *Curr. Drug Targets* **2015**, *16*, 843–867. [CrossRef] [PubMed]
5. Evans, C.W.; Iyer, K.S.; Hool, L.C. The potential for nanotechnology to improve delivery of therapy to the acute ischemic heart. *Nanomedicine* **2016**, *11*, 817–832. [CrossRef] [PubMed]
6. Penna, C.; Bassino, E.; Alloatti, G. Platelet activating factor: The good and the bad in the ischemic/reperfused heart. *Exp. Biol. Med. (Maywood)* **2011**, *236*, 390–401. [CrossRef] [PubMed]

7. Sluijter, J.P.; Condorelli, G.; Davidson, S.M.; Engel, F.B.; Ferdinandy, P.; Hausenloy, D.J.; Lecour, S.; Madonna, R.; Ovize, M.; Ruiz-Meana, M.; et al. Novel therapeutic strategies for cardioprotection. *Pharmacol. Ther.* **2014**, *144*, 60–70. [CrossRef] [PubMed]

8. Petrosillo, G.; Di Venosa, N.; Ruggiero, F.M.; Pistolese, M.; D'Agostino, D.; Tiravanti, E.; Fiore, T.; Paradies, G. Mitochondrial dysfunction associated with cardiac ischemia/reperfusion can be attenuated by oxygen tension control. Role of oxygen-free radicals and cardiolipin. *Biochim. Biophys. Acta* **2005**, *1710*, 78–86. [CrossRef] [PubMed]

9. Penna, C.; Rastaldo, R.; Mancardi, D.; Raimondo, S.; Cappello, S.; Gattullo, D.; Losano, G.; Pagliaro, P. Post–conditioning induced cardioprotection requires signaling through a redox-sensitive mechanism, mitochondrial ATP–sensitive K$^+$ channel and protein kinase C activation. *Basic Res. Cardiol.* **2006**, *101*, 180–189. [CrossRef] [PubMed]

10. Tsutsumi, Y.M.; Yokoyama, T.; Horikawa, Y.; Roth, D.M.; Patel, H.H. Reactive oxygen species trigger ischemic and pharmacological postconditioning: In vivo and in vitro characterization. *Life Sci.* **2007**, *81*, 1223–1227. [CrossRef] [PubMed]

11. Vinten-Johansen, J.; Granfeldt, A.; Mykytenko, J.; Undyala, V.V.; Dong, Y.; Przyklenk, K. The multidimensional physiological responses to postconditioning. *Antioxid. Redox Signal.* **2011**, *14*, 791–810. [CrossRef] [PubMed]

12. Hausenloy, D.J.; Barrabes, J.A.; Bøtker, H.E.; Davidson, S.M.; Di Lisa, F.; Downey, J.; Engstrom, T.; Ferdinandy, P.; Carbrera-Fuentes, H.A.; Heusch, G.; et al. Ischaemic conditioning and targeting reperfusion injury: A 30 year voyage of discovery. *Basic Res. Cardiol.* **2016**, *111*, 70. [CrossRef] [PubMed]

13. Irie, T.; Uekama, K. Pharmaceutical applications of cyclodextrins. III. Toxicological issues and safety evaluation. *J. Pharm. Sci.* **1997**, *86*, 147–162. [CrossRef] [PubMed]

14. Trotta, F.; Zanetti, M.; Cavalli, R. Cyclodextrin-based nanosponges as drug carriers. *Beilstein J. Org. Chem.* **2012**, *8*, 2091. [CrossRef] [PubMed]

15. Shende, P.; Kulkarni, Y.A.; Gaud, R.S.; Deshmukh, K.; Cavalli, R.; Trotta, F.; Caldera, F. Acute and repeated dose toxicity studies of different β-cyclodextrin-based nanosponge formulations. *J. Pharm. Sci.* **2015**, *104*, 1856–1863. [CrossRef] [PubMed]

16. Torne, S.J.; Ansari, K.A.; Vavia, P.R.; Trotta, F.; Cavalli, R. Enhanced oral paclitaxel bioavailability after administration of paclitaxel-loaded nanosponges. *Drug Deliv.* **2010**, *17*, 419–425. [CrossRef] [PubMed]

17. Gigliotti, C.L.; Ferrara, B.; Occhipinti, S.; Boggio, E.; Barrera, G.; Pizzimenti, S.; Giovarelli, M.; Fantozzi, R.; Chiocchetti, A.; Argenziano, M.; et al. Enhanced cytotoxic effect of camptothecin nanosponges in anaplastic thyroid cancer cells *in vitro* and *in vivo* on orthotopic xenograft tumors. *Drug Deliv.* **2017**, *24*, 670–680. [CrossRef] [PubMed]

18. Loftsson, T.; Leeves, N.; Bjornsdottir, B.; Duffy, L.; Masson, M. Effect of cyclodextrins and polymers on triclosan availability and substantivity in toothpastes in vivo. *J. Pharm. Sci.* **1999**, *88*, 1254–1258. [CrossRef] [PubMed]

19. Shityakov, S.; Puskás, I.; Pápai, K.; Salvador, E.; Roewer, N.; Förster, C.; Broscheit, J.A. Sevoflurane-sulfobutylether-β-cyclodextrin complex: Preparation, characterization, cellular toxicity, molecular modeling and blood-brain barrier transport studies. *Molecules* **2015**, *20*, 10264–10279. [CrossRef] [PubMed]

20. Shityakov, S.; Salmas, R.E.; Salvador, E.; Roewer, N.; Broscheit, J.; Förster, C. Evaluation of the potential toxicity of unmodified and modified cyclodextrins on murine blood-brain barrier endothelial cells. *J. Toxic. Sci.* **2016**, *41*, 175–184. [CrossRef] [PubMed]

21. Cavalli, R.; Akhter, A.K.; Bisazza, A.; Giustetto, P.; Trotta, F.; Vavia, P. Nanosponge formulations as oxygen delivery systems. *Int. J. Pharm.* **2010**, *402*, 254–257. [CrossRef] [PubMed]

22. Trotta, F.; Caldera, F.; Cavalli, R.; Mele, A.; Punta, C.; Melone, L.; Castiglione, F.; Rossi, B.; Ferro, M.; Crupi, V.; et al. Synthesis and characterization of a hyper-branched water-soluble β-cyclodextrin polymer. *Beilstein J. Org. Chem.* **2014**, *10*, 2586–2593. [CrossRef] [PubMed]

23. Trotta, F.; Tumiatti, W. Cross-Linked Polymers Based on Cyclodextrins for Removing Polluting Agents. WO 03/085002 A1, 16 October 2003.

24. Pasqua, T.; Tota, B.; Penna, C.; Corti, A.; Cerra, M.C.; Loh, Y.P.; Angelone, T. pGlu-serpinin protects the normotensive and hypertensive heart from ischemic injury. *J. Endocrinol* **2015**, *227*, 167–178. [CrossRef] [PubMed]

25. Yin, Y.; Guan, Y.; Duan, J.; Wei, G.; Zhu, Y.; Quan, W.; Guo, C.; Zhou, D.; Wang, Y.; Xi, M.; et al. Cardioprotective effect of Danshensu against myocardial ischemia/reperfusion injury and inhibits apoptosis of H9c2 cardiomyocytes via Akt and ERK1/2 phosphorylation. *Eur. J. Pharmacol.* **2013**, *699*, 219–226. [CrossRef] [PubMed]

26. Penna, C.; Perrelli, M.G.; Karam, J.P.; Angotti, C.; Muscari, C.; Montero-Menei, C.N.; Pagliaro, P. Pharmacologically active microcarriers influence VEGF-A effects on mesenchymal stem cell survival. *J. Cell. Mol. Med.* **2013**, *17*, 192–204. [CrossRef] [PubMed]

27. Motoyama, K.; Arima, H.; Toyodome, H.; Irie, T.; Hirayama, F.; Uekama, K. Effect of 2,6-di-O-methyl-alpha-cyclodextrin on hemolysis and morphological change in rabbit's red blood cells. *Eur. J. Pharm. Sci.* **2006**, *29*, 111–119. [CrossRef] [PubMed]

28. Alvarez-Lorenzo, C.; García-González, C.A.; Concheiro, A. Cyclodextrins as versatile building blocks for regenerative medicine. *J. Control. Release* **2017**, *268*, 269–281. [CrossRef] [PubMed]

29. Bisazza, A.; Giustetto, P.; Rolfo, A.; Caniggia, I.; Balbis, S.; Guiot, C.; Cavalli, R. Microbubble-mediated oxygen delivery to hypoxic tissues as a new therapeutic device. In Proceedings of the 30th Annual International Conference of the IEEE Engineering in Medicine and Biology Society EMBS 2008, Vancouver, BC, Canada, 20–25 August 2008; pp. 2067–2070.

30. Cavalli, R.; Bisazza, A.; Giustetto, P.; Civra, A.; Lembo, D.; Trotta, G.; Guiot, C.; Trotta, M. Preparation and characterization of dextran nanobubbles for oxygen delivery. *Int. J. Pharm.* **2009**, *381*, 160–165. [CrossRef] [PubMed]

31. Cavalli, R.; Bisazza, A.; Rolfo, A.; Balbis, S.; Madonnaripa, D.; Caniggia, I.; Guiot, C. Ultrasound-mediated oxygen delivery from chitosan nanobubbles. *Int. J. Pharm.* **2009**, *378*, 215–217. [CrossRef] [PubMed]

32. Magnetto, C.; Prato, M.; Khadjavi, A.; Giribaldi, G.; Fenoglio, I.; Jose Gulino, G.; Cavallo, F.; Quaglino, E.; Benintende, E.; Varetto, G.; et al. Ultrasound-activated decafluoropentane-cored and chitosan-shelled nanodroplets for oxygen delivery to hypoxic cutaneous tissues. *RCS Adv.* **2014**, *4*, 38433–38441. [CrossRef]

33. Prato, M.; Magnetto, C.; Jose, J.; Khadjavi, A.; Cavallo, F.; Quaglino, E.; Panariti, A.; Rivolta, I.; Benintende, E.; Varetto, G.; et al. 2H, 3H-decafluoropentane-based nanodroplets: New perspectives for oxygen delivery to hypoxic cutaneous tissues. *PLoS ONE* **2015**, *10*, e0119769. [CrossRef] [PubMed]

34. Cramer, F.; Henglein, F.M. Einschlussverbindungen der cyclodextrine mit gasen. *Angew. Chem.* **1956**, *68*, 649. [CrossRef]

35. Trotta, F.; Cavalli, R.; Martina, K.; Biasizzo, M.; Vitillo, J.; Bordiga, S.; Vavia, P.; Ansari, K. Cyclodextrin nanosponges as effective gas carriers. *J. Incl. Phenom. Macrocycl. Chem.* **2011**, *71*, 189–194. [CrossRef]

36. Dufort, S.; Sancey, L.; Coll, J.L. Physico-chemical parameters that govern nanoparticles fate also dictate rules for their molecular evolution. *Adv. Drug Deliv. Rev.* **2012**, *64*, 179–189. [CrossRef] [PubMed]

37. Daga, M.; Ullio, C.; Argenziano, M.; Dianzani, C.; Cavalli, R.; Trotta, F.; Ferretti, C.; Zara, G.P.; Gigliotti, L.C.; Ciamporcero, E.; et al. GSH-targeted nanosponges increase doxorubicin-induced toxicity "in vitro" and "in vivo" in cancer cells with high antioxidant defenses. *Free Radic. Biol. Med.* **2016**, *97*, 24–37. [CrossRef] [PubMed]

38. Swaminathan, S.; Pastero, L.; Serpe, L.; Trotta, F.; Vavia, P.; Aquilano, D.; Trotta, M.; Zara, G.P.; Cavalli, R. Cyclodextrin-based nanosponges encapsulating camptothecin: Physicochemical characterization, stability and cytotoxicity. *Eur. J. Pharm. Biopharm.* **2010**, *74*, 193–201. [CrossRef] [PubMed]

polymers

MDPI

Article

Reusable Xerogel Containing Quantum Dots with High Fluorescence Retention

Xiang-Yong Liang [1,2], Lu Wang [1,2], Zhi-Yi Chang [1,2], Li-Sheng Ding [1], Bang-Jing Li [1,*] and Sheng Zhang [3,*]

[1]   Key Laboratory of Mountain Ecological Restoration and Bioresource Utilization, Chengdu Institute of Biology, Chinese Academy of Sciences, Chengdu 610041, China; liangxy1216@163.com (X.-Y.L.); luwangbest@163.com (L.W.); changzy1@cib.ac.cn (Z.-Y.C.); lsding@cib.ac.cn (L.-S.D.)
[2]   College of Life Sciences, University of Chinese Academy of Sciences, Beijing 100049, China
[3]   State Key Laboratory of Polymer Materials Engineering, Polymer Research Institute of Sichuan University, Sichuan University, Chengdu 610065, China
*   Correspondence: libj@cib.ac.cn (B.-J.L.); zslbj@163.com (S.Z.)

Received: 27 November 2017; Accepted: 7 March 2018; Published: 13 March 2018

Abstract: Although various analytical methods have been established based on quantum dots (QDs), most were conducted in solution, which is inadequate for storage/transportation and rapid analysis. Moreover, the potential environmental problems caused by abandoned QDs cannot be ignored. In this paper, a reusable xerogel containing CdTe with strong emission is established by introducing host–guest interactions between QDs and polymer matrix. This xerogel shows high QDs loading capacity without decrease or redshift in fluorescence (the maximum of loading is 50 wt % of the final xerogel), which benefits from the steric hindrance of β-cyclodextrin (βCD) molecules. Host–guest interactions immobilize QDs firmly, resulting in the excellent fluorescence retention of the xerogel. The good detecting performance and reusability mean this xerogel could be employed as a versatile analysis platform (for quantitative and qualitative analyses). In addition, the xerogel can be self-healed by the aid of water.

Keywords: quantum dots; host–guest interactions; xerogel

1. Introduction

Quantum dots (QDs), mainly referring to nanocrystals of II-VI or IV-VI semiconductors, have been studied as one of the hottest research fields worldwide in recent decades due to their unique optical and electronic properties. They have the potential to bring revolutions to many technologies such as bio-imaging, (bio)sensors, solar-energy transformation, etc. [1–4]. Among the advantages of QDs, analytical chemists are most interested in the strong affinity between fluorescence property and the surface chemical conditions. Researchers have established a large trove of analytical methods based on "turn-on" or "turn-off" the fluorescence of the QDs [5–8]. However, it is difficult to find one commercial product whose operation mechanism is based on the reported methods. One of the main reasons is that most of the proposed analytical methods are based on QDs in solution, which does not benefit storage, transportation and rapid analysis, limiting their practical applications [9–11]. Therefore, the immobilization of QDs has attracted considerable attention in recent years.

Thus far, QD-based hydrogels have been in the center of the arena, marrying the unique photoelectrical properties of QDs with construction features of hydrogels [12–14]. Sol-gel method has proven that it is a powerful approach to prepare QD hydrogels. QD nanoparticles could assemble into gels spontaneously after the partial disengagement of the stabilizers [10,15–17]. However, the added oxidants or photochemical treatment may inactivate biomolecules, narrowing the QD applications in the field of biological detection. Even though the Eychmüller group improved the traditional sol-gel method and entrapped tyrosinase with high activity, it was still time-consuming [18]. Some other strategies based on the QDs inter-particle assembly have been developed to promote sol-gel transition, in which the elaborate design on the QDs surface was necessary [8,19–24]. A better choice to simplify the preparation procedure is constructing polymer–QDs structure. Emerging technologies, for example, layer-by-layer assembly [25], electrospun/electrospray [26,27], micropatterning [28] and inkjet/contact printing [29,30], have been employed to fabricate QDs-based sensors. Among these methods, in situ polymerization is more convenient, suitable and cost-effective for industrial manufacture, but the issues of loading capacity, stability and the size distribution of QDs should be considered primarily. Otherwise, they would lead to fluorescent properties of the final materials being uncontrollable [31,32]. Jiang et al. developed a kind of supramolecular hydrogels with QDs incorporated by host–guest interactions, in which the QDs with narrow size distribution kept their satisfactory photoluminescence, but the low QDs load (the decline of fluorescence intensity was observed when the CdS concentration exceeded 5 mg/mL) limited the practical applications of this method [33–35]. One might notice that almost all the above-mentioned QD immobilization proposals stopped at the hydrogel form, requiring strict storage conditions, which is unfavorable for transportation. Hydrogels could be dried at ambient atmosphere or by supercritical fluid drying system to form xerogels or aerogels, respectively [36]. Typically, aerogels dried by supercritical fluid drying system showed strong emissions (>90% of the original fluorescence intensity), while the fluorescence intensity decreased significantly (~50–60% of the original fluorescence intensity) if hydrogels had been dried at ambient atmosphere (i.e., xerogels), due to the enhanced inter-particle interactions caused by palpable volume shrinkage [9,15].

Another barrier to commercialize QDs-based products is the lack of environmentally friendly programs to treat the abandoned ones. Some research groups have raised the alarm that the huge risks posed by QDs should be considered seriously [37–40]. Meanwhile, the properties of each kind of QDs differs greatly from another. There is a long way to establish complete systems to dispose of the abandoned QDs.

In this paper, we developed a new strategy to prepare reusable xerogels containing QDs (cadmium telluride (CdTe), one of the typical QDs) with high fluorescence retention, which may pave the way to overcome those obstacles with following advantages: (a) the CdTe nanocrystals are immobilized in the polymer matrix via common polymerization procedure and dried at ambient atmosphere; (b) the xerogels could load a large amount of CdTe (ideally, the maximum is 50 wt % of the final xerogel) and exhibited strong emission, maintaining 85% of the QD solution fluorescence intensity; (c) the xerogels could be reused, alleviating the potential hazards caused by abandoned QDs; and (d) the presence of βCD molecules on the surface of CdTe make the xerogels potential treasure houses to be excavated (isomers, chiral distinctions, etc.) [41–45]. In addition, the xerogels showed the potential to act as biosensors also by entrapping corresponding enzymes.

2. Experimental Section

2.1. Materials

Cadmium chloride ($\check{C}dCl_2$, anhydrous), tellurium (Te), 1-adamantanecarboxylic acid (Ad-COOH), thionyl chloride, potassium persulfate ($K_2S_2O_8$, KPS) and *N,N,N′,N′*-tetramethylethylenediamine (TEA) were purchased from Aladdin Reagent Co., Ltd. (Shanghai, China). 2-hydroxyethyl methacrylate (HEMA), dopamine, 3,4-dimethoxyphenethylamine (DMPA) and 3-mercaptopropionic acid (MPA) were purchased from J&K Chemical Technology (Beijing, China). *N*-(hydroxymethyl)acrylamide (HMAAm) was obtained from Tokyo Chemical Industry Co., Ltd. (Tokyo, Japan) β-cyclodextrin (βCD) was purchased from Chengdu Kelong Chemical Reagent Factory (Chengdu, China). Sodium dihydrogen phosphate (NaH_2PO_4), thiocarbamide, sodium borohydride ($NaBH_4$), and vanillin (Van) were purchased from Tianjin Kemiou Chemical Reagent Co., Ltd. (Tianjin, China) Sodium hydroxide (NaOH) and disodium phosphate dodecahydrate ($Na_2HPO_4 \cdot 12H_2O$) were purchased from Guangdong Guanghua Sci-Tech Co., Ltd. (JHD) (Shantou, China). Tyrosinase (TYR, 570 $U \cdot mg^{-1}$) was obtained from Beijing Solarbio Science & Technology Co., Ltd. (Beijing, China). Ultrapure water (UP water, resistivity > 18.0 MΩ·cm, 25 °C) was used in this work. All other reagents and solvents were of analytical grade and were used directly without further purification. HEMA-Ad was synthesized according to our previous works [46,47].

2.2. Characterization

^{1}H/^{13}C NMR and 2D NOESY spectra were recorded on a Bruker Avance-III 400 NMR spectrometer (Bruker, Billerica, MA, USA). Mass spectroscopy was performed using Finnigan LCQ^{DECA} mass spectrometer (Thermo Fisher Scientific, Waltham, MA, USA). FT-IR spectra of sample pellets with KBr were recorded on a Fourier transform infrared (FTIR) spectrometer Spectrum 100 (PerkinElmer, Waltham, MA, USA). SEM images of the surface of the xerogel were obtained under scanning electron microscope (JSM-7500F, JEOL, Tokyo, Japan). The UV/Vis and fluorescence spectra were detected using Varioskan Flash (Thermo). Please note that the optical properties of solutions and xerogels are different, which stemmed from the essences of different physical forms. To ensure the comparability of the collected fluorescence spectra, all controllable parameters were set as the same. The thermos-gravimetric analysis (TGA) was characterized using Thermal Analyzer EXSTAR 6000 (25–600 °C, 10 °C min^{-1}, N$_2$ atmosphere, Seiko Instruments Inc., Chiba, Japan).

2.3. Synthesis of Mono-6-thio-β-cyclodextrin (mSH-CD)

First, mono-deoxy-6-(*p*-tolylsulfonyl)-β-cyclodextrin (6-OTs-β-CD) was prepared according to the reported protocol [46] Then 6-OTs-β-CD (9.68 g, 7.51 mmol) and thiourea (11.41 g, 0.15 mol) were refluxed in 80% (*v/v*) methanol-water solution (400 mL) for 60 h. After the solvent was removed under vacuum, the residue was stirred in methanol for 5 h. The solid was collected by filtration and stirred in 10 wt % NaOH solution for 8 h at 50 °C. Then, the pH of the solution was adjusted to 2 with 10 wt % HCl after cooling to ambient temperature. The crud product was obtained by filtration after adding 20 mL trichloroethylene and stirred for 12 h. Finally, the crud product was stirred in methanol for 12 h and collected, and then dried under vacuum (yield: 48%).

^{1}H NMR (400 MHz, DMSO) δ = 5.86–5.57 (m, 14H, OH–2, OH–3), 4.84 (dd, *J* = 16.6, 3.4 Hz, 7H, H–1), 4.64–4.38 (m, 6H, OH–6), 3.86–3.24 (m, overlapping with HDO, H-2/3/4/5/6), 3.02–2.91 (m, 1H, H–6a which bonding with –SH), 2.81–2.69 (m, 1H, H–6b which bonding with –SH), 2.05 (dd, *J* = 10.2, 6.1 Hz, 1H, SH); MS (ESI) *m/z*: 1173.98 [M + Na]$^+$.

2.4. Preparation of mSH-CD Capped CdTe (βCD-CdTe)

βCD-CdTe was synthesized according to Zhang's study with minor changes [17]. Firstly, tellurium (40 mg, 0.3 mmol) and NaBH4 (115 mg, 3.0 mmol) were stirred in 5 mL UP water under argon atmosphere for 6 h in ice bath. Thus, colorless NaHTe solution was obtained. Secondly, MPA (52 μL, 0.6 mmol) was added in CdCl$_2$ solution (110 mg, 0.6 mmol in 200 mL UP water), and then mSH-CD (345 mg, 0.3 mmol) was added. The pH was adjusted to 8 by adding NaOH solution (1 mol L^{-1}). Finally, NaHTe (2.0 mL) was added to the second solution and refluxed for different times to obtain different particle size of βCD-CdTe. Argon bubbling was needed throughout the above process.

CdTe was also prepared by following the above processes but without adding mSH-CD.

2.5. Preparation of HEMA-Ad@βCD-CdTe Complex

The βCD-CdTe solution was concentrated to about 20 mL and stirred with HEMA-Ad (88 mg, 0.3 mmol) for 7 days in darkness. Then, the solution was filtered (microporous membrane, 0.4 μm) and dried by vacuum freeze dryer, thus HEMA-Ad@βCD-CdTe complex was obtained.

2.6. Preparation of the Xerogel

HMAAm (1.01 g, 10 mmol), KPS (11 mg, 0.04 mmol) and HEMA-Ad@βCD-CdTe (101 mg) were stirred in 5 mL UP water with argon bubbling for 30 min under ice bath. Then, TEA (11 μL, 0.07 mmol) was added and the mixture was transferred to a lid of 96-well plate, covered by a piece of glass. The circular grooves (diameter: 8.71 mm, depth: 0.26 mm) in the lid were used as molds to shape the HEMA-Ad@βCD-CdTe/PHMAAm hydrogel. The polymerization was conducted at room temperature for 24 h. The obtained hydrogel was rinsed with water three times and dried at room temperature for 48 h to obtain xerogels of HEMA-Ad@βCD-CdTe/PHMAAm. For recycling tests, the circular hydrogel was peeled off from the lid and dried under the same conditions.

The xerogel of HEMA-Ad@βCD-CdTe/PHMAAm loading TYR (TYR/HEMA-Ad@βCD-CdTe/PHMAAm) was prepared the same way, but displacing the solvent with PBS (pH = 6.53, 50 mM) and adding 1 mg TYR along with HMAAm.

2.7. Detection of Analytes

Xerogels' fluorescence spectroscopy (λ_{ex} = 380 nm) were obtained before detecting analytes. According to the fluorescence spectroscopy, the median fluorescence intensity around the maximum emission wavelength (MFI, λ_{em} ± 5 nm) was used as the quantitative standard.

In the case of detecting Van, 5 μL of Van PBS solution (pH = 8.00, 50 mM) was added to the xerogel of HEMA-Ad-βCD-CdTe/PHMAAm. After 5 min, the median fluorescence intensity (MFI) was evaluated to calculate the concentration of the sample. For recycling test, the xerogel of HEMA-Ad-βCD-CdTe/PHMAAm was immersed in UP water for 5 min, then oscillated in dichloromethane for 10 min. Finally, the hydrogel dried at room temperature for 24 h to obtain xerogel for the next cycle of detection.

In the case of detecting dopamine, 5 μL of Van PBS solution (pH = 6.53, 50 mM) was added to the xerogel of TYR/HEMA-Ad@βCD-CdTe/PHMAAm. After incubation at 30 °C for 5 min, the MFI was measured to calculate the concentration of the sample.

2.8. Estimation of the Number of βCD Molecules Bound to One CdTe Particle and Calculating Inclusion Rate of HEMA-Ad@βCD-CdTe

The number of βCD molecules bound to one CdTe particle could be estimated based on TGA and TEM results.

In the range 200–400 °C, the mass loss was 42.10 wt % for βCD-CdTe, which resulted from the decomposition of MPA (x wt %) and βCD molecules (y wt %). Taking the molar ratio (n_{MPA}:n_{CD} = 2:1) into account, the content of βCD in βCD-CdTe could be calculated through following formula:

$$x + y = 42.10 \tag{1}$$

$$\frac{x}{M_{MPA}} = 2 \times \frac{y}{M_{CD}} \tag{2}$$

where M_{MPA} represents the molar mass of MPA (M_{MPA} = 106.14 g/mol) and M_{CD} represents the molar mass of βCD molecule (M_{CD} = 1151.05 g/mol). The calculate results are that x = 6.66 and y = 35.54, suggesting that 35.54 wt % weight loss of βCD-CdTe between 200 and 400 °C was due to βCD molecules decomposition. In other words, βCD-CdTe consists of about 0.36 g/g of βCD molecules.

The number of βCD molecules bound to one CdTe particle (n) could be calculated according to the following equation (assumed that 1 g of βCD-CdTe was taken):

$$n = \frac{\frac{4}{3} \times \pi \times r^3 \times \rho}{W_{CdTe}} \times \frac{W_{CD}}{M_{CD}} \times N_A \tag{3}$$

where r represents the radius of single CdTe nanoparticle, r = 2 nm (according to the TEM); ρ represents the density of CdTe, ρ = 5.85 g/cm^3; W_{CdTe} represents the weight of CdTe in 1 g of βCD-CdTe particle (W_{CdTe} = 0.58 g); W_{CD} represents the weight of βCD in 1 g of βCD-CdTe particle (W_{CD} = 0.36 g); and N_A represents Avogadro constant, N_A = 6.02 × 10^{23}. The calculated result is that $n \approx 65$.

Then, the inclusion rate of HEMA-Ad@βCD-CdTe could be calculated according to the above results and the TGA result of HEMA-Ad@βCD-CdTe.

During the same temperature range (200–400 °C), the mass loss of HEMA-Ad@βCD-CdTe was 49.11 wt %, thus the content of CdTe in HEMA-Ad@βCD-CdTe is 50.89 wt %. For HEMA-Ad@ βCD-CdTe and βCD-CdTe, the mass ratio among CdTe, MPA and βCD is constant. Therefore, HEMA-Ad@βCD-CdTe consists of 5.85 MPA, 31.24 βCD, 12.02 HEMA-Ad and 50.89 wt % CdTe. The inclusion rate of HEMA-Ad@βCD-CdTe (ϕ) could be calculated according the following equation (assuming that 1 g of HEMA-Ad@βCD-CdTe was taken):

$$\phi\ (\%) = \frac{\frac{W_{HEMA\text{-}Ad}}{M_{HEMA\text{-}Ad}}}{\frac{W_{CD}}{M_{CD}}} \times 100\% \tag{4}$$

where $W_{HEMA\text{-}Ad}$ represents the weight of HEMA-Ad in 1 g of HEMA-Ad@βCD-CdTe, $W_{HEMA\text{-}Ad}$ = 0.12 g and $M_{HEMA\text{-}Ad}$ represents the molar mass of HEMA-Ad molecule, $M_{HEMA\text{-}Ad}$ = 292.38 g/mol; W_{CD} = 0.31 g. The calculated inclusion rate of HEMA-Ad@βCD-CdTe is 152%, which is higher than 100%. The main reason is the formation of hydrogen bonds between MPA and HEMA-Ad. According to the calculated result, almost all of the βCD molecules on the CdTe surface are occupied by HEMA-Ad.

3. Results and Discussion

3.1. Preparation of the Xerogel

The fabrication process of the reusable xerogel is illustrated in Scheme 1, including four steps in sequence: modifying β-cyclodextrin (βCD) molecules on the CdTe surface (termed as βCD-CdTe); introducing guest molecules (HEMA-Ad, bearing polymerization groups) via host–guest interactions (termed as HEMA-Ad@βCD-CdTe); initiating copolymerization process with KPS to obtain hydrogel (termed as HEMA-Ad@βCD-CdTe/PHMMAAm); and drying the hydrogel at ambient atmosphere to obtain the final xerogel.

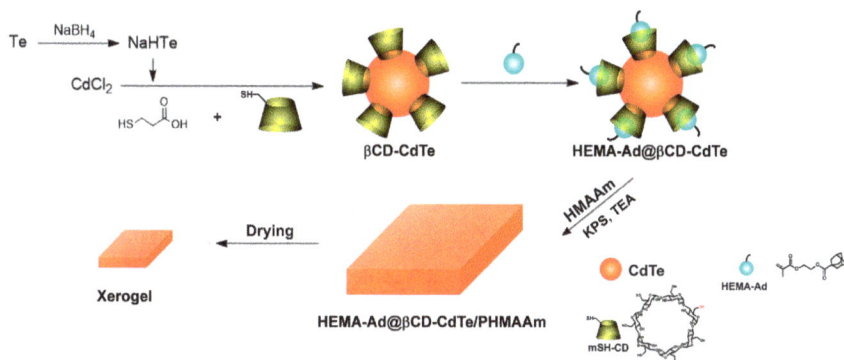

Scheme 1. The fabrication process of HEMA-Ad@βCD-CdTe/PHMAAm xerogel.

Mono-6-thio-β-cyclodextrin (mSH-CD), acting as one of the stabilizers, is introduced during the synthesis of CdTe firstly [48]. The successful immobilization of βCD is confirmed by the results of Fourier transform infrared spectroscopy (FT-IR) and thermos-gravimetric analysis (TGA). In the FT-IR spectrum, βCD-CdTe shows much stronger peaks than CdTe at 1151 and 1032 cm^{-1}, which stemmed from the C–O stretch and C–O–C antisymmetric vibrational modes in βCD molecules, respectively (Figure S1a, Supplementary Materials) [49]. The TGA curves show that the weight of βCD-CdTe lost 42.11% during the span of 200–400 °C, which is resulted from the decomposition of the stabilizers on the CdTe surface, including 3-mercaptopropionic acid (MPA) and mSH-CD (Figure S1b, Supplementary Materials) [50]. The amount of βCD on the surface of CdTe is about 0.31 mmol/g (about 65 βCD molecules bound to one CdTe nanoparticle), calculated from TGA results (in Experimental Section).

Figure 1a shows the fluorescence spectra of βCD-CdTe at different reflux time. It can be seen that the fluorescence intensity of βCD-CdTe increases significantly with the prolongation of reflux time in the first stage, and then declines slowly (Figure 1a). The increase in fluorescence intensity in the first stage is due to the crystallization of βCD-CdTe. Then, as the reflux time increases, the emission peak shifts to longer wavelength and the fluorescence intensity weakens gradually due to quantum-confined size effect [51]. Thus, the reflux process is optimized to 50 min to achieve the strongest fluorescence intensity of βCD-CdTe. The diameter of βCD-CdTe is about 4 nm, as shown in TEM images (Figure 1c,d). The fluorescence intensity of βCD-CdTe varies with different excitation wavelengths (λ_{ex}). In the condition of λ_{ex} = 380 nm, βCD-CdTe exhibits the highest fluorescence intensity (Figure S2, Supplementary Materials). The fluorescence property of βCD-Cd is further investigated under different pH conditions (Figure 1b). The median fluorescence intensity (MFI) around the emission wavelength (($\lambda_{em} \pm 5$) nm) is used to quantify the fluorescence intensity of βCD-CdTe. The strongest fluorescence intensity of βCD-CdTe, which is set as 100%, is obtained under the condition of pH = 7.97. In the acidic environment, especially when the pH value is less than 4, protonation of the thiol ligands on the surface of βCD-CdTe leads to the formation of aggregates, resulting in a sharp drop-off in fluorescence intensity due to the mentioned quantum-confined size effect [52,53]. It could be observed that alkaline environment is more suitable for βCD-CdTe, as about 80% of the fluorescence intensity is still maintained even when the pH value is up to 12 (Figure 1b).

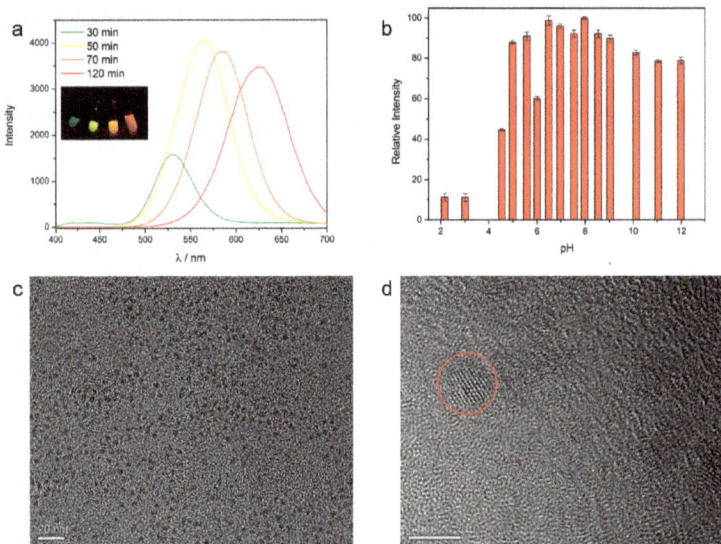

Figure 1. (**a**) Fluorescence spectra of βCD-CdTe at different reflux time (λ_{ex} = 380 nm). Insert is the photograph of the corresponding βCD-CdTe solutions (a small amount of the reaction solution is taken out at specific times during the reflux process) under ultraviolet lamp (λ_{ex} = 365 nm). From left to right: 30, 50, 70, 120 min. (**b**) Relative fluorescence intensity of βCD-CdTe under different pH conditions (PBS, 50 mM, βCD-CdTe: 5 mg/mL). The fluorescence intensity of βCD-CdTe at pH = 7.97 is set as 100%. (**c**) TEM (scale bar: 20 nm); and (**d**) high-resolution TEM (HRTEM, scale bar: 5 nm) images of βCD-CdTe dispersed in ultrapure water. One of the βCD-CdTe nanocrystals is marked by a red circle in the HRTEM image.

HEMA-Ad was prepared through the esterification reaction between 1-adamantanecarboxylic acid (Ad–COOH) and hydroxyethyl methacrylate (HEMA) according to our previous works [46,47]. The successful preparation of HEMA-Ad is confirmed by the results of ^{1}H NMR and MS (Figure S3, Supplementary Materials).

It is well known that βCD could interact with adamantane and its derivatives to form stable inclusion complexes in various environment. The immobilized βCD molecules on the surface of βCD-CdTe could also include HEMA-Ad in their cavities to form HEMA-Ad@βCD-CdTe complex, acting as crosslinking agent in the next polymerization process. Nuclear overhauser effect spectroscopy (NOESY) is one of the most convincing means to authenticate the occurrence of inclusion between βCDs and Ad groups [46,54]. The corresponding correlation signals could be observed clearly in the NOESY spectra of HEMA-Ad@βCD-CdTe, suggesting the successful formation of assemblies (Figure S4, Supplementary Materials). Almost all of the βCD molecules on the CdTe surface are assembled with HEMA-Ad, according to the TGA results (Figure S5, Supplementary Materials, the detailed calculation process is shown in the Experimental Section).

To optimize the dosage of HEMA-Ad@βCD-CdTe, we investigated their fluorescence properties with different concentrations. Typically, the significant decrease in the overall fluorescence intensity or redshift could be observed when the QDs concentration exceeded a certain threshold due to electronic coupling or exciton energy transfer between nanocrystals [55–57]. According to Jiang's report, the threshold of concentration is about 5 mg/mL [34]. Surprisingly, in this work, no decrease in the overall fluorescence intensity and only a little redshift occurred even when the concentration of HEMA-Ad@βCD-CdTe is up to 200 mg/mL (Figure 2a,b), suggesting its excellent dispersity and stability, laying the foundation of the high loading and strong fluorescence properties of the final xerogel. We owe this prominent fluorescence retention to the introduction of βCD on the βCD-CdTe surface. βCD molecules possess cyclic structure, which could suppress the interactions between nanocrystals effectively due to steric hindrance. The solid state βCD-Cd could be re-dispersed in water easily after freeze drying. For CdTe, which is stabilized by 3-mercaptopropionic acid only, it is difficult to achieve a homogeneous dispersion solution again, since severe aggregation formed during the freeze drying (Figure S6, Supplementary Materials).

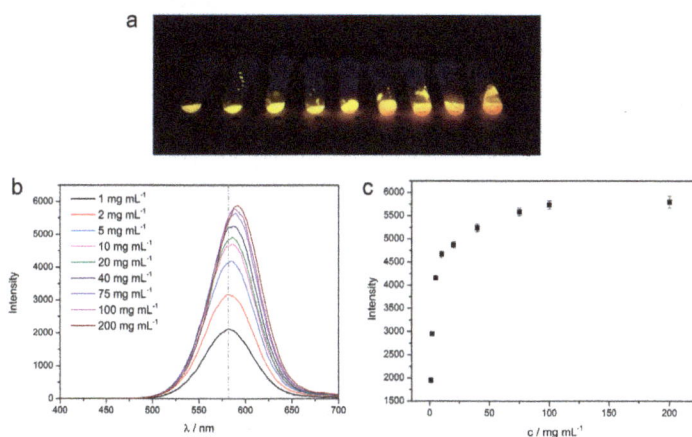

Figure 2. The relationship between fluorescence property and its concentration of HEMA-Ad/βCD-CdTe. (**a**) Photograph of HEMA-Ad@βCD-CdTe under ultraviolet light (λ_{ex} = 365 nm) with different concentrations (from left to right: 1/2/5/10/20/40/75/100/200 mg/mL); (**b**) fluorescence spectra of HEMA-Ad@βCD-CdTe with different concentrations (λ_{ex} = 380 nm); and (**c**) MFI scatterplot of HEMA-Ad@βCD-CdTe with different concentrations.

As shown in Figure 2c, in the first stage, the fluorescence intensity of HEMA-Ad@βCD-CdTe increases sharply as its concentration increases. However, the increase of fluorescence intensity slows down significantly when the concentration is higher than 20 mg/mL. Therefore, the content of HEMA-Ad@βCD-CdTe in the polymer formulation is set as 20 mg/mL. In addition, the fluorescence intensity of HEMA-Ad@βCD-CdTe is a little stronger than βCD-CdTe at the same concentration (Figure 3b,c), which may be attributed to the increased spatial distance between nanocrystals.

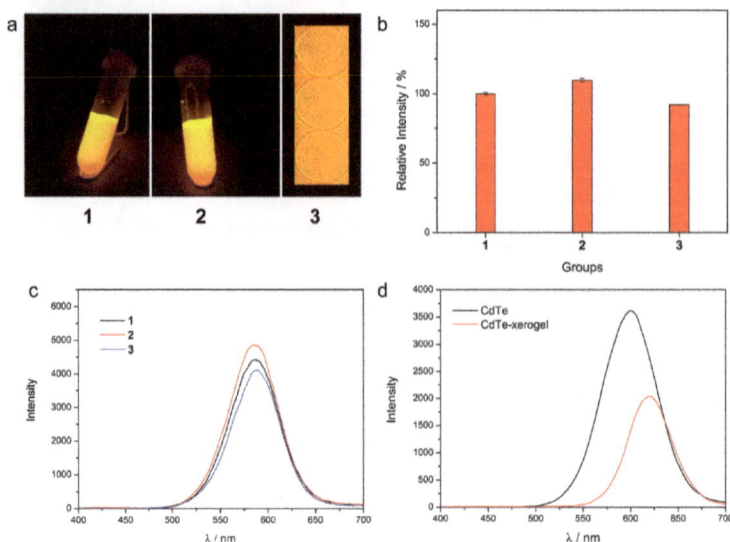

Figure 3. The fluorescence property of the xerogels. (**a**) photographs of βCD-CdTe (**1**, 20 mg/mL), HEMA-Ad@βCD-CdTe (**2**, 20 mg/mL) and the corresponding xerogel (**3**) under ultraviolet light (λ_{ex} = 365 nm); (**b**) the relative fluorescence intensity of **1**, **2** and **3**, in which the fluorescence intensity of **1** is set as 100%; (**c**) fluorescence spectra of **1**, **2** and **3**; and (**d**) fluorescence spectra of CdTe (20 mg/mL) and the corresponding xerogel (CdTe-xerogel). For all the fluorescence spectra, the excitation wavelength is 380 nm.

HEMA-Ad@βCD-CdTe/PHMAAm hydrogel was prepared by copolymerization HEMA-Ad@βCD-CdTe and HMAAm at room temperature. The final xerogel is obtained after drying the prepared hydrogel at ambient atmosphere. Predictably, shrinkage occurred during the drying process, and the surface of the xerogel is plicate (Figure S7, Supplementary Materials).

3.2. Fluorescence Property of the Xerogel

Ordinarily, as mentioned, QDs hydrogels dried at ambient atmosphere will result in violent decrease of fluorescence intensity, as the structure is shrunken and the density of QDs is increased (the spatial distance of adjacent nanocrystals is compressed) [55,56]. Unexpectedly, the xerogel (**3**) we prepared exhibits strong fluorescence property, whose intensity is 90% of βCD-CdTe solution (**1**), and 85% of HEMA-Ad@βCD-CdTe solution (**2**, Figure 3a,b), which is much better than reported studies (the reported fluorescence retention is about 50–60%) [9,15]. In addition, compared with the fluorescence curves of βCD-CdTe and HEMA-Ad@βCD-CdTe solution, no redshift occurs for the xerogel, suggesting that nanocrystals kept homogeneous dispersion throughout the drying process [55]. As a comparison, we also measured the fluorescence property of xerogel of CdTe/PHMAAm (CdTe-xerogel), which lacks the constraint of host–guest interactions. It could be seen that the control sample only remained 56% fluorescence intensity of the corresponding solution, along with the obvious redshift (Figure 3d), suggesting that some of the nanocrystals aggregated during the drying process. The prominent fluorescence retention of xerogel (**3**) stems from the introduction of βCD on the βCD-CdTe surface. Firstly, as we mentioned before, the βCD molecules on the particles' surface suppressed their interactions effectively due to steric hindrance. Secondly, the quantum dot nanoparticles are immobilized in the polymer matrix through host–guest interactions, preventing the aggregations even during the drying process of the hydrogel at ambient atmosphere (the emission wavelength is almost same as the corresponding solution state, Figure 3c).

Although our strategy is quite similar to Jiang's, there are two very different points. Firstly, the "supramolecular cross-linker" preparation is different. Jiang employed perthiolated βCD as the sole stabilizer on the surface of CdS and the dryness process was conducted in a vacuum drier at room temperature. Aggregation of CdS was expected during the dryness process as hydrogen bonds between hydroxyl groups of βCD and the high surface energy of CdS nanocrystals. In this work, the inter-nanocrystals aggregation is suppressed effectively by employing freeze dryer. Secondly, rather than stop at the successful preparation of QDs–polymer hydrogel, we further explore the fluorescence property and versatility of the xerogel.

3.3. The Detecting Performance and Reusable Ability of the Xerogel

Vanillin (Van), one of the most immensely popular flavor components, having close relations to our daily life [58–60], is chosen as a model molecule to evaluate the detecting performance and the reusable property of the xerogels.

A small amount of sample (5 μL) is required for the detection process. After adding the sample (20 mg/L Van in 50 mM PBS, pH = 8.01), the fluorescence intensity of the xerogel is monitored every 5 min. After the first 5 min, the fluorescence intensity keeps almost constant (Figure S8, Supplementary Materials). Thus, in the following experiments, the responding time between Van and the xerogel is set as 5 min. It can be seen that there is a good linear relationship between I_0/I and the Van concentration (10–2000 mg/L): $I_0/I = 1.3209 + 0.0029$ c ($R^2 = 0.999$) (Figure 4a,b), in which I_0 and I are the MFI of the xerogel in the absence and presence of Van, respectively, and the value of I_0/I reflects the degree of fluorescence quenching; and c is the concentration of Van. Typically, there are two widely accepted quenching mechanisms: fluorescence resonance energy transfer (FRET) or electron transfer. As the fluorescence spectrum of xerogel does not overlap with UV–Vis spectra of Van (Figure S9, Supplementary Materials), electron transfer accounts for the fluorescence quenching phenomenon [61].

The spent xerogel could be recycled by following steps easily (Figure 4c). Firstly, the used xerogel was immersed in water for 5 min, and transformed into hydrogel due to water absorption. Secondly, the "hydrogel" was oscillated in dichloromethane for 10 min to clear Van molecules in the polymer matrix due to extraction effect. Finally, recycled xerogel was achieved after drying the "hydrogel" at room temperature for 48 h in darkness. The fluorescence of the first time recycled xerogel restored almost completely (98% of the original state, Figure 4d). Even after three cycles, the recycled xerogel kept strong fluorescence property (82% of the original state). The synergistic effect of the first two steps is necessary to achieve the extraordinary restoration of fluorescence. Soaking the utilized xerogel in water in the first step removed the remaining phosphate and relaxed the polymer chains. The host–guest interactions between βCD molecules on the CdTe surface and Ad groups linked with the polymer chains prevent quantum dots from leaking, providing a solid foundation for the fluorescence restoration. The loose polymer network structure is beneficial to clean up Van molecules owing to extraction effect in the second step.

As shown in Figure 5a, various aromatic molecules, not only Van, could quench the fluorescence of xerogel with different degrees, including protocatechualdehyde (PCA), hydroquinone (HQ), *p*-nitrophenol (*p*-NP) and *o*-nitrophenol (*o*-NP, Figure 5c), indicating that the xerogels have great potential for acting as a versatile platform for quantitative analyze many aromatic molecules. One may notice that the quenching efficiency of *p*-NP (91.75%) is much higher than *o*-NP (31.53%) with the same concentration, suggesting that the xerogel could also be used to identify isomers (i.e., qualitative analysis) (Figure 5b) [48].

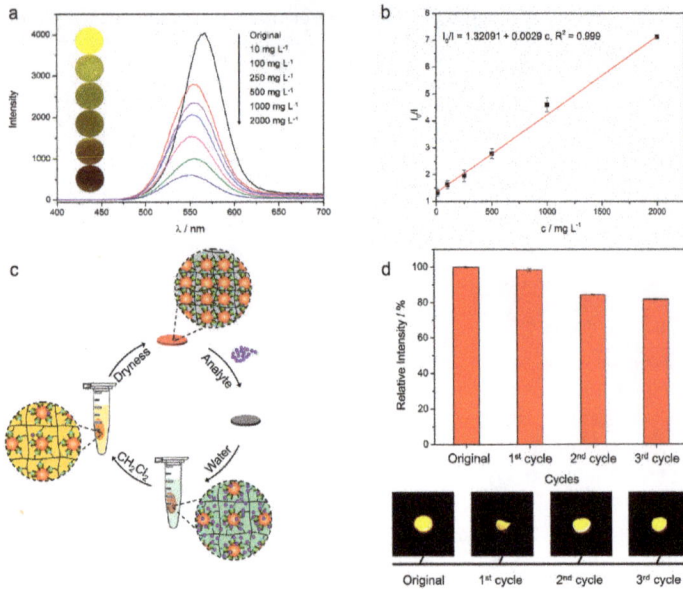

Figure 4. The detecting performance and reusable property of the xerogel. (**a**) Fluorescence spectra of the xerogel with different concentration Van (λ_{ex} = 380 nm). At the left is the corresponding photographs under ultraviolet light (λ_{ex} = 365 nm); (**b**) Calibration plot between I_0/I and the concentration of Van (c, mg/L); (**c**) Schematic illustration of the reclamation process; (**d**) The relative fluorescence intensity between the original xerogel and the recycled ones. The fluorescence intensity of the original xerogel is set as 100%. Below this are photographs of the corresponding xerogels under ultraviolet light (λ_{ex} = 365 nm). For the above test samples, Van is dissolved in PBS (pH = 8.01, 50 mM).

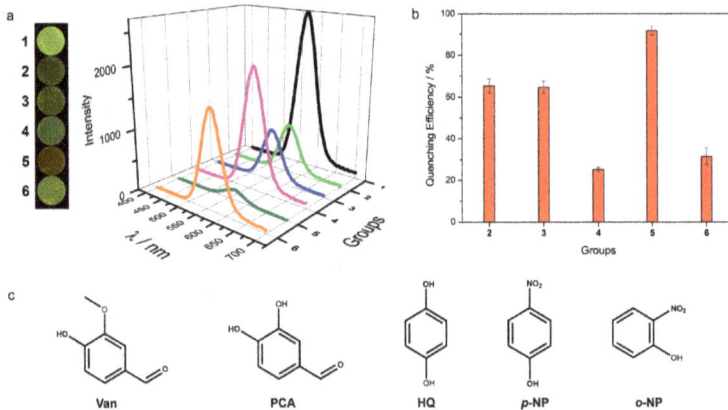

Figure 5. The interactions between the xerogel and various aromatic molecules. (**a**) The fluorescence spectra (λ_{ex} = 380 nm) of the xerogel after interacting with PBS (**1**, blank control, pH = 8.00), Van (**2**), PCA (**3**), HQ (**4**), *p*-NP (**5**) and *o*-NP (**6**); (**b**) The calculated quenching efficiency and (**c**) chemical structures of the detected aromatic molecules.

According to the mentioned experiment results, the prepared βCD-CdTe shows strong fluorescence in a wide pH range, which overlap the optimal pH range of various enzymes. By the way, polymerization process is taken place at room temperature. All of these provide an opportunity to embed enzymes in the xerogel to expand its applications. To bring this concept to reality, we attempted to encapsulate tyrosinase (TYR) in the xerogel to analyze dopamine, one of the most important neurotransmitters [62]. The ratio I_0/I is proportional to the concentration of dopamine: $I_0/I = 1.25677 + 0.00063$ c ($R^2 = 0.997$), in which I_0 and I are the MFI of the xerogel in the absence and presence of dopamine, respectively, c is the concentration of dopamine (Figure S10a,b, Supplementary Materials). The excellent linearity between I_0/I and c results from the oxidation of dopamine to dopaquinone, which is catalyzed by TYR [63]. This could be verified through control experiments: Case 1, utilizing common xerogels (without TYR) to detect dopamine; Case 2, utilizing TYR contained xerogels to detect 3,4-dimethoxyphenethylamine (DMPA), whose chemical structure is very similar with dopamine (Figure S10d, Supplementary Materials); and Case 3, utilizing TYR contained xerogels to detect dopamine. The concentrations of all analytes are the same. As shown in Figure S10c, no significant decrease of fluorescence intensity is observed in both Case 1 and Case 2, which is quite different from Case 3. These results also indicate the specificity of the established platform to detect dopamine, certificate the fact that the xerogel is a suitable and promising candidate to fabricate various enzyme-based biosensors.

Finally, as the xerogels could be used circularly, the intrinsic self-healing property of the xerogels guarantees the long-time service. As shown in Figure S11 (Supplementary Materials), the fractured xerogel could combine again under the assistance of very little water in 5 min. The self-healing mechanism is similar to other self-healing materials containing host–guest groups [47]: the added water softens polymer network, enhancing the mobility of polymer chains; when the fractured surface contacted, the exposed βCD molecules on the CdTe surface and Ad groups linked to the polymer chain on both sides reassociate together autonomously across the interface due to host–guest interactions; at the same time, polymer chains entanglement occurred. Under the synergy of host–guest interactions and polymer chains entanglement, the fractured xerogel is restored to an entirety.

4. Conclusions

In summary, we fabricated a reusable xerogel containing QDs with high fluorescence retention and versatility. βCD molecules on the QDs surface are the protagonists in this strategy. Their cyclic structure suppresses interactions between nanocrystals (electronic coupling or exciton energy transfer) effectively at high concentration (up to 200 mg/mL), ensuring the high QDs loading capacity of the xerogel. Host–guest interactions between βCD molecules on the CdTe surface and Ad groups on the polymer chains immobilize QDs firmly, resulting in the excellent fluorescence retention of the xerogel, as well as self-healing capacity. The prepared xerogel could serve as not only a platform of quantitative analysis (detecting Van, for example) but also a platform of qualitative analysis (distinguishing between *p*-NP and *o*-NP, for example). The used xerogels could be recycled through three steps easily. Moreover, the xerogel could also entrap enzymes to act as biosensors.

Supplementary Materials: The following are available online at www.mdpi.com/2073-4360/10/3/310/s1. Including the structure characterization of βCD-CdTe and supplementary experimental results serving for conclusions of the main text.

Acknowledgments: This work was funded by the National Natural Science Foundation of China (Grant Nos. 51373174 and 51573187), and the West Light Foundation of CAS.

Author Contributions: Xiang-Yong Liang designed and performed the experiments, and drafted the manuscript; Lu Wang and Zhi-Yi Chang assisted with data collection and analysis; Li-Sheng Ding contributed analysis tools; and Bang-Jing Li and Sheng Zhang conceived the experiments and revised the manuscript.

Conflicts of Interest: The authors declare no conflict of interest.

References

1. Medintz, I.L.; Uyeda, H.T.; Goldman, E.R.; Mattoussi, H. Quantum dot bioconjugates for imaging, labelling and sensing. *Nat. Mater.* **2005**, *4*, 435–446. [CrossRef] [PubMed]

2. Carey, G.H.; Abdelhady, A.L.; Ning, Z.; Thon, S.M.; Bakr, O.M.; Sargent, E.H. Colloidal Quantum Dot Solar Cells. *Chem. Rev.* **2015**, *115*, 12732–12763. [CrossRef] [PubMed]

3. Silvi, S.; Credi, A. Luminescent sensors based on quantum dot-molecule conjugates. *Chem. Soc. Rev.* **2015**, *44*, 4275–4289. [CrossRef] [PubMed]

4. Hildebrandt, N.; Spillmann, C.M.; Algar, W.R.; Pons, T.; Stewart, M.H.; Oh, E.; Susumu, K.; Díaz, S.A.; Delehanty, J.B.; Medintz, I.L. Energy Transfer with Semiconductor Quantum Dot Bioconjugates: A Versatile Platform for Biosensing, Energy Harvesting, and Other Developing Applications. *Chem. Rev.* **2017**, *117*, 536–711. [CrossRef] [PubMed]

5. Smith, A.M.; Nie, S. Chemical analysis and cellular imaging with quantum dots. *Analyst* **2004**, *129*, 672–677. [CrossRef] [PubMed]

6. Wang, J.-H.; Wang, H.-Q.; Zhang, H.-L.; Li, X.-Q.; Hua, X.-F.; Cao, Y.-C.; Huang, Z.-L.; Zhao, Y.-D. Purification of denatured bovine serum albumin coated CdTe quantum dots for sensitive detection of silver(I) ions. *Anal. Bioanal. Chem.* **2007**, *388*, 969–974. [CrossRef] [PubMed]

7. Tang, C.; Qian, Z.; Huang, Y.; Xu, J.; Ao, H.; Zhao, M.; Zhou, J.; Chen, J.; Feng, H. A fluorometric assay for alkaline phosphatase activity based on β-cyclodextrin-modified carbon quantum dots through host–guest recognition. *Biosens. Bioelectron.* **2016**, *83*, 274–280. [CrossRef] [PubMed]

8. Zhang, X.; Ding, S.; Cao, S.; Zhu, A.; Shi, G. Functional surface engineering of quantum dot hydrogels for selective fluorescence imaging of extracellular lactate release. *Biosens. Bioelectron.* **2016**, *80*, 315–322. [CrossRef] [PubMed]

9. Arachchige, I.U.; Brock, S.L. Highly Luminescent Quantum-Dot Monoliths. *J. Am. Chem. Soc.* **2007**, *129*, 1840–1841. [CrossRef] [PubMed]

10. Arachchige, I.U.; Brock, S.L. Sol–Gel Methods for the Assembly of Metal Chalcogenide Quantum Dots. *Acc. Chem. Res.* **2007**, *40*, 801–809. [CrossRef] [PubMed]

11. Li, X.; Zhou, Y.; Zheng, Z.; Yue, X.; Dai, Z.; Liu, S.; Tang, Z. Glucose Biosensor Based on Nanocomposite Films of CdTe Quantum Dots and Glucose Oxidase. *Langmuir* **2009**, *25*, 6580–6586. [CrossRef] [PubMed]

12. Mamedov, A.A.; Belov, A.; Giersig, M.; Mamedova, N.N.; Kotov, N.A. Nanorainbows: Graded Semiconductor Films from Quantum Dots. *J. Am. Chem. Soc.* **2001**, *123*, 7738–7739. [CrossRef] [PubMed]

13. Gattás-Asfura, K.M.; Zheng, Y.; Micic, M.; Snedaker, M.J.; Ji, X.; Sui, G.; Orbulescu, J.; Andreopoulos, F.M.; Pham, S.M.; Wang, C.; et al. Immobilization of Quantum Dots in the Photo-Cross-Linked Poly(ethylene glycol)-Based Hydrogel. *J. Phys. Chem. B* **2003**, *107*, 10464–10469. [CrossRef]

14. Zhao, F.; Yao, D.; Guo, R.; Deng, L.; Dong, A.; Zhang, J. Composites of Polymer Hydrogels and Nanoparticulate Systems for Biomedical and Pharmaceutical Applications. *Nanomaterials* **2015**, *5*, 2054–2130. [CrossRef] [PubMed]

15. Mohanan, J.L.; Arachchige, I.U.; Brock, S.L. Porous Semiconductor Chalcogenide Aerogels. *Science* **2005**, *307*, 397–400. [PubMed]

16. Hendel, T.; Lesnyak, V.; Kuehn, L.; Herrmann, A.-K.; Bigall, N.C.; Borchardt, L.; Kaskel, S.; Gaponik, N.; Eychmueller, A. Mixed Aerogels from Au and CdTe Nanoparticles. *Adv. Funct. Mater.* **2013**, *23*, 1903–1911. [CrossRef]

17. Nahar, L.; Esteves, R.J.A.; Hafiz, S.; Özgür, Ü.; Arachchige, I.U. Metal–Semiconductor Hybrid Aerogels: Evolution of Optoelectronic Properties in a Low-Dimensional CdSe/Ag Nanoparticle Assembly. *ACS Nano* **2015**, *9*, 9810–9821. [CrossRef] [PubMed]

18. Yuan, J.; Wen, D.; Gaponik, N.; Eychmüller, A. Enzyme-Encapsulating Quantum Dot Hydrogels and Xerogels as Biosensors: Multifunctional Platforms for Both Biocatalysis and Fluorescent Probing. *Angew. Chem. Int. Ed.* **2013**, *52*, 976–979. [CrossRef] [PubMed]

19. Wolf, A.; Lesnyak, V.; Gaponik, N.; Eychmueller, A. Quantum-Dot-Based (Aero)gels: Control of the Optical Properties. *J. Phys. Chem. Lett.* **2012**, *3*, 2188–2193. [CrossRef] [PubMed]

20. Yan, J.-J.; Wang, H.; Zhou, Q.-H.; You, Y.-Z. Reversible and Multisensitive Quantum Dot Gels. *Macromolecules* **2011**, *44*, 4306–4312. [CrossRef]

21. Cayuela, A.; Kennedy, S.R.; Soriano, M.L.; Jones, C.D.; Valcarcel, M.; Steed, J.W. Fluorescent carbon dot-molecular salt hydrogels. *Chem. Sci.* **2015**, *6*, 6139–6146. [CrossRef]

22. Xie, X.; Ma, D.; Zhang, L.-M. Fabrication and properties of a supramolecular hybrid hydrogel doped with CdTe quantum dots. *RSC Adv.* **2015**, *5*, 58746–58754. [CrossRef]

23. Cayuela, A.; Soriano, M.L.; Kennedy, S.R.; Steed, J.W.; Valcárcel, M. Fluorescent carbon quantum dot hydrogels for direct determination of silver ions. *Talanta* **2016**, *151*, 100–105. [CrossRef] [PubMed]

24. Chatterjee, S.; Maitra, U. A novel strategy towards designing a CdSe quantum dot-metallohydrogel composite material. *Nanoscale* **2016**, *8*, 14979–14985. [CrossRef] [PubMed]

25. Liu, W.; Xu, S.; Li, Z.; Liang, R.; Wei, M.; Evans, D.G.; Duan, X. Layer-by-Layer Assembly of Carbon Dots-Based Ultrathin Films with Enhanced Quantum Yield and Temperature Sensing Performance. *Chem. Mater.* **2016**, *28*, 5426–5431. [CrossRef]

26. Park, H.-Y.; Oh, J.H.; Hong, S.P.; Do, Y.R.; Jang, S.-Y. Optical Properties Enhancement of Electrosprayed Quantum Dot/Polymer Nanohybrid Films by a Solvent Vapor Treatment. *Sci. Adv. Mater.* **2016**, *8*, 224–230. [CrossRef]

27. Raja, S.N.; Luong, A.J.; Zhang, W.; Lin, L.; Ritchie, R.O.; Alivisatos, A.P. Cavitation-Induced Stiffness Reductions in Quantum Dot–Polymer Nanocomposites. *Chem. Mater.* **2016**, *28*, 2540–2549. [CrossRef]

28. Jang, E.; Son, K.J.; Kim, B.; Koh, W.-G. Phenol biosensor based on hydrogel microarrays entrapping tyrosinase and quantum dots. *Analyst* **2010**, *135*, 2871–2878. [CrossRef] [PubMed]

29. Kim, C.-H.; Bang, J.-H.; Hong, K.B.; Park, M.-H. Fabrication of highly photoluminescent quantum dot-polymer composite micropatterned surface using thiol-ene chemistry. *RSC Adv.* **2016**, *6*, 96700–96705. [CrossRef]

30. Schnee, V.P.; Bright, C.J.; Nallon, E.C.; Polcha, M.P. Contact printing of a quantum dot and polymer cross-reactive array sensor. *Sens. Actuators B* **2016**, *236*, 506–511. [CrossRef]

31. Jańczewski, D.; Tomczak, N.; Han, M.-Y.; Vancso, G.J. Introduction of Quantum Dots into PNIPAM microspheres by precipitation polymerization above LCST. *Eur. Polym. J.* **2009**, *45*, 1912–1917. [CrossRef]

32. Jańczewski, D.; Tomczak, N.; Han, M.-Y.; Vancso, G.J. Stimulus Responsive PNIPAM/QD Hybrid Microspheres by Copolymerization with Surface Engineered QDs. *Macromolecules* **2009**, *42*, 1801–1804. [CrossRef]

33. Guo, M.; Jiang, M. Supramolecular Hydrogels with CdS Quantum Dots Incorporated by Host–Guest Interactions. *Macromol. Rapid Commun.* **2010**, *31*, 1736–1739. [CrossRef] [PubMed]

34. Du, P.; Chen, G.; Jiang, M. Electrochemically sensitive supra-crosslink and its corresponding hydrogel. *Sci. China Chem.* **2012**, *55*, 836–843. [CrossRef]

35. Liu, J.; Chen, G.; Guo, M.; Jiang, M. Dual Stimuli-Responsive Supramolecular Hydrogel Based on Hybrid Inclusion Complex (HIC). *Macromolecules* **2010**, *43*, 8086–8093. [CrossRef]

36. Cheng, W.; Rechberger, F.; Niederberger, M. Three-Dimensional Assembly of Yttrium Oxide Nanosheets into Luminescent Aerogel Monoliths with Outstanding Adsorption Properties. *ACS Nano* **2016**, *10*, 2467–2475. [CrossRef] [PubMed]

37. Derfus, A.M.; Chan, W.C.W.; Bhatia, S.N. Probing the Cytotoxicity of Semiconductor Quantum Dots. *Nano Lett.* **2004**, *4*, 11–18. [CrossRef] [PubMed]

38. Hardman, R. A Toxicologic Review of Quantum Dots: Toxicity Depends on Physicochemical and Environmental Factors. *Environ. Health Perspect.* **2006**, *114*, 165–172. [CrossRef] [PubMed]

39. Lewinski, N.; Colvin, V.; Drezek, R. Cytotoxicity of Nanoparticles. *Small* **2008**, *4*, 26–49. [CrossRef] [PubMed]

40. Wang, W.; He, Y.; Yu, G.; Li, B.; Sexton, D.W.; Wileman, T.; Roberts, A.A.; Hamilton, C.J.; Liu, R.; Chao, Y.; et al. Sulforaphane Protects the Liver against CdSe Quantum Dot-Induced Cytotoxicity. *PLoS ONE* **2015**, *10*, e0138771. [CrossRef] [PubMed]

41. Han, C.; Li, H. Chiral recognition of amino acids based on cyclodextrin-capped quantum dots. *Small* **2008**, *4*, 1344–1350. [CrossRef] [PubMed]

42. Li, H.; Han, C. Sonochemical Synthesis of Cyclodextrin-Coated Quantum Dots for Optical Detection of Pollutant Phenols in Water. *Chem. Mater.* **2008**, *20*, 6053–6059. [CrossRef]

43. Cao, Y.; Wu, S.; Liang, Y.; Yu, Y. The molecular recognition of beta-cyclodextrin modified CdSe quantum dots with tyrosine enantiomers: Theoretical calculation and experimental study. *J. Mol. Struct.* **2013**, *1031*, 9–13. [CrossRef]

44. Wei, Y.; Hao, H.; Zhang, J.; Hao, X.; Dong, C. Sensitive and selective detection of L-tryptophan using Mn-ZnS QDs as the ratiometric emission probe. *Anal. Methods* **2014**, *6*, 3227–3230. [CrossRef]

45. Zhou, J.; Liu, Y.; Zhang, Z.; Yang, S.; Tang, J.; Liu, W.; Tang, W. Cyclodextrin-clicked silica/CdTe fluorescent nanoparticles for enantioselective recognition of amino acids. *Nanoscale* **2016**, *8*, 5621–5626. [CrossRef] [PubMed]

46. Guo, K.; Zhang, D.-L.; Zhang, X.-M.; Zhang, J.; Ding, L.-S.; Li, B.-J.; Zhang, S. Conductive Elastomers with Autonomic Self-Healing Properties. *Angew. Chem. Int. Ed.* **2015**, *54*, 12127–12133. [CrossRef] [PubMed]

47. Liang, X.-Y.; Wang, L.; Wang, Y.-M.; Ding, L.-S.; Li, B.-J.; Zhang, S. UV-Blocking Coating with Self-Healing Capacity. *Macromol. Chem. Phys.* **2017**, *218*. [CrossRef]

48. Zhang, Z.; Zhou, J.; Liu, Y.; Tang, J.; Tang, W. Cyclodextrin capped CdTe quantum dots as versatile fluorescence sensors for nitrophenol isomers. *Nanoscale* **2015**, *7*, 19540–19546. [CrossRef] [PubMed]

49. Yoon, S.; Nichols, W.T. Cyclodextrin directed self-assembly of TiO$_2$ nanoparticles. *Appl. Surf. Sci.* **2013**, *285*, 517–523. [CrossRef]

50. Guo, Y.; Guo, S.; Ren, J.; Zhai, Y.; Dong, S.; Wang, E. Cyclodextrin Functionalized Graphene Nanosheets with High Supramolecular Recognition Capability: Synthesis and Host–Guest Inclusion for Enhanced Electrochemical Performance. *ACS Nano* **2010**, *4*, 4001–4010. [CrossRef] [PubMed]

51. Yang, M.; Wang, Y.; Wang, H. beta-cyclodextrin functionalized CdTe quantum dots for electrochemiluminescent detection of benzo[a]pyrene. *Electrochim. Acta* **2015**, *169*, 7–12. [CrossRef]

52. Aldana, J.; Lavelle, N.; Wang, Y.; Peng, X. Size-Dependent Dissociation pH of Thiolate Ligands from Cadmium Chalcogenide Nanocrystals. *J. Am. Chem. Soc.* **2005**, *127*, 2496–2504. [CrossRef] [PubMed]

53. Zhang, Y.; Mi, L.; Wang, P.-N.; Ma, J.; Chen, J.-Y. pH-dependent aggregation and photoluminescence behavior of thiol-capped CdTe quantum dots in aqueous solutions. *J. Lumin.* **2008**, *128*, 1948–1951. [CrossRef]

54. Dong, Z.-Q.; Cao, Y.; Yuan, Q.-J.; Wang, Y.-F.; Li, J.-H.; Li, B.-J.; Zhang, S. Redox- and Glucose-Induced Shape-Memory Polymers. *Macromol. Rapid Commun.* **2013**, *34*, 867–872. [CrossRef] [PubMed]

55. Kagan, C.R.; Murray, C.B.; Bawendi, M.G. Long-range resonance transfer of electronic excitations in close-packed CdSe quantum-dot solids. *Phys. Rev. B* **1996**, *54*, 8633–8643. [CrossRef]

56. Artemyev, M.V.; Bibik, A.I.; Gurinovich, L.I.; Gaponenko, S.V.; Woggon, U. Evolution from individual to collective electron states in a dense quantum dot ensemble. *Phys. Rev. B* **1999**, *60*, 1504–1506. [CrossRef]

57. Noh, M.; Kim, T.; Lee, H.; Kim, C.-K.; Joo, S.-W.; Lee, K. Fluorescence quenching caused by aggregation of water-soluble CdSe quantum dots. *Colloids Surf. A* **2010**, *359*, 39–44. [CrossRef]

58. Karathanos, V.T.; Mourtzinos, I.; Yannakopoulou, K.; Andrikopoulos, N.K. Study of the solubility, antioxidant activity and structure of inclusion complex of vanillin with β-cyclodextrin. *Food Chem.* **2007**, *101*, 652–658. [CrossRef]

59. Kayaci, F.; Uyar, T. Encapsulation of vanillin/cyclodextrin inclusion complex in electrospun polyvinyl alcohol (PVA) nanowebs: Prolonged shelf-life and high temperature stability of vanillin. *Food Chem.* **2012**, *133*, 641–649. [CrossRef]

60. Duran, G.M.; Contento, A.M.; Rios, A. beta-Cyclodextrin coated CdSe/ZnS quantum dots for vanillin sensoring in food samples. *Talanta* **2015**, *131*, 286–291. [CrossRef] [PubMed]

61. Jia, L.; Xu, J.-P.; Li, D.; Pang, S.-P.; Fang, Y.; Song, Z.-G.; Ji, J. Fluorescence detection of alkaline phosphatase activity with β-cyclodextrin-modified quantum dots. *Chem. Commun.* **2010**, *46*, 7166–7168. [CrossRef] [PubMed]

62. Wen, D.; Liu, W.; Herrmann, A.-K.; Haubold, D.; Holzschuh, M.; Simon, F.; Eychmüller, A. Simple and Sensitive Colorimetric Detection of Dopamine Based on Assembly of Cyclodextrin-Modified Au Nanoparticles. *Small* **2016**, *12*, 2439–2442. [CrossRef] [PubMed]

63. Zhang, F.; Xu, L.; Zhao, Q.; Sun, Y.; Wang, X.; Ma, P.; Song, D. Dopamine-modified Mn-doped ZnS quantum dots fluorescence probe for the sensitive detection of tyrosinase in serum samples and living cells imaging. *Sens. Actuators B* **2018**, *256*, 1069–1077. [CrossRef]

![Polymers logo] *polymers*

MDPI

Article

Supramolecular Hydrogel Based on pNIPAm Microgels Connected via Host–Guest Interactions

Iurii Antoniuk [1], Daria Kaczmarek [2], Attila Kardos [2,3], Imre Varga [2,3,*] and Catherine Amiel [1,*]

[1] University Paris Est, ICMPE (UMR 7182), CNRS, UPEC, F-94320 Thiais, France; iurii.antoniuk@gmail.com
[2] Institute of Chemistry, Eötvös Loránd University, Pázmány s. 1/A, 1117 Budapest, Hungary; daria.kaczmarek@nanos3.eu (D.K.); kardosattila87@gmail.com (A.K.)
[3] Department of Chemistry, University J. Selyeho, 945 01 Komárno, Slovakia
*   Correspondence: imo@chem.elte.hu (I.V.); amiel@icmpe.cnrs.fr (C.A.); Tel.: +33-1-4978-1219 (C.A.); Fax: +33-1-4978-1208 (C.A.)

Received: 19 March 2018; Accepted: 18 May 2018; Published: 23 May 2018

Abstract: In this work, host–guest supramolecular hydrogels were prepared from poly(*N*-isopropylacrylamide) (pNIPAm) microgels utilizing electrostatic and host/guest self-assembly. First, pNIPAm microgels bearing a poly(acrylic acid) (pAAc) shell were coated with positively charged β-cyclodextrin polymers. Addition of adamantane-substituted dextrans (Dex-Ada) allowed us to establish interparticle connections through β-cyclodextrin-adamantane (βCD-Ada) inclusion complex formation, and thus to prepare hierarchical hydrogels. Under the conditions of hydrogel formation, close contact between the microgels was ensured. To the best of our knowledge, this is the first example of doubly crosslinked microgels prepared by noncovalent crosslinking via host–guest interactions. The prepared macrogels were studied with rheology, and fast mechanical response to temperature variation was found. Furthermore, the hydrogels exhibit fully reversible temperature-induced gel–sol transition at the physiological temperature range (37–41 °C), due to the synergetic effect between shrinking of the microgels and dissociation of βCD-Ada crosslinks at higher temperatures. This opens up attractive prospects of their potential use in biomedical applications.

Keywords: host–guest polymer complex; microgel; hydrogel; β-cyclodextrin polymer; rheology; temperature-induced sol–gel transition

---

1. Introduction

Hydrogels are crosslinked 3D networks composed of hydrophilic polymeric chains or low-molecular-mass gelators [1]. Due to their structural and functional resemblance to extracellular matrix, hydrogels are widely used in biomedical industry as scaffolds in regenerative medicine, and depos for sustained release of therapeutics [2–4]. Incorporation of particulate drug-delivery vehicles, and thermoresponsive microgels in particular, in hydrogel matrices might increase the pharmacological performance of both components [4]. The resulting composite or "plum puddinG" hydrogels often show better mechanical properties, and the biocompatibility of microgels/nanoparticles is increased by integrating the latter in the biocompatible hydrogel network. The hydrogel matrix can act as a diffusion barrier aimed at diminishing the "burst release" of drug load that is an issue observed for nanoscale carriers [5]. At the same time, fixation of the nanoparticles in the matrix prevents their migration or washing away from the site of action in vivo. Finally, multiple types of microgels carrying different payloads can be entrapped in the same gel matrix and simultaneously delivered [6].

Richtering and Saunders distinguish three types of composite hydrogels containing microgels [7]: (i) microgel-filled hydrogels; (ii) microgel-reinforced hydrogels and (iii) doubly crosslinked microgels (DX microgels). All these types of hydrogels exhibit structural hierarchy with two or more distinct length scales between the adjacent crosslinks represented by the correlation lengths. In the first two cases, microgels are either mechanically entrapped or covalently linked to the bulk hydrogel matrix. In the case of DX microgels, however, the particles are directly interlinked and no additional hydrogel matrix is required.

Sivakumaran et al. have recently reported nanocomposite hydrogels composed of thermoresponsive pNIPAm-*co*-pAAc microgels either entrapped or covalently linked to inert carboxymethyl cellulose hydrogel matrix [8]. The hydrogels showed capacity to facilitate temperature-triggered long-term release of bupivacaine hydrochloride, a cationic drug. Interestingly, both increase of internal crosslink densities and covalent linking of microgels to the matrix led to lower overall rates of drug release, which was ascribed to reduced kinetics of microgel deswelling at higher temperatures (37 °C). Similar microgel-reinforced systems were reported, composed of hyaluronic acid–glycidyl methacrylate (HA–GMA) matrix and covalently linked ethylene-oxide (EO)-based microgels [9], and HA hydrogel network and covalently integrated surface-modified HA-microgels [10,11]. However, the mechanical properties of the hydrogels did not benefit from the presence of microgels in these cases, which is apparently related to the quite high distance between the embedded particles.

Temperature-induced change in the mechanical properties was found for polyacrylamide (PAAm)-based composite hydrogels noncovalently filled with pNIPAm microgels [12] and for pNIPAM-based composite hydrogels covalently filled with pNIPAm nanogels [13]. At high temperatures ($t >$ volume phase-transition temperature—VPTT—of the microgels), the collapse of the filler microgel particles led to a change in their nature from a soft to a hard filler.

As mentioned above, DX microgels are formed via direct interaction between the microgel particles in a swollen state, provided that their surface groups and dangling polymer chains are close enough to establish a link [7]. DX microgels were first described in 2000 by Hu et al. [14]. Their system was composed of epichlorhydrin-crosslinked pNIPAm-*g*-pAAc microgels. In this case and in many later works, secondary crosslinking of the microgels was done covalently, by functionalizing them with vinyl groups and using free-radical chemistry [15,16]. Such DX microgels were used as injectable formulations to reinforce degenerated intervertebral disc tissue [16]. The microgel crosslinking was performed directly in vivo and the resulting hydrogels had remarkably high toughness ($G' = 0.7$–1.3×10^5 Pa). However, because of the permanent character of the secondary crosslinking, such DX microgels lack reversibility and self-healing capacity. The aim of the present work is to design reversible DX microgels using noncovalent crosslinking based on host–guest interactions.

The introduction of cyclodextrins (CDs) as structural units of both chemical and physical hydrogels proved to be beneficial given the ease and versatility of CD functionalization (via multiple primary and secondary hydroxyls), their high biocompatibility and the opportunity to exploit the cyclodextrin–hydrophobic guest inclusion chemistry [17]. Over the last decade, a large variety of complex supramolecular hydrogels containing CDs have been developed [3]. In some of the recent examples, βCD-Ada host–guest chemistry (HG) was used as a means to crosslink soft nanoscale objects into noncovalent 3D hydrogel networks. Indeed, adamantyl moieties are very often used as guests because of their strong affinity toward βCD hosts (schematic structures in Figure S1). For instance, Himmelein et al. used hydrophobically self-assembled cyclodextrin vesicles as multivalent 3D joints which were converted to a gel by addition of adamantane-modified hydroxyethylcellulose [18]. Due to its reversible nature, their system showed shear-thinning/self-healing properties and was injectable through a syringe. Recently, pNIPAM-based microgels have been surface modified with host and guest moieties, and it has been shown that the microgels were able to selectively interact with each other via a lock-and-key mechanism leading to the design of self-sorting colloidal systems [19,20].

Kardos et al. have recently developed [21] a novel approach to prepare core-shell pNIPAm microgels with unrestricted shell composition, for example, microgel particles with pNIPAm core and 100% AAc shell. The pure polyelectrolyte shell composition was aimed at facilitating the strong interaction and complex formation with various cationic agents and at making the pH-induced changes in phase behavior more pronounced.

In this paper, we aimed at preparing microgel networks, which are crosslinked via host–guest interactions. Electrostatic interactions between pNIPAm–shell–pAAc microgels and positively charged β-cyclodextrin polymers (pCD) allowed the formation of pNIPAm/pCD core-shell microgel complexes. These βCD-coated microgels were subsequently used as a building block for hierarchical DX microgel formation by noncovalent crosslinking with adamantane-substituted dextrans (Dex-Ada). The Dex-Ada polymer was expected to establish interparticle connections between the microgels through the βCD-Ada inclusion complexation (Figure 1).

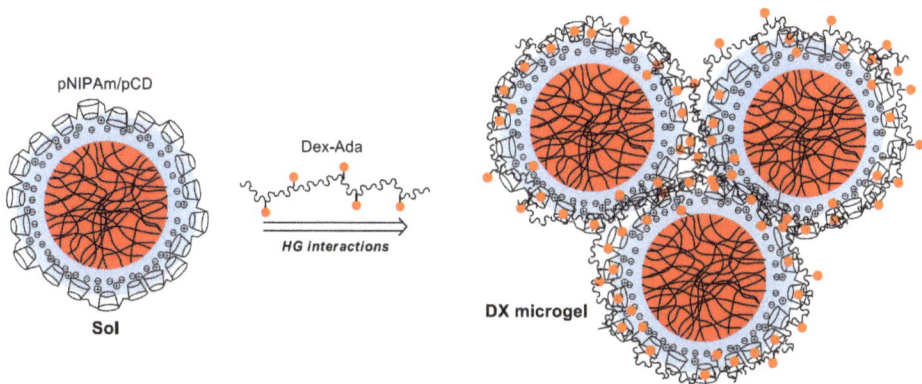

Figure 1. Schematic illustration of the strategy for host–guest-driven crosslinking of pCD-coated pNIPAm core (red)/AAc shell (blue) microgels into a 3D DX microgel network.

2. Experimental Section

2.1. Materials and Reagents

N-isopropylacrylamide (NIPAm), methylenebisacrylamide (BA), ammonium persulfate (APS), acrylic acid (AAc), 4-dimethylaminopyridine (DMAP), pyridine, 1-adamantanecarbonyl chloride, anhydrous grade *N,N*-dimethylformamide (DMF) and sodium dodecyl sulfate (SDS) were purchased from Sigma-Aldrich. *N*-isopropylacrylamide was recrystallized from hexane, methylenebisacrylamide was recrystallized from acetone and kept in a freezer usually for a few days before they were used for the synthesis of the microgel particle. To remove the inhibitor from the acrylic acid it was vacuum distilled and used immediately. Lithium chloride (Sigma-Aldrich, Lyon, France) and dextrans (M_w 115 kDa (Dex$_{115}$), 500 kDa (Dex$_{500}$), Amersham, Umeå, Sweden) were dried overnight under vacuum at 80 °C. All other materials were used as received. All solutions were prepared in ultraclean Milli-Q water (total organic content = 4 ppb; resistivity = 18 mΩ·cm, filtered through a 0.2 μm membrane filter to remove particulate impurities).

2.2. Microgel Synthesis

Microgels composed of crosslinked poly(*N*-isopropylacrylamide) (pNIPAm) core and 100% poly(acrylic acid) (pAAc) shell were prepared using a semibatch precipitation polymerization technique developed by Kardos et al. [21]. In a typical polymerization reaction, a calculated amount of NIPAm was dissolved in Milli-Q water. Calculated volumes of a BA crosslinker stock solution and SDS stock solution were added to the NIPAm solution and the total volume of the mixture was adjusted to reach the desired final volume. The solution was introduced in a double-wall Pyrex glass reactor and it was stirred vigorously. To keep the temperature of the reaction mixture at constant 80 °C, the outer shell of the reactor was connected to a temperature bath and controlled temperature water was circulated in it. The reaction mixture was degassed by purging it with nitrogen for 60 min. Then, the reaction was initiated by adding a small volume of aqueous APS solution to the reactor. After reaching 95% conversion of the NIPAm monomer, the AAc monomer was fed into the reaction mixture (using a calculated volume of degassed aqueous AAc stock solution) to form the pure AAc-shell on the crosslinked pNIPAm microgel core. The final product, pNIPAm–shel–100%pAAc microgel, was purified from unreacted monomers and polymeric byproducts by ultracentrifugation (Beckman Optima XPN ultracentrifuge, 362,000 g), decantation, and redispersion. The centrifuged microgels were redispersed in Milli-Q water and the cycle was repeated up to 5 times. Finally, the purified microgels were freeze-dried and stored in a freezer before further use. The acrylic acid content of the prepared microgel particles was determined by conductometric titration in nitrogen atmosphere.

2.3. Cationic Poly(β-cyclodextrin) (pCD) Synthesis

pCD was prepared in a two-step procedure according to a previously described method [22,23]. Briefly, neutral epichlorohydrin-βCD polymers were first prepared by reacting β-cyclodextrin and epichlorohydrin in alkaline media. Then, these polymers were cationized by reaction with glycidyltrimethylammonium chloride. The success of the reaction was confirmed by ^{1}H NMR (Bruker, Champs sur Marne, France, Avance II Ultrashield Plus 400 MHz NMR spectrometer). ^{1}H NMR was also used to determine the βCD content and the number of cationic groups per cyclodextrin (N+/CD) in the prepared pCD polymer, using the approach previously described by our group [24]. The molecular weight (M_w) was determined by size-exclusion chromatography. Size-exclusion chromatography coupled to multi-angle laser light scattering (SEC-MALLS) was performed in deionized water with $0.1\ mol\cdot L^{-1}$ $LiNO_3$ (0.05% NaN_3) on TSK-gel type SW4000-3000 columns and detection by a Wyatt Dawn 8+ light-scattering detector and a Wyatt Optilab Rex refractive index detector.

2.4. Adamantane-Modified Guest Polymers

Adamantane grafted dextrans (Dex-Ada) were prepared via esterification reaction according to the procedure described elsewhere [25]. In a typical procedure, calculated amounts of dextrans with molecular weights of either Dex_{115} or Dex_{500} and lithium chloride were dissolved under stirring at 80 °C in anhydrous DMF. After the addition of calculated amounts of DMAP, pyridine and 1-adamantanecarbonyl chloride, the reaction mixture was left for 3 h at 80 °C and 15 h at room temperature. The polymers were isolated by precipitation into 2-propanol and filtration on sintered-glass funnel. The pure Dex-Ada samples were obtained by dialysis of the polymers' concentrated water solutions against water, followed by freeze-drying. The degree of substitution by adamantyl groups in mole % per glucose unit was determined by ^{1}H NMR in deuterated dimethylsulfoxide from the ratio of the integration of the protons of adamantyl groups and of the integration of anomeric and hydroxylic protons, as described in our earlier work [26].

2.5. Preparation of βCD-Coated Microgel Particles

(pNIPAm–shell–pAAc/pCD core-shell microgel complexes): In a typical procedure, 2 mL of pNIPAm–shell–pAAc microgel solution at 0.145 or 0.29 mM of negative charges (acrylic acid monomer concentration) were mixed with 2 mL of pCD solutions with calculated concentration of positive charges. The pCD solution was added to the stirred microgel solution by an automatic pipette and the mixture was stirred for an additional 2 min at 600 rpm at room temperature (pH = 7.0).

2.6. Supramolecular Hydrogel Preparation

Aqueous dispersions of pNIPAm–shell–pAAc/pCD (c_{AAc} = 6 mM, c(+)/c(−) = 0.8–2.0) coated microgels were mixed with aqueous solutions of Dex-Ada at pH 7 at room temperature. The obtained mixtures were then concentrated under vacuum at 55 °C (for further details see Discussion). Transition from turbid liquids to transparent gels occurred upon cooling the concentrated samples to 25 °C.

2.7. Methods and Instrumentation

2.7.1. Rheology Measurements

Rheological characterization of the hydrogels was performed with a DHR-2 Rheometer (TA Instruments, Guyancourt, France), equipped with a 1° cone geometry of 40 mm diameter. An insulated ring was placed around the geometry to prevent water evaporation. The hydrogel sample under investigation (V = 400–500 µL) was placed on the center of the Peltier plate with a spatula or with a pipette (in the case of weak gels or viscous liquids) and the geometry was applied. Oscillatory frequency sweep measurements at 20 °C and 0.1% strain were performed in order to evaluate storage G′ and loss G″ moduli as a function of frequency in the range from 0.01 to 100 Hz. Limits of linear viscoelasticity regime were studied by strain amplitude sweep experiments with the strain amplitudes varying from 0.001% to 1000% (f = 10 Hz). Temperature sweep experiments from 25 to 62 °C were performed at a heating rate of 2 °C/min (f = 10 Hz, γ = 0.1%). In heating–cooling cycles experiment G′ and G″ were monitored as a function of time at two alternately changing temperatures, 25 and 50 °C (f = 10 Hz, γ = 0.1%).

2.7.2. Dynamic Light Scattering (DLS)

Particle size and polydispersity were determined by DLS in pure water, microgel concentration of 10–20 ppm and temperature 25 °C. The measurements were performed with a Brookhaven Instruments (Holtsville, NY, USA) device, which consists of a BI-200SM goniometer and a BI-9000AT digital autocorrelator. A Coherent Genesis MX488-1000 STM laser was used as a light source. The laser was used at a wavelength of 488.0 nm and it emitted vertically polarized light. The autocorrelator was set in a "multi τ" mode; that is, the time axis was logarithmically spaced to span the required correlation time range. The autocorrelation functions were measured at a detection angle of 90° with a 100 µm pinhole size. The obtained autocorrelation functions were then analyzed by a second-order cumulant and the CONTIN methods. The extracted diffusion coefficients (D) were converted into hydrodynamic diameters d_h using the Stokes–Einstein equation.

2.7.3. Electrophoretic Mobility

The electrophoretic mobility measurements were performed using a Malvern Zetasizer NanoZ instrument. The data were analyzed using the M3-PALS technique. The standard error in the values of the electrophoretic mobility was around 10%, and measurements were always performed on freshly mixed samples.

2.7.4. Cryo-Transmission Electron Microscopy (Cryo-TEM)

Cryo-TEM experiments were performed in Sorbonne University, Institut de Minéralogie, de Physique des Matériaux et de Cosmochimie, with a kind assistance by Jean-Michel Guigner. The images were recorded on an Ultrascan 2k CCD camera (Gatan, Pleasanton, CA, USA), using a LaB6 JEOL JEM 2100 (JEOL, Akishima, Tokyo, Japan) cryo-microscope operating at 200 kV with a JEOL low-dose system (Minimum Dose System, MDS) to protect the thin-ice film from any irradiation before imaging and to reduce the irradiation during the image capture. The images were recorded at 93 K. The samples were prepared as follows. A drop of the microgel complexes solution at pH 2 was deposited on a Quantifoil grid (Micro Tools GmbH, Jena, Germany). The grids were previously treated by glow-discharge in the presence of argon to increase their surface hydrophilicity. The excess of solution was then blotted out with a filter paper, and before evaporation the grid was quench-frozen in liquid ethane to form a thin vitreous ice film. The grid was mounted in a Gatan 626 cryo-holder cooled with liquid nitrogen and transferred in the microscope.

3. Results and Discussion

3.1. Preparation of Host-Molecule Functionalized Microgels

To facilitate the formation of 3D host/guest macrogels using pNIPAm microgels as the building blocks we prepared host-molecule-decorated microgels and guest-molecule-grafted polymers. We hypothesized that in the mixture of these materials, the host-molecule-decorated microgels can act as multivalent crosslinkers and enable the formation of a 3D gel matrix. However, it could be expected that efficient gel formation can take place only if the surfaces of the microgel beads are functionalized by the host molecules in high-enough density. To achieve high βCD coverage on the pNIPAm microgel surface, we used polyelectrolyte complex formation between pNIPAm–shell–AAc microgels and cationic β-cyclodextrin polymers (pCD, M_w = 238 kDa, with 1.62 positive charge per βCD). The characteristics of these compounds are summarized in Table 1 where c_{MG}, c_{AAc}, c_{CD}, c_{N+}, c_{Ada} are the concentrations in microgel, AAC, CD, N+ and Ada respectively and n_{N+}/n_{CD} is the molar ratio of N+ over CD. The main advantage of this approach is that the interaction of the highly charged non-crosslinked acrylic acid shell of the microgel and the small positively charged pCD coils leads to the accumulation of the pCD on the surface of the microgel and exposes the βCD moieties for host–guest interactions.

Table 1. Characteristics of the synthetized polymers and microgels.

Microgel (MG)	c_{AAc} in c_{MG}	d_h (nm) (pH 7, 25 °C)		
pNIPAm-shell-100%pAAc	1 mM in 0.40 g·L^{-1}	450		
Host polymers	c_{N+} in c_{pCD}	M_w (kDa)	c_{CD} (wt %)	n_{N+}/n_{CD}
pCD(1.6N+)	1 mM in 1.20 g·L^{-1}	238	58.2	1.62
pCD(3.2N+)	1 mM in 0.68 g·L^{-1}	268	51.8	3.24
Guest polymers	c_{Ada} in c_{DexAda}	M_w (kDa)	c_{Ada} (mol %)	
Dex$_{115}$-Ada$_5$	1 mM in 3.68 g·L^{-1}	115	4.7	
Dex$_{500}$-Ada$_6$	1 mM in 2.66 g·L^{-1}	533	6.5	

In order to utilize the pCD-covered microgels in host–guest gel formation, first we had to understand the phase behavior of the microgel/pCD mixtures. We found that upon mixing the microgels and pCD dilute solutions (60 ppm for the microgel) in stoichiometric ratio (1:1 ratio of negative charges of the pAAc monomers to the positive charges of the pCD), fast aggregation of the components took place resulting in a high-turbidity mixture. The turbidity of the samples decreased dramatically upon aging, as precipitate settled to the bottom of the container leaving a polymer-free transparent supernatant. Electrophoretic mobility measurements indicated that the fresh aggregates formed during mixing were charge–neutral, confirming the formation of colloidally unstable complexes.

However, when pCD was added to the microgel solution either in excess or in lower-than-stoichiometric positive-to-negative charge ($c(+)/c(-)$) ratios, the turbidity quickly decreased and transparent, stable mixtures formed. In the case of microgel excess, the $c(+)/c(-)$ ratio had to be decreased to 0.8 or below to avoid precipitation. These samples had negative electrophoretic mobility indicating the excess of polyacrylic acid chains in the shell of the microgel/pCD complexes. In the case of pCD excess, the ($c(+)/c(-)$) ratio had to be increased to 1.5 to avoid precipitation. The microgel/pCD complexes had a positive electrophoretic mobility (~1×10^{-8} m^2·V^{-1}·s^{-1}) showing that pCD is in excess of pAAc in the microgel shell. However, if the positive-to-negative ratio was further increased to 2, the electrophoretic mobility could be increased significantly further (~2×10^{-8} m^2·V^{-1}·s^{-1}), demonstrating the formation of a more robust pCD shell on the microgel particles. To confirm that in excess pCD the formed microgel/pCD complexes are present as individual particles, cryo-transmission electron microscopy (cryo-TEM) experiments were performed. As shown in Figure 2, in the case of excess pCD (2:1 positive-to-negative charge ratio), no sign of aggregation can be detected, but well-separated individual particles are present in the sample.

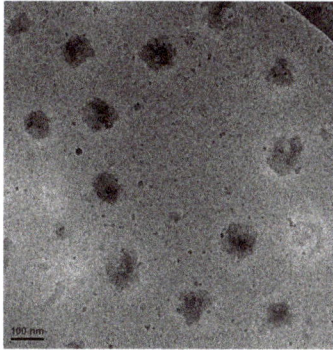

Figure 2. Cryo-transmission electron microscopy (cryo-TEM) image of microgel/pCD complexes prepared using pCD excess (c(+)/c(−) = 2). The scale bar indicates 100 nm.

3.2. Hydrogel Formation

Having established that stable microgel/pCD complexes can be prepared, we aimed at using these complexes for the preparation of 3D host–guest gels. We aimed at using the pCD-coated microgels as multivalent crosslinkers of adamantane-grafted dextran (Table 1). It can be expected that gels could form only if the adamantane-grafted dextran can bridge the gap between the neighboring microgels and strong (multiple) bonds can form between the microgel particles. To meet these requirements, the volume fraction of the microgel particles has to be high enough to ensure the close proximity of the microgels. According to Senff and Richtering, the rheological properties and phase behavior of pNIPAm microgel suspensions are strongly dependent on their effective volume fraction (ϕ_{ef}) [27]. For instance, by measuring zero shear viscosity of microgels at different temperatures below lower critical solution temperature (LCST) as a function of ϕ_{ef}, they observed a sharp increase in viscosity at $\phi_{ef} > 0.5$ [28]. The phenomenon was ascribed to the microgel soft spheres being close to contact and interacting with each other at these concentrations. Hence, we decided to use microgel samples for the 3D gel formation with this effective volume fraction ($\phi_{ef} = 0.5$).

Previously it was found by dynamic light-scattering measurements [29] that the molar weight of microgel particles (MG) prepared with similar collapsed size (d_h = 150 nm determined by DLS at 40 °C) always exceeded the value of M_{MG} ~10^8 g·mol^{-1}. Using this molecular weight as a lower limit and the hydrodynamic diameter of the pCD-coated microgels measured in their swollen state (305 nm at 25 °C), we could calculate a lower limit for the microgel concentration in terms of dry weight per solution volume required to reach the desired effective volume fraction (ϕ_{ef} = 0.5). Using this simple estimation, we found that 6.0 g/L microgel concentration (15 mM for acrylic acid monomer concentration) should be sufficient in the final mixture to ensure the close proximity of the microgel beads required for gel formation.

To connect the neighboring microgel particles by multiple host–guest interactions, preferably large-molecular-weight Dex-Ada polymer has to be used with high-enough graft density of adamantane in high-enough concentration. To meet these requirements, we used dextran molecules with a molecular weight of 115 kDa (~700 glucose units) and 4.7% graft density (ca. 30 adamantane moieties on each polymer chain), and the concentration of the Dex-Ada polymer was chosen to ensure at least 1:1 adamantane:CD ratio in the samples.

Unfortunately, when we tried to prepare the 3D gel network by the direct mixing of the concentrated components at room temperature, the loss of colloid stability of the system resulted in precipitation. To circumvent this limitation, an indirect gelation approach was applied. In the first step, stable β-cyclodextrin polymer-coated microgel particles (MG/pCD) were prepared at lower concentrations (typically c_{AAc} = 6 mM, c(+) = 4.8–12 mM in the final mixture) at room temperature. These coated microgels were mixed with the Dex-Ada polymer solution at close to stoichiometric Ada/CD molar ratios. Then, the resulting homogenous mixtures were concentrated in vacuum at 50 °C to reach the desired 6.0 g/L microgel concentration. After cooling down the concentrated mixtures to room temperature, the pNIPAm core of the MG/pCD complexes could reswell, leading to the formation of final samples.

First, we investigated the effect of the pCD coverage of microgel particles on the gel formation. We used the three different surface coverages providing colloidally stable MG/pCD complexes: a low pCD coverage that provides stable complexes with acrylic acid excess in the shell (c(+)/c(−) = 0.8); a pCD coverage that provides stable complexes with a slight pCD excess in the shell (c(+)/c(−) = 1.5); and a full pCD excess where electrophoretic mobility reaches the plateau for the MG/pCD complexes (c(+)/c(−) = 2). In each case we used the same final microgel concentration (c_{tot} ~6 g/L), while the ratio of the CD and the adamantane moieties was kept at a fixed value (n_{Ada}/n_{CD} ~1.3). For further details see Table 2. These experiments indicated that the pCD coverage of the microgel particles has a profound effect on the gel formation. In the case of the lowest pCD coverage (*G1*), precipitation occurred during the sample concentration. This can be explained by the loss of the colloid stability of the complex particles, whose charge could not counteract the increased van der Waals interaction among the collapsed complex particles (T > LCST). In the case of the overcharged complexes (excess pCD), no precipitation was observed and stable concentrated samples could be produced. At the same time, the lower pCD coverage (*G2*: c(+)/c(−) = 1.5) resulted in a viscous liquid sample, while the sample with the largest pCD coverage (*G3*) behaved as a physical gel (see insets in Figure 3). This indicates that although the MG/pCD complexes preserved their colloid stability in both cases during sample concentration, connectivity among the microgel particles could reach a percolation threshold only in the latter case due to the larger number of links performed by the adamantyl units, increasing the amount of Dex-Ada polymer being needed to ensure the constant adamantane/cyclodextrin ratio in the samples.

Table 2. Molar ratios and weight concentrations of the components in supramolecular gel samples prepared from microgels (MG), pCD host polymers and Dex-Ada guest polymers.

Sample Code	Components	c(+)/c(−)	n_{Ada}/n_{CD}	c_{MG} (wt %)	c_{pCD} (wt %)	$c_{Dex-Ada}$ (wt %)	c_{tot} (wt %)	Observation
G1	MG/pCD(1.6N+)/Dex$_{110}$Ada$_5$	0.8	1.25	0.6	1.4	3.4	5.4	phase separation
G2	MG/pCD(1.6N+)/Dex$_{110}$Ada$_5$	1.5	1.25	0.6	2.6	6.3	9.4	viscous liquid
G3	MG/pCD(1.6N+)/Dex$_{110}$Ada$_5$	2.0	1.25	0.5	3.3	8.1	11.9	gel
R1	MG	-	-	10.1	0	0	10.1	liquid
R2	MG/pCD(1.6N+)	2.0	-	1.5	9.0	0	10.5	phase separation
G4	MG/pCD(3.2N+)/Dex$_{110}$Ada$_5$	2.0	1.0	1.5	5.0	8.4	14.9	liquid
G5	MG/pCD(3.2N+)/Dex$_{500}$Ada$_6$	2.0	1.0	1.3	4.4	5.3	11.0	gel
R3	pCD(3.2N+)/Dex$_{500}$Ada$_6$	-	1.0	0	4.3	5.2	9.5	gel

To confirm that the gel formation is indeed related to the host–guest complex formation, we prepared two control samples without the addition of the guest polymer (Table 2). Since it has been reported in the literature that uniform pNIPAm microgel dispersions with concentrations corresponding to close-packing (~5 wt %) undergo sol–gel transition below the LCST [30], we also concentrated the uncoated pNIPAm–shell–pAAc microgels to test if they can give a sol/gel transition in the investigated concentration range. This experiment indicated that the core/shell microgel particles did not form a gel even at concentrations as high as 10.1 wt % (*R1*), where the acrylic acid shells of the microgel particles strongly interpenetrate each other (semidilute systems). This can be explained by the strong electrostatic repulsion of the highly charged acrylic acid chains in the shell of the microgel particles (pH = 7).

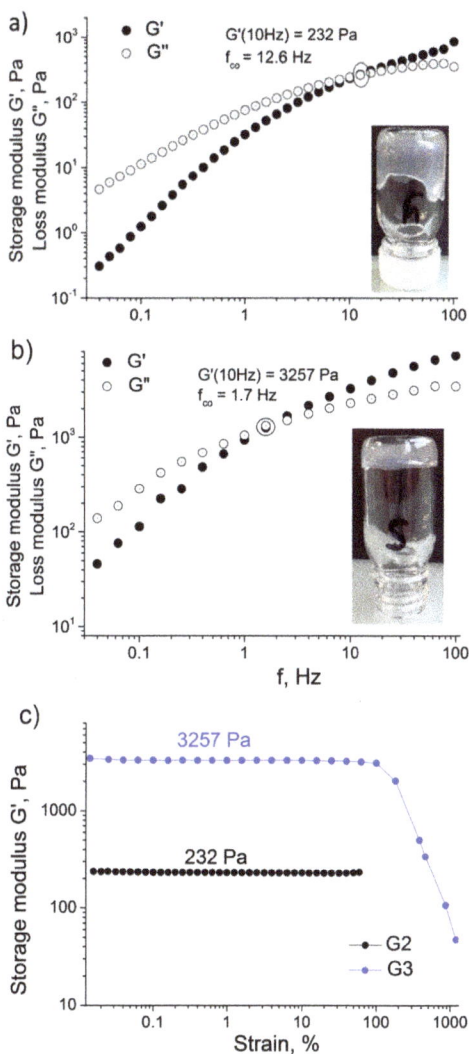

Figure 3. Oscillatory rheological measurements of hydrogel samples with varying pCD coverage of the microgel particles. Storage G′ and loss G″ moduli obtained from frequency sweep performed at 0.1% strain for (**a**) *G2*; (**b**) *G3*; (**c**) Storage modulus G′ of *G2* and *G3* samples obtained from amplitude-strain sweep performed at $f = 10$ Hz. All measurements were performed at 25 °C.

In addition, the bare MG/pCD complexes (c(+)/c(−) = 2) were also concentrated to 10.5 wt % (*R2*), which resulted in the precipitation of the samples. The observed precipitation is a consequence of the loss of the colloid stability of the collapsed MG/pCD complexes. This clearly indicates that the colloid stability of the microgel complexes observed during the concentration of *G2* and *G3* samples is provided by the adamantane-grafted dextran chains and not exclusively by the charge of the MG/pCD complexes. Thus we can conclude that the host–guest complex formation indeed took place, and guest-polymer-wrapped hierarchical complexes formed upon mixing.

To get a better insight into the mechanical characteristics of the prepared samples, the viscoelastic properties of G2 and G3 were studied by oscillatory rheological measurements. Frequency sweep measurements of storage G′ and loss G″ moduli are shown in the two upper panels of Figure 3. The G′ and G″ values prove to be frequency dependent in both cases, which was previously reported as a typical behavior for associating systems [31–33]. At higher frequencies and shorter observation times, materials show gel character with the elastic component G′ dominating the viscous component G″, while at lower frequencies the system starts to behave as a viscous liquid with G″ > G′. In a number of earlier studies on host–guest polymeric hydrogels, such behavior is explained by the weak and reversible character of the host–guest links, which are being broken and reformed under applied stress [32,34,35].

The finite lifetimes of the interpolymer crosslinks are due to the fluctuations in the relative kinetic energy of the interacting adamantane and βCD-groups. These lifetimes (τ_{co}), also known as relaxation times, may be estimated from the frequency (f_{co}) of the crossover point of G′ and G″ in the frequency sweeps:

$$\tau_{co} = (2\pi \cdot f_{co})^{-1}. \tag{1}$$

f_{co} should decrease with the number of host–guest links between individual microgels. Indeed, the G2 sample, which behaves as a viscous liquid (Figure 2a), has the higher f_{co} value of 12.6 Hz (τ_{co} = 0.013 s), whereas in the case of G3, which behaves as a gel, the f_{co} is shifted down to 1.7 Hz (τ_{co} = 0.094 s) (Figure 2b). The strength of the hydrogel structures was also studied by strain-amplitude sweep performed at f = 10 Hz (Figure 2c). The plateau G′ moduli vary from 232 Pa for G2, behaving as a viscous liquid, to 3257 Pa for the G3 hydrogel. G2 and G3 exhibit quite high strain resistance and maintain their structure up to γ ~100% (Figure 2c). Similar values of elastic storage modulus were found for recently described cyclodextrin-based hierarchical hydrogels such as βCD-coated quantum dots crosslinked with azobenzene-modified thermoresponsive copolymers (G′ = 200–400 Pa, c_{tot} = 12.5 wt % and T > 40 °C) [36] or βCD vesicles interconnected with Ada-modified cellulose polymer (G′ = 200–400 Pa, c_{tot} = 2–3 wt %) [18].

We also attempted to fit the frequency sweep data of the samples by the Maxwell model. This basic viscoelasticity model was widely used for describing the behavior of entangled physically bonded polymer networks [37,38] and polymeric host–guest hydrogels [35]. The Maxwell model implies that storage G and loss G″ moduli of a viscoelastic body vary as a function of the strain angular frequency ω according to the following equations:

$$G'(\omega) = \frac{G_0(\omega\tau)^2}{1 + (\omega\tau)^2}, \tag{2}$$

$$G''(\omega) = \frac{G_0\omega\tau}{1 + (\omega\tau)^2}, \tag{3}$$

where ω is the angular frequency [rad·s^{-1}], τ is the relaxation time and G_0 is the limiting value of G′ at high frequencies where it typically reaches a plateau. From the equations above, it also follows that $G'/G'' = \omega\tau$. In our case, τ was obtained from the G′ and G″ crossover frequency value and was set as a fixed parameter during the fitting. Since G′ did not reach the plateau in the available range of angular frequencies, G_0 was set as a free optimized parameter. Despite the viscoelastic character of the samples evidenced by the frequency sweep data (Figure 3), rather poor fits to the Maxwell model were obtained. An example of such a fit for the G3 sample is illustrated in Figure S3. Only the low-frequency part (0.1–1.6 rad·s^{-1}) of the loss modulus G″ curve showed a reasonable agreement with the experimental data. The failure of the fitting might be related to the oversimplifications of the Maxwell model, which is suitable only for the description of the viscoelastic materials with a single and well-defined relaxation time, τ. For instance, van de Manakker et al. obtained relatively good fits for host–guest hydrogel networks composed of linear and four-arm star PEG polymers end-functionalized with βCD and cholesterol groups [35]. However, they found that the fitting to the Maxwell model was no longer possible for structurally more complex hydrogels composed of eight-arm PEGs with the same end-functionalities. They ascribed this discrepancy to the presence of a broader range of relaxation mechanisms with different relaxation times involved in the stress relaxation in the latter case. We assume that the same arguments hold for the structurally rather complex pNIPAm/pCD/DexAda doubly crosslinked microgels. Indeed, a good fit of the frequency sweep data has been obtained using a generalized Maxwell model using a distribution of relaxation times (Figure S3).

3.3. Influence of Host and Guest Polymer Characteristics on the Gel Formation

To develop a better understanding of the key parameters controlling the supramolecular gel formation, we also tested the effect of the host polymer (pCD) charge density on the gel formation. We used a host polymer to prepare the MG/pCD complexes, which had twice-as-large charge density (pCD3.2N+) as the host polymer used in the previous experiments (pCD1.6N+). This meant that while we could prepare the MG/pCD complexes with the same overall charge and colloid stability, the number of the CD groups present in the shell could be halved. The prepared complexes indeed preserved their colloid stability, however, when the system was concentrated in the presence of stoichiometric amount of guest polymer ($n_{Ada}/n_{CD} = 1$), gel formation could not be observed. To facilitate the gel formation, we repeated the same experiment with doubling the MG concentration (*G4* in Table 2), thus ensuring that the same amount of adamantane moieties were present in the sample as in the case of the successfully gelled *G3* sample, while the increased microgel concentration resulted in tighter contact of the microgel particles. Surprisingly, the mixture remained a viscous liquid even at as high a total polymer content as $c_{tot} = 14.8$ wt %. Taking into account that doubling the charge density of the pCD polymer while keeping the charge ratio constant ($c(+)/c(−) = 2$) decreased the surface concentration of the pCD coverage by half, we hypothesized that the lack of gel formation was related to the decreased number of host/guest interactions of the guest polymers bridging the microgel particles.

In an attempt to overcome this problem, we used a guest polymer with increased number of grafted adamantane groups. This was achieved by using a larger-molecular-weight and graft density guest polymer (see $Dex_{500}Ada_6$, in Table 1) as a crosslinker. This polymer had ~200 grafted adamantane groups compared to the 32 groups of the previously used polymer ($Dex_{110}Ada_5$). This approach proved successful: A stable transparent gel was formed at $c_{tot} = 11.0$ wt % (*G5* in Table 2), confirming that the number of host–guest interactions established by the bridging polymer chains by the connected microgel particles is a key parameter in gel formation.

We also made frequency sweep rheological measurements on the prepared gel sample (*G5*). The measured curves (Figure S4) had the same qualitative features as measured previously for the *G3* sample (Figure 3), but we found a lower crossover frequency ($f_{co} = 1.1$ Hz compared to the 1.7 Hz observed for *G3*), which represents a slower relaxation time ($\tau_{co} = 0.145$ s vs. $\tau_{co} = 0.094$ s for *G3*). This observation is in good agreement with the expected increased cooperativity of the interactions.

3.4. Thermoresponsive Behavior of the Supramolecular Gels

Since the obtained supramolecular hydrogels are composed of crosslinked thermoresponsive pNIPAm beads, their swelling and mechanical properties are expected to be temperature dependent. Previously, it has been found that hydrogels filled either noncovalently or covalently with pNIPAm microgels [12,13] became mechanically more robust at high temperatures ($t >$ volume phase-transition temperature) due to changing the nature of the microgel particles from a soft to a hard filler. It has also been shown that the nanostructured gels formed by chemically crosslinked pNIPAm microgels or by metal–ligand interactions [39] shrunk with increasing temperature, and their volume deswelling closely mimicked the volume deswelling of the microgel building blocks.

Interestingly, the supramolecular gel samples we prepared did not shrink as a uniform body on temperature increase. Instead they remained a single-phase system with practically constant volume, and the gel sample turned into a liquid system at high temperatures. In order to follow the gel–sol transition of the supramolecular hydrogels, the storage and loss moduli were recorded as a function of temperature between 25 and 60 °C for both *G3* and *G5* (Figure 4). The measurements provided very similar results for the two gel samples. Both G′ and G″ decreased monotonously with increasing temperature. Furthermore, in the temperature range where the volume phase transition of the microgel particles occurs (~30 °C < t < 35 °C), the decrease of the storage and loss moduli accelerated then leveled off again, giving rise to a step-like decrease of the moduli. However, it is interesting to note this step-like decrease was significantly larger for G′ than for G″, decreasing the gap between the two curves. Since G′ decreased faster than G″, at some point G″ became larger (crossover temperature—T_{co}), and the gel-to-sol phase transition occurred. It should also be noted that in the case of the *G3* sample, the crossover temperature was in a good agreement with the end of the temperature range where the pNIPAm/pCD microgels collapsed (VPTT: 37 °C), however, in the case of the *G5* sample, the crossover temperature was shifted to a higher value (41 °C) and it seemed independent of the microgel collapse. Finally, when the samples were gradually cooled back from 60 to 25 °C, the moduli showed full reversibility; hysteresis could not be observed neither for G′ nor for G″.

Contrary to previous investigations [12,13,39], our measurements clearly showed that the prepared supramolecular gels became softer with increasing temperature, and finally turned into a liquid at the crossover temperature. Since the main difference between the previously investigated samples and our supramolecular gels was that we used host–guest interaction for the crosslinking of the microgel particles, we assumed that the observed temperature dependence was related to the characteristics of the host–guest complex formation. Indeed, the group of B. Ravoo has also reported decreasing mechanical strength in the case of host–guest supramolecular hydrogels connected by vesicles [18].

To confirm this interpretation, we prepared a reference sample that had the same concentrations of the host and guest polymers as *G5* but did not contain microgel particles (*R3*). The storage and loss moduli recorded for this sample as a function of temperature are plotted in panel (c) of Figure 4. Similarly to the microgel-containing samples, G′ and G″ decreased monotonously with increasing temperature. However, the step-like steep decrease of the moduli in the temperature range of the microgel collapse was missing in this case, and as a consequence, the crossover temperature shifted to a much higher value (T_{co} = 51 °C).

Based on these experimental results, it can be concluded that two independent processes take place in the supramolecular gel samples on temperature increase. On the one hand, as the volume phase-transition temperature is approached, the microgel building blocks shrink and become fully collapsed when the VPTT is reached. This should lead to the uniform shrinking of the gel samples and give rise to mechanically more-robust gels. However, this is counteracted by the fact that the host–guest association is an equilibrium process that has a negative enthalpy change (H). Thus, as the temperature is increased, the host–guest complex formation is shifted towards dissociation. This decreases the number of host–guest complexes in the system, and as a consequence, the average strength of the bonds bridging the microgel particles (the average number of host–guest complexes formed by a guest polymer) also decreases. This is evidenced by the continuously decreasing moduli and the gel–sol transition taking place even in the pure mixture of the host and guest polymers. At the same time, in the presence of the microgels, an osmotic effect attributed to the Dex-Ada chains opposes the deswelling of the gel sample. When the microgels shrink, the interstitial volume between the microgels is increased and the Dex-Ada chains accommodate this space to oppose the deswelling. This is the reason why there is no phase separation for these gels, contrary to the covalently crosslinked DX hydrogels [12,13,39]. As a consequence, the local Dex-Ada concentration decreases, inducing a decrease of the number of host–guest complexes. In a previous study of host–guest hydrogels, it has been shown that moduli were decreasing with the concentrations of host and/or guest moieties [29]. As a result, the average number of host/guest complexes formed by a guest polymer connecting the neighboring microgel beads decreases, and the supramolecular gel becomes softer, giving rise to the step-like decrease of moduli as the VPTT is approached.

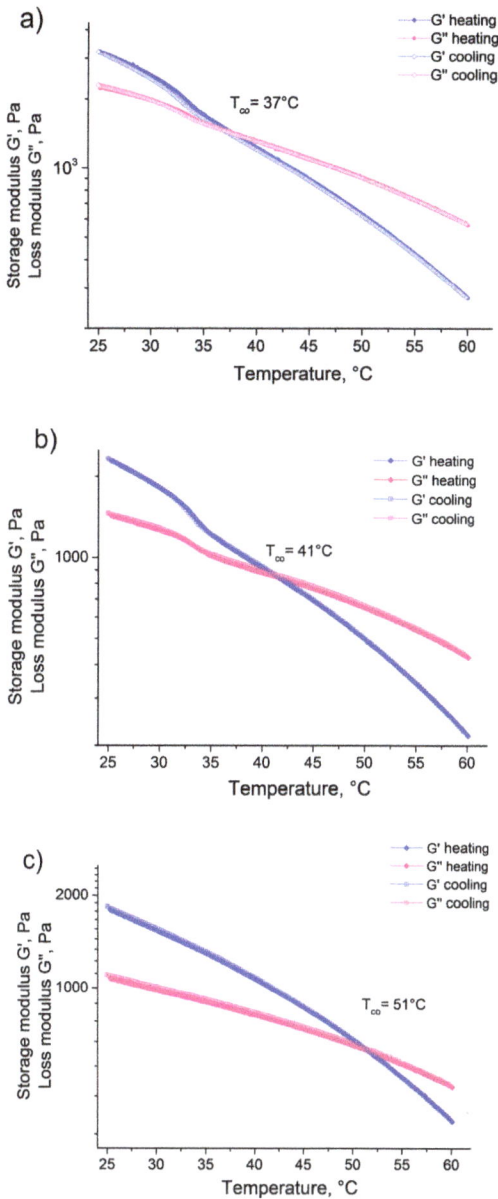

Figure 4. Evolution of storage G′ and loss G″ moduli of supramolecular gels under heating and cooling temperature ramps (temperature change rate = 2 °C/min, f = 10 Hz, γ = 0.1%) for (**a**) *G3*; (**b**) *G5*; (**c**) *R3* hydrogel samples.

The gel-to-sol transition temperature (T_{CO}) corresponds to a temperature at which a percolation threshold in the connected microgel beads is attained. Increasing the number of adamantyl groups on the guest polymer should result in a larger number of host–guest complexes at any temperature, and thus to a shift of T_{CO} to larger temperature. This explains why the T_{CO} of *G3*, made with

Dex$_{110}$-Ada$_5$ (32 adamantyl groups per chain), is lower by 4 °C than T_{CO} of G5 made with Dex$_{500}$-Ada$_6$ (200 adamantyl groups per chain). An important message of this conclusion is that the gel–sol transition could be tuned within a wide temperature range by varying the size and graft density of the guest molecules, as well as by tuning the VPTT of the microgel beads.

To test the response dynamics of the supramolecular gels, the sample was subjected to several heating–cooling cycles, where its temperature was alternated between 25 and 50 °C (Figure 5). As indicated by the figure, the temperature jumps required only a few seconds. G′ and G″ were recorded at a constant frequency (10 Hz). Both moduli changed promptly when the temperature jump was triggered, significant delay in their response to the temperature change could not be observed on the applied timescale. G″ was larger than G′ at 50 °C, whereas the inverse situation was obtained at 25 °C, meaning that a gel-to-sol transition occurred. The cycle was repeated several times, and the initial G′ and G″ values were fully restored in each case, proving the reversibility of the temperature-induced phase transition. One should note that the sample in Figure 5 has the same nominal concentration as G3, but the final concentrated sample was a softer gel at the end due to the uncertainties of the sample preparation. We are working on a more robust and simple sample-preparation method.

The observed fast response of the supramolecular hydrogel to the temperature change is in contrast with the observations made for the chemically crosslinked pNIPAM hydrogels, which have characteristic relaxation times on the order of 100 s under the same geometry conditions [40]. This can be explained by the fact that in this case, the overall volume of the gel does not change, thus the transport of the liquid out of and into the macroscopic gel structure is not required. Instead, the liquid transport is localized to the close neighborhood of the microgel beads and confined to a few-hundred-nanometer length scale. This renders the response time of the system to the subsecond timescale.

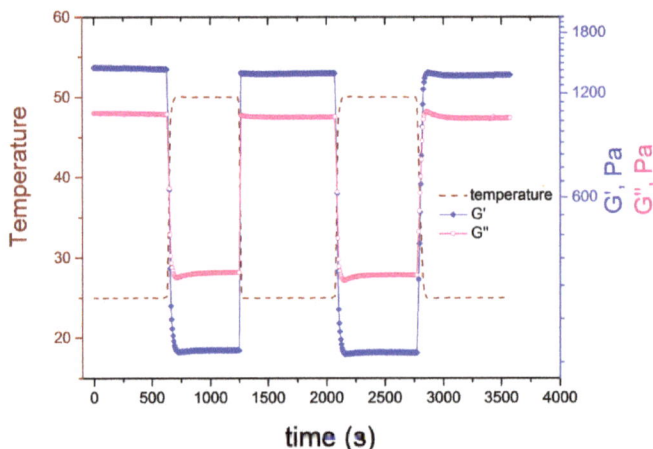

Figure 5. Oscillatory rheological measurements of *G3* under temperature ramps (f = 10 Hz, γ = 0.1%).

4. Conclusions

In summary, novel host–guest supramolecular hydrogels were prepared, containing pNIPAm–shell–pAAc microgels as responsive building blocks. We were able to define the conditions of supramolecular hydrogel formation in terms of the concentrations of the three components: The close contact between the microgel particles has to be ensured ($\phi_{ef} \geq 0.5$), pCD concentrations to give rise to a compact pCD shell formation is required (corresponding to twice the pCD concentration needed for stoichiometric MG/pCD complex formation), and close-to-stoichiometric βCD/Ada ratio is needed.

Polymers **2018**, *10*, 566

To the best of our knowledge, this is the first example of DX microgels noncovalently crosslinked via host–guest interactions.

The material properties were studied with rheology and showed fast mechanical response to temperature variation. An important feature of these materials is their low deswelling sensitivity to temperature increase, contrary to most pNIPAM-based materials. This is due to the osmotic effect exerted by the Dex-Ada chains. Furthermore, the hydrogels exhibit fully reversible temperature-induced gel–sol transition due to the synergetic effect between shrinking of the microgels and dissociation of βCD-Ada crosslinks at higher temperatures. Due to the presence of microgels, the gel–sol transition temperature of pNIPAm/pCD/Dex-Ada is significantly shifted down to the physiological temperature range (37–41 °C) as compared to uniform pβCDN/DT-Ada host–guest hydrogels (51 °C). It opens up attractive prospects of their potential use in biomedical applications.

Supplementary Materials: The following are available online at http://www.mdpi.com/2073-4360/10/6/566/s1, Figure S1: Schematic structure of β-cyclodextrin and adamantyl moieties. Figure S2: Hydrodynamic diameter as a function of temperature for positively overcharged microgels. Figure S3: Frequency sweep data of *G3* fitted with a generalized Maxwell model. Figure S4: Oscillatory rheological measurements of *G5* sample.

Author Contributions: C.A. and I.V. conceived and designed the experiments; A.K. synthetized the core shell microgel, while I.A. prepared the host and a guest polymers. I.A. and D.K. performed all other experiments and analyzed the data.

Acknowledgments: The authors would like to thank Clémence Le Coeur for her help in rheology study and Jean-Michel Guignier for cryo-TEM experiment. This research has received funding from the People Programme (Marie Curie Actions) of the European Union's Seventh Framework Programme FP7/2007–2013/under REA grant agreement N.290251 and from the Hungarian National Research, Development and Innovation Office (NKFIH K116629), which is gratefully acknowledged.

Conflicts of Interest: The authors declare no conflict of interest.

References

1. Lee, K.Y.; Mooney, D.J. Hydrogels for tissue engineering. *Chem. Rev.* **2001**, *101*, 1869–1880. [CrossRef] [PubMed]
2. Balakrishnan, B.; Banerjee, R. Biopolymer-based hydrogels for cartilage tissue engineering. *Chem. Rev.* **2011**, *111*, 4453–4474. [CrossRef] [PubMed]
3. Tan, S.; Ladewig, K.; Fu, Q.; Blencowe, A.; Qiao, G.G. Cyclodextrin-Based Supramolecular Assemblies and Hydrogels: Recent Advances and Future Perspectives. *Macromol. Rapid Commun.* **2014**, *35*, 1166–1184. [CrossRef] [PubMed]
4. Hoare, T.R.; Kohane, D.S. Hydrogels in drug delivery: Progress and challenges. *Polymer* **2008**, *49*, 1993–2007. [CrossRef]
5. Chen, P.C.; Kohane, D.S.; Park, Y.J.; Bartlett, R.H.; Langer, R.; Yang, V.C. Injectable microparticle–gel system for prolonged and localized lidocaine release. II. In vivo anesthetic effects. *J. Biomed. Mater. Res. Part A* **2004**, *70*, 459–466. [CrossRef] [PubMed]
6. Lynch, I.; de Gregorio, P.; Dawson, K. Simultaneous release of hydrophobic and cationic solutes from thin-film "plum-puddinG" gels: A multifunctional platform for surface drug delivery? *J. Phys. Chem. B* **2005**, *109*, 6257–6261. [CrossRef] [PubMed]
7. Richtering, W.; Saunders, B.R. Gel architectures and their complexity. *Soft Matter* **2014**, *10*, 3695–3702. [CrossRef] [PubMed]
8. Sivakumaran, D.; Maitland, D.; Oszustowicz, T.; Hoare, T. Tuning drug release from smart microgel–hydrogel composites via crosslinking. *J. Colloid Interface Sci.* **2013**, *392*, 422–430. [CrossRef] [PubMed]
9. Bencherif, S.A.; Siegwart, D.J.; Srinivasan, A.; Horkay, F.; Hollinger, J.O.; Washburn, N.R.; Matyjaszewski, K. Nanostructured hybrid hydrogels prepared by a combination of atom transfer radical polymerization and free radical polymerization. *Biomaterials* **2009**, *30*, 5270–5278. [CrossRef] [PubMed]
10. Jha, A.K.; Malik, M.S.; Farach-Carson, M.C.; Duncan, R.L.; Jia, X. Hierarchically structured, hyaluronic acid-based hydrogel matrices via the covalent integration of microgels into macroscopic networks. *Soft Matter* **2010**, *6*, 5045–5055. [CrossRef] [PubMed]

11. Jia, X.; Yeo, Y.; Clifton, R.J.; Jiao, T.; Kohane, D.S.; Kobler, J.B.; Zeitels, S.M.; Langer, R. Hyaluronic acid-based microgels and microgel networks for vocal fold regeneration. *Biomacromolecules* **2006**, *7*, 3336–3344. [CrossRef] [PubMed]

12. Meid, J.; Dierkes, F.; Cui, J.; Messing, R.; Crosby, A.J.; Schmidt, A.; Richtering, W. Mechanical properties of temperature sensitive microgel/polyacrylamide composite hydrogels—From soft to hard fillers. *Soft Matter* **2012**, *8*, 4254–4263. [CrossRef]

13. Xia, L.-W.; Xie, R.; Ju, X.-J.; Wang, W.; Chen, Q.; Chu, L.-Y. Nano-structured smart hydrogels with rapid response and high elasticity. *Nat. Commun.* **2013**, *4*, 2226. [CrossRef] [PubMed]

14. Hu, Z.; Lu, X.; Gao, J.; Wang, C. Polymer gel nanoparticle networks. *Adv. Mater.* **2000**, *12*, 1173–1176. [CrossRef]

15. Liu, R.; Milani, A.H.; Freemont, T.J.; Saunders, B.R. Doubly crosslinked pH-responsive microgels prepared by particle inter-penetration: Swelling and mechanical properties. *Soft Matter* **2011**, *7*, 4696–4704. [CrossRef]

16. Milani, A.H.; Freemont, A.J.; Hoyland, J.A.; Adlam, D.J.; Saunders, B.R. Injectable doubly crosslinked microgels for improving the mechanical properties of degenerated intervertebral discs. *Biomacromolecules* **2012**, *13*, 2793–2801. [CrossRef] [PubMed]

17. Liao, X.; Chen, G.; Liu, X.; Chen, W.; Chen, F.; Jiang, M. Photoresponsive pseudopolyrotaxane hydrogels based on competition of host–guest interactions. *Angew. Chem.* **2010**, *122*, 4511–4515. [CrossRef]

18. Himmelein, S.; Lewe, V.; Stuart, M.C.; Ravoo, B.J. A carbohydrate-based hydrogel containing vesicles as responsive noncovalent crosslinkers. *Chem. Sci.* **2014**, *5*, 1054–1058. [CrossRef]

19. Han, K.; Go, D.; Tigges, T.; Rahimi, K.; Kuehne, A.J.; Walther, A. Social Self-Sorting of Colloidal Families in Co-Assembling Microgel Systems. *Angew. Chem. Int. Ed.* **2017**, *56*, 2176–2182. [CrossRef] [PubMed]

20. Han, K.; Go, D.; Hoenders, D.; Kuehne, A.J.; Walther, A. Switchable Supracolloidal Coassembly of Microgels Mediated by Host/Guest Interactions. *ACS Macro Lett.* **2017**, *6*, 310–314. [CrossRef]

21. Kardos, A.; Varga, I. Core-shell pNIPAm microgels with unrestricted shell composition. *Langmuir* **2018**, Submitted.

22. Blomberg, E.; Kumpulainen, A.; David, C.; Amiel, C. Polymer bilayer formation due to specific interactions between beta-cyclodextrin and adamantane: A surface force study. *Langmuir* **2004**, *20*, 10449–10454. [CrossRef] [PubMed]

23. Renard, E.; Deratani, A.; Volet, G.; Sebille, B. Preparation and characterization of water soluble high molecular weight beta-cyclodextrin-epichlorohydrin polymers. *Eur. Polym. J.* **1997**, *33*, 49–57. [CrossRef]

24. El Fagui, A.; Wintgens, V.; Gaillet, C.; Dubot, P.; Amiel, C. Layer-by-Layer Coated PLA Nanoparticles with Oppositely Charged β-Cyclodextrin Polymer for Controlled Delivery of Lipophilic Molecules. *Macromol. Chem. Phys.* **2014**, *215*, 555–565. [CrossRef]

25. Layre, A.-M.; Wintgens, V.; Gosselet, N.-M.; Dalmas, F.; Amiel, C. Tuning the interactions in cyclodextrin polymer nanoassemblies. *Eur. Polym. J.* **2009**, *45*, 3016–3026. [CrossRef]

26. Layre, A.-M.; Volet, G.; Wintgens, V.; Amiel, C. Associative Network Based on Cyclodextrin Polymer: A Model System for Drug Delivery. *Biomacromolecules* **2009**, *10*, 3283–3289. [CrossRef] [PubMed]

27. Senff, H.; Richtering, W. Influence of crosslink density on rheological properties of temperature-sensitive microgel suspensions. *Colloid Polym. Sci.* **2000**, *278*, 830–840. [CrossRef]

28. Senff, H.; Richtering, W. Temperature sensitive microgel suspensions: Colloidal phase behavior and rheology of soft spheres. *J. Chem. Phys.* **1999**, *111*, 1705–1711. [CrossRef]

29. Varga, I.; Gilányi, T.; Mészáros, R.; Filipcsei, G.; Zrínyi, M. Effect of Cross-Link Density on the Internal Structure of Poly(N-isopropylacrylamide) Microgels. *J. Phys. Chem. B* **2001**, *105*, 9071–9076. [CrossRef]

30. Zhao, Y.; Cao, Y.; Yang, Y.; Wu, C. Rheological study of the sol-gel transition of hybrid gels. *Macromolecules* **2003**, *36*, 855–859. [CrossRef]

31. Charbonneau, C.; Chassenieux, C.; Colombani, O.; Nicolai, T. Controlling the dynamics of self-assembled triblock copolymer networks via the pH. *Macromolecules* **2011**, *44*, 4487–4495. [CrossRef]

32. Charlot, A.; Auzely-Velty, R.; Rinaudo, M. Specific interactions in model charged polysaccharide systems. *J. Phys. Chem. B* **2003**, *107*, 8248–8254. [CrossRef]

33. Wintgens, V.; Daoud-Mahammed, S.; Gref, R.; Bouteiller, L.; Amiel, C. Aqueous polysaccharide associations mediated by β-cyclodextrin polymers. *Biomacromolecules* **2008**, *9*, 1434–1442. [CrossRef] [PubMed]

34. Li, L.; Guo, X.; Wang, J.; Liu, P.; Prud'homme, R.K.; May, B.L.; Lincoln, S.F. Polymer networks assembled by host−guest inclusion between adamantyl and β-cyclodextrin substituents on poly(acrylic acid) in aqueous solution. *Macromolecules* **2008**, *41*, 8677–8681. [CrossRef]

35. Van de Manakker, F.; Vermonden, T.; el Morabit, N.; van Nostrum, C.F.; Hennink, W.E. Rheological behavior of self-assembling PEG-β-cyclodextrin/PEG-cholesterol hydrogels. *Langmuir* **2008**, *24*, 12559–12567. [CrossRef] [PubMed]

36. Liu, J.; Chen, G.; Guo, M.; Jiang, M. Dual stimuli-responsive supramolecular hydrogel based on hybrid inclusion complex (HIC). *Macromolecules* **2010**, *43*, 8086–8093. [CrossRef]

37. Cates, M. Reptation of living polymers: Dynamics of entangled polymers in the presence of reversible chain-scission reactions. *Macromolecules* **1987**, *20*, 2289–2296. [CrossRef]

38. Vermonden, T.; van Steenbergen, M.J.; Besseling, N.A.; Marcelis, A.T.; Hennink, W.E.; Sudhölter, E.J.; Cohen Stuart, M.A. Linear rheology of water-soluble reversible neodymium (III) coordination polymers. *J. Am. Chem. Soc.* **2004**, *126*, 15802–15808. [CrossRef] [PubMed]

39. Lee, J.; Choi, E.J.; Varga, I.; Claesson, P.M.; Yun, S.-H.; Song, C. Terpyridine-functionalized stimuli-responsive microgels and their assembly through metal-ligand interactions. *Polym. Chem.* **2018**, *9*, 1032–1039. [CrossRef]

40. Tanaka, T.; Sato, E.; Hirokawa, Y.; Hirotsu, S.; Peetermans, J. Critical kinetics of volume phase transition of gels. *Phys. Rev. Lett.* **1985**, *55*, 2455. [CrossRef] [PubMed]

MDPI

St. Alban-Anlage 66

4052 Basel

Switzerland

Tel. +41 61 683 77 34

Fax +41 61 302 89 18

www.mdpi.com

Polymers Editorial Office

E-mail: polymers@mdpi.com

www.mdpi.com/journal/polymers

www.ingramcontent.com/pod-product-compliance
Lightning Source LLC
Chambersburg PA
CBHW051726210326
41597CB00032B/5621